Hydraulic Systems Volume 5
Safety and Maintenance

Dr. Medhat Kamel Bahr Khalil, Ph.D, CFPHS, CFPAI.
Director of Professional Education and Research Development,
Applied Technology Center, Milwaukee School of Engineering,
Milwaukee, WI, USA.

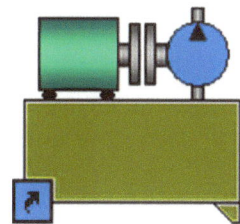

CompuDraulic LLC
www.CompuDraulic.com

CompuDraulic LLC

Hydraulic System Volume 5

Safety and Maintenance

ISBN: 978-0-9977816-5-6

Printed in the United States of America
First Published by June 2022
First Revision by   -----

All rights reserved for CompuDraulic LLC.
3850 Scenic Way, Franksville, WI, 53126 USA.
www.compudraulic.com

Disclaimer

It is always advisable to review the relevant standards and the recommendations from the system manufacturer. However, the content of this book provides guidelines based on the author's experience.

Any portion of information presented in this book might not be suitable for some applications due to various reasons. Since errors can occur in circuits, tables, and text, the author/publisher assumes no liability for the safe and/or satisfactory operation of any system designed based on the information in this book.

The author/publisher does not endorse or recommend any brand name product by including such brand name products in this book. Conversely the author/publisher does not disapprove any brand name product not included in this book. The publisher obtained data from catalogs, literatures, and material from hydraulic components and systems manufacturers based on their permissions. The author/publisher welcomes additional data from other sources for future editions. This disclaimer is applicable for the workbook (if available) for this textbook.

Hydraulic Systems Volume 5
Best Practices for Safety and Reliability

Chapter 3: Hydraulic Measuring Instruments, 128

Chapter 6: Maintenance of Cylinders, 211

6.1-BP-Cylinders-01-Selection and Replacement
6.2-BP-Cylinders-02-Maintenance Scheduling
6.3-BP-Cylinders-03-Installation and Maintenance
  6.3.1- Proper Cylinder Disassembly
  6.3.2- Proper Cylinder Inspection and Maintenance
  6.3.3- Proper Seal Replacement and Installation
  6.3.4- Proper Cylinder Assembly
  6.3.5- Proper Mounting with the Machine Structure
  6.3.6- Proper Alignment with the Load
  6.3.6- Proper Alignment with the Load
  6.3.7- Proper Air Bleeding
  6.3.8- Proper Connection with Transmission Lines
  6.3.9- Proper Installation of External Limit Switches
  6.3.10- Protect End Caps Against Impact Load
  6.3.11- Protect Cylinder from External Hazard
  6.3.12- Protect Cylinder Rod from Corrosion
  6.3.13- Protect Air Chamber of a Single-Acting Cylinder from the Environment
  6.3.14- Review Range of Allowable and Maximum Working Conditions
6.4-BP-Cylinders-04-Standard Tests and Calibration
6.5-BP-Cylinders-05-Transportation and Storage

Chapter 7: Maintenance of Valves, 243

7.1-BP-Valves-01-Selection and Replacement
7.2-BP-Valves-02-Maintenance Scheduling
7.3-BP-Valves-03-Installation and Maintenance
  7.3.1- Proper Valve Disassembly
  7.3.2- Proper Valve Inspection and Maintenance
  7.3.3- Proper Valve Assembly
  7.3.4- Proper Hydraulic Connections
  7.3.5- Proper Electrical Connections
  7.3.6- Proper Valve Adjustment
  7.4-BP-Valves-04-Standard Tests and Calibration
  7.4.1- Flow Control Valve Test
  7.4.2- Test for Pressure Relief Valves
  7.4.3- Test for Pressure Reducing Valves
  7.4.4- Test for Directional Control Valve
  7.4.5- Test for Proportional and Servo Valve
7.4-BP-Valves-04-Standard Tests and Calibration
7.5-BP-Valves-05-Transportation and Storage

Chapter 12: Maintenance of Filters, 321

APPENDIXES, 352

INDEX, 377

PREFACE

Safety and maintenance are very important experience to whoever involved in designing, commissioning, operating, and servicing hydraulic systems. Gaining such experience help to avoid future unexpected shutdowns, hence improve system reliability and safety of work environment. This book introduces the concept and best practices of hydraulic components and system maintenance.

The book presents five sets of best practices for each of the following components: pumps, motors, cylinders valves, accumulators, reservoirs, transmission lines, heat exchangers, and filters. The five sets of best practices are as follows:

- BP-X Components-01-Selection and Replacement
- BP-X Components-02-Maintenance Scheduling
- BP-X Components-03-Installation and Maintenance Practices
- BP-X Components-04-Standard Tests and Calibration
- BP-X Components-05-Transportation and Storage

This book is targeting industry professionals who are in charge for operating, maintaining, and troubleshooting hydraulic systems. This book is also a great resource for mechanical engineers and service manuals technical writers.

The author is working hard to finish his goal of supporting fluid power professional education by developing the following series of volumes and relevant software:

- Hydraulic Systems Volume 1: Introduction to Hydraulics for Industry Professionals.
- Hydraulic Systems Volume 2: Electro-Hydraulic Components and Systems.
- Hydraulic Systems Volume 3: Hydraulic Fluids and Contamination Control.
- Hydraulic Systems Volume 4: Hydraulic Fluids Conditioning. Under Development
- Hydraulic Systems Volume 5: Safety and Maintenance.
- Hydraulic Systems Volume 6: Troubleshooting and Failure Analysis.
- Hydraulic Systems Volume 7: Modeling and Simulation for Application Engineers.
- Hydraulic Systems Volume 8: Design Strategies of Hydraulic Systems. (Under Development).
- Hydraulic Systems Volume 9: Design Strategies of Electro-Hydraulic Systems. (Under Development).
- Hydraulic Systems Volume 10: Hydraulic Components Modeling and Simulation. (Under Development).

Dr. Medhat Kamel Bahr Khalil

ACKNOWLEDGEMENT

All praises are to Allah who granted me the knowledge, resources and health to finish this work

To the soul of my parents who taught me the values of believe in God

To my wife who offered me all the best she can to make this work completed

To my family: wife, sons, daughters in law, and grandson "Adam"

To my best teachers and supervisors

The author also thanks the following gentlemen for their effective support in developing this book:

- Thomas Bray, former Dean of Applied Technology Center at Milwauke School of Eng.
- Kamara Sheku, Dean of Applied Researches at Milwaukee School of Engineering.
- Tom Wanke, CFPE, Director of Fluid Power Industrial Consortium and Industry Relations at Milwaukee School of Engineering.

The author also thanks the following companies (listed alphabetically) for granting the permission to use portions of their copyrighted literatures in this book.

- International Hydraulic Safety Authority
- American Technical Publishers
- Assofluid
- Bosch Rexroth
- C.C. Jensen Inc
- CFC Industrial Training
- Conco Systems
- Danfoss
- Donaldson
- ENERPAC
- Flo-Tech
- Fluid Power Safety Institute
- Fluid Power Training Institute
- Gates Corporation
- Hedland
- Hydraulic and Pneumatic Magazine
- Hydraforce
- Hydac
- International Fluid Power Society
- Max

- Mcmillan Engineering Group
- MPFiltri
- Moog
- MSOE
- NFPA
- Noria Corporation
- Pall Corporation
- Parker Hannifin
- Spectro Scientific
- Trelleborg
- Turck
- Vickers
- Webtec
- Womack

Lastly, the author extends his thanks to the following sources of public information used to enrich the contents of the book.

- www.dgdfluidpower.com
- https://www.noshok.com
- www.flender.com
- wastewater101.net
- www. degelman.com
- www.ame.com
- www.machinerylubrication.com
- www.stahlbus.com
- www.hydrauliccylindersinc.com
- dietzautomation.com
- www.bondfluidaire.com
- mac-hyd.com
- www.new-line.com
- www.Flaretite.com
- www.aircraftsystemstech.com
- hcheattransfer.com/fouling1.html

ABOUT THE BOOK

Book Description:

This book is targeting industry professionals who are in charge of operating, maintaining, and troubleshooting hydraulic systems. This book is also a great resource for mechanical engineers and service manuals technical writers. The books start by introducing best practices for safe design, commissioning, operating and servicing hydraulic components. The book then presents the maintenance concepts and measuring instruments commonly applicable for hydraulic system maintenance. The book follows that by presenting five sets of best practices for each of the following components: pumps, motors, cylinders valves, accumulators, reservoirs, transmission lines, heat exchangers, and filters. It is to be noted that, the listed best practices and maintenance actions are based on the author's experience and may not be applicable in each case. Therefore, the author is highly recommending reviewing the instructions from component and system manufacturers as considered the main and first source of information. This book is colored and has the size of standard A4. The book contains a total of twelve chapters distributed over 350 pages with very demonstrative figures and tables. The contents of the book are brand non-biased and intends to introduce the latest technologies related to the subject of the book. The book is the fifth in a series that the author plans to publish to offer a complete and comprehensive teaching curriculum for fluid power industry. The book is associated with a separate colored workbook. The workbook contains printed power point slides, chapter reviews and assignments.

Book Objectives:

Chapter 01: Hydraulic System Safety

Hydraulic equipment is widely used daily at almost all industrial sectors. Hydraulic power units range from a simple fraction of horsepower such as in a hydraulic hand tool to a large 500 horsepower machine. Lack of training, understanding how it works, and awareness of the associated hazards are reflected on increasing number of annual related injuries. The objective of this chapter is to increase the awareness about the hydraulic system safety during different phases including system design, startup, normal operation, and servicing. This chapter also explores the safety of the individuals, workplace, equipment, and the public.

Chapter 02: Basic Concepts of Hydraulic System Maintenance

This chapter covers basic rules of hydraulic system maintenance and skill set required for service workers. Impact of maintenance on system reliability and various maintenance techniques are presented. Common mistakes and reasons to void warranty are discussed. Best practices of maintaining a specific hydraulic component will be presented in the relevant chapter including guidelines for selection, replacement, installation, storage, maintenance scheduling, and standard testing.

Chapter 03: Hydraulic Measuring Instruments

This chapter provides an overview of the common measuring devices used in hydraulic systems including devices for measure pressure, flow, temperature, oil level, and load cells. The chapter introduces the difference between a meter, a switch, and a sensor. The chapter also discusses the best practices for measuring devices selection & replacement, maintenance scheduling, installation & maintenance, and standard tests & calibration.

Chapter 04: Maintenance of Pumps

This chapter provides guidelines for **pumps** selection, replacement, maintenance scheduling, installation, testing, storage and transportation. This chapter is supported by examples and figures provided by leading fluid power manufacturers.

Chapter 05: Maintenance of Motors

This chapter provides guidelines for **motors** selection, replacement, maintenance scheduling, installation, testing, storage and transportation. This chapter is supported by examples and figures granted by leading fluid power manufacturers.

Chapter 06: Maintenance of Cylinders

This chapter provides guidelines for **cylinders** selection, replacement, maintenance scheduling, installation, testing, storage and transportation. This chapter is supported by examples and figures granted by leading fluid power manufacturers.

Chapter 07: Maintenance of Valves

This chapter provides guidelines for **valves** selection, replacement, maintenance scheduling, installation, testing, storage and transportation. This chapter is supported by examples and figures granted by leading fluid power manufacturers.

Chapter 08: Maintenance of Accumulators

This chapter provides guidelines for **accumulator's** selection, replacement, maintenance scheduling, installation, testing, storage and transportation. This chapter is supported by examples and figures granted by leading fluid power manufacturers.

Chapter 09: Maintenance of Reservoirs

This chapter provides guidelines for **reservoirs** selection, replacement, maintenance scheduling, installation, testing, storage and transportation. This chapter is supported by examples and figures granted by leading fluid power manufacturers.

Chapter 10: Maintenance of Transmission Lines

This chapter provides guidelines for **transmission lines** selection, replacement, maintenance scheduling, installation, testing, storage and transportation. This chapter is supported by examples and figures granted by leading fluid power manufacturers.

Chapter 11: Maintenance of Heat Exchange

This chapter provides guidelines for **heat exchangers** selection, replacement, maintenance scheduling, installation, testing, storage and transportation. This chapter is supported by examples and figures granted by leading fluid power manufacturers.

Chapter 12: Maintenance of Filters

This chapter provides guidelines for **Filters** selection, replacement, maintenance scheduling, installation, testing, storage and transportation. This chapter is supported by examples and figures granted by leading fluid power manufacturers.

Book Statistics:

Chapter #	Pages	Figures	Tables	Words	Editing Time (Hours)
Chapter 1	97	116	2	890	88
Chapter2	17	17	1	3350	108
Chapter 3	34	42	1	4784	118
Chapter 4	42	37	4	8341	330
Chapter 5	7	4	1	1037	91
Chapter 6	32	36	2	5397	121
Chapter 7	20	19	2	3458	119
Chapter 8	15	8	1	2962	106
Chapter 9	6	3	1	1469	95
Chapter 10	26	24	4	4150	121
Chapter 11	11	10	1	2249	107
Chapter 12	31	25	5	4845	120
Total	338	341	25	42932	1,524 Hour = 63 Days

ABOUT THE AUTHOR

Medhat Khalil, Ph.D. is Director of Professional Education & Research Development at the Applied Technology Center, Milwaukee School of Engineering, Milwaukee, WI, USA. Medhat has consistently been working on his academic development through the years, starting from bachelor's and master's Degrees in Mechanical Engineering in Cairo Egypt and proceeding with his Ph.D. in Mechanical Engineering and Post-Doctoral Industrial Research Fellowship at Concordia University in Montreal, Quebec, Canada. He has been certified and is a member of many institutions such as: Certified Fluid Power Hydraulic Specialist (CFPHS) by the International Fluid Power Society (IFPS); Certified Fluid Power Accredited Instructor (CFPAI) by the International Fluid Power Society (IFPS); Member of Center for Compact and Efficient Fluid Power Engineering Research Center (CCEFP); Listed Fluid Power Consultant by the National Fluid Power Association (NFPA); and Listed Professional Instructor by the American Society of Mechanical Engineers (ASME). Medhat has balanced academic and industrial experience. Medhat has vast working experience in Fluid Power teaching courses for industry professionals. Being quite aware of the technological developments in the field of fluid power,

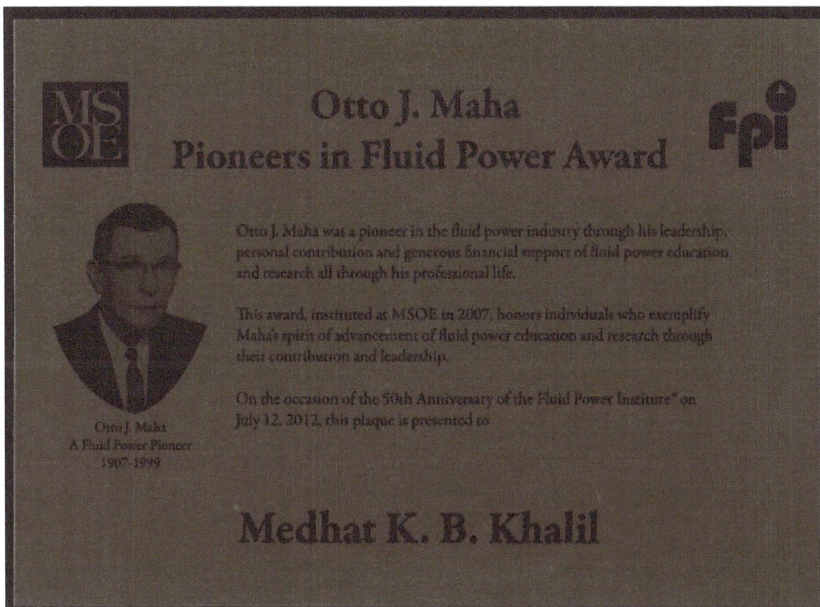

Medhat had worked for several world-wide recognized industrial organizations such as Rexroth in Egypt and CAE in Canada. Medhat had designed several hydraulic systems and developed several analytical and educational software. Medhat also has considerable experience in modeling and simulation of dynamic systems using Matlab-Simulink. Medhat has been selected among the inductees for Pioneers in fluid Power by NFPA (2012) and Hall of Fam in fluid Power by IFPS (2021).

Chapter 1
Hydraulic System Safety

Objectives

Hydraulic equipment is widely used daily at almost all industrial sectors. Hydraulic power units range from a simple fraction of horsepower such as in a hydraulic hand tool to a large 500 horsepower machine. Lack of training, understanding how it works, and awareness of the associated hazards are reflected on increasing number of annual related injuries. The objective of this chapter is to increase the awareness about the hydraulic system safety during different phases including system design, startup, normal operation, and servicing. This chapter also explores the safety of the individuals, workplace, equipment, and the public.

Brief Contents

1.1- Introduction
1.2- Why Hydraulic Systems Safety is Important?
1.3- Who is Responsible for Hydraulic Systems Safety?
1.4- Where to Find General Industry Safety and Health Standards?
1.5- What are the Sources of Best Practices for Hydraulic Systems Safety?
1.6- When to Apply Hydraulic Systems Safety Best Practices?
1.7- BP-Safety-01: Design for Safe and Reliable Hydraulic Systems
1.8- BP-Safety-02: Safety of Hydraulic System Operators
1.9- BP-Safety-03: Safety of Hydraulic System Work Environment
1.10- BP-Safety-04: Safety of Hydraulic System Workspace
1.11- BP-Safety-05: Safe Startup of Hydraulic Systems
1.12- BP-Safety-06: Safe Operation of Hydraulic Systems
1.13- BP-Safety-07: Safe Servicing of Hydraulic Systems
1.14- BP-Safety-08: Oil Injection Avoidance and Treatment
1.15- BP-Safety-09: Safe usage of Hydraulic Powered Tools
1.16- BP-Safety-10: Safe Storage and Transportation of Hydraulic Systems

Chapter 1: Basic Concepts of Hydraulic System Maintenance

1.1- Introduction

As it has been stated previously in the book disclaimer, information provided in this book cannot cover every situation and is not intended to do so. It is highly recommended and always advisable to review the safety precautions provided by the components, systems, and machines manufacturers. Hydraulic-driven machines are used widely in all industrial sectors. Figure 1.1 reveals the fact that, like electrical systems, hydraulics systems must be treated properly to maintain safe machine operation. As shown in Fig. 1.2, hazard from electrical and hydraulic transmission lines are equal. Electrical shock from a transmission line can cause loss of life. Also, blow out of a hydraulic transmission line can cause loss of life.

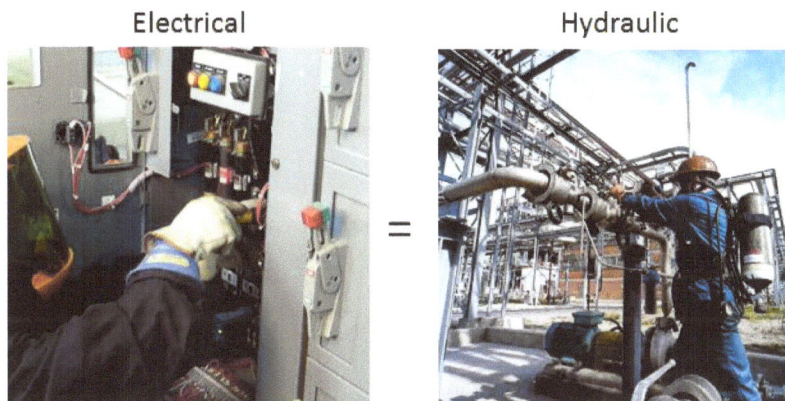

Electrical   Hydraulic

=

Fig. 1.1 – Safety of a Hydraulics System is as Important as that of an Electrical System

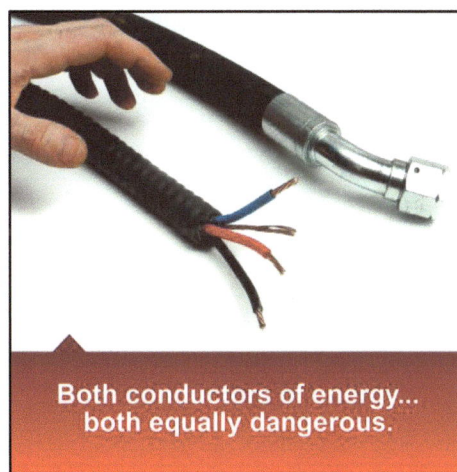

Both conductors of energy...
both equally dangerous.

**Fig. 1.2 – Hazard from Electrical and Hydraulic Transmission Lines are Equal
(Courtesy of the International Hydraulic Safety Authority)**

After brain storming of hydraulic system safety, as shown in Fig. 1.3, the following questions need answers:

- **Why** hydraulic system safety is important?
- **Who** is responsible for the safety of hydraulic systems?
- **Where** to find general industry safety and health standards?
- **What** are the sources of best practices for hydraulic systems safety?
- **When** to apply hydraulic system safety best practices?

The following sections present answers to these questions with examples to explore the ideas.

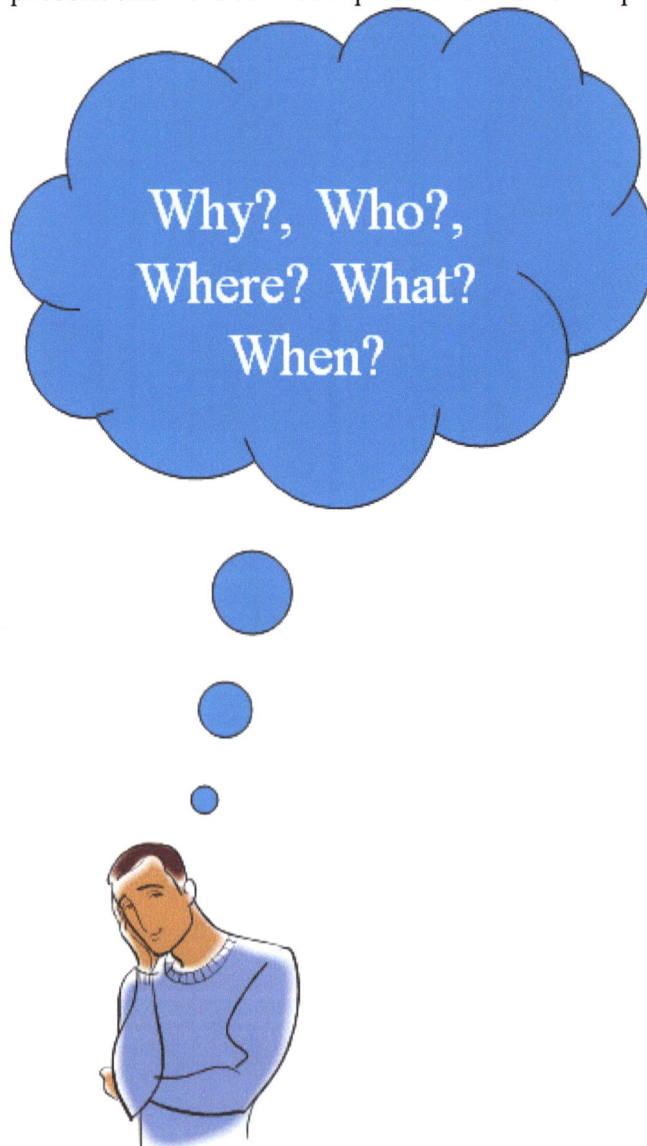

Fig. 1.3 – Hydraulics System Safety Related Questions

1.2- Why Hydraulic System Safety is Important?

As shown in Fig. 1.4, a statement has been made in one of the famous movies was found very applicable to industrial and mobile hydraulic driven machines. The statement was "With great power, there should be great responsibility".

Fig. 1.4 – Hydraulic Power Associated with Large Loads

As shown in Fig. 1.5, the main goal of enforcing the rules for hydraulic system safety is to prevent injury or death to the human, damage to the environment, surroundings and the machine itself. The next section provides detailed interpretation of the previous statement.

-- HUMAN (1): Using hydraulic power usually associated with quite large loads. Underestimating or un-securing such loads may cause major injuries or even loss of life. An example of this, as shown in the figure, is oil injection. Employees recognize that when the employer takes care of them, they take care of their job and that leverages productivity.

-- ENVIRONMENT (2): Machine failure may result in damaging the environment. Repair of such damage results in additional cost and liabilities. An example of this, as shown in the figure, is oil leakage.

-- SURROUNDINGS (3): Unexpected or unsecured actuator movement my damage the surroundings including the final products that might be of a considerable value. An example if this, as shown in the figure, is falling of heavy loads.

-- MACHINE (4): Unsafe operation of a hydraulic system may cause machine failure. The plant/machine shutdown cost is much more than the cost of repairing the machine. An example of this, as shown in the figure, pump failure.

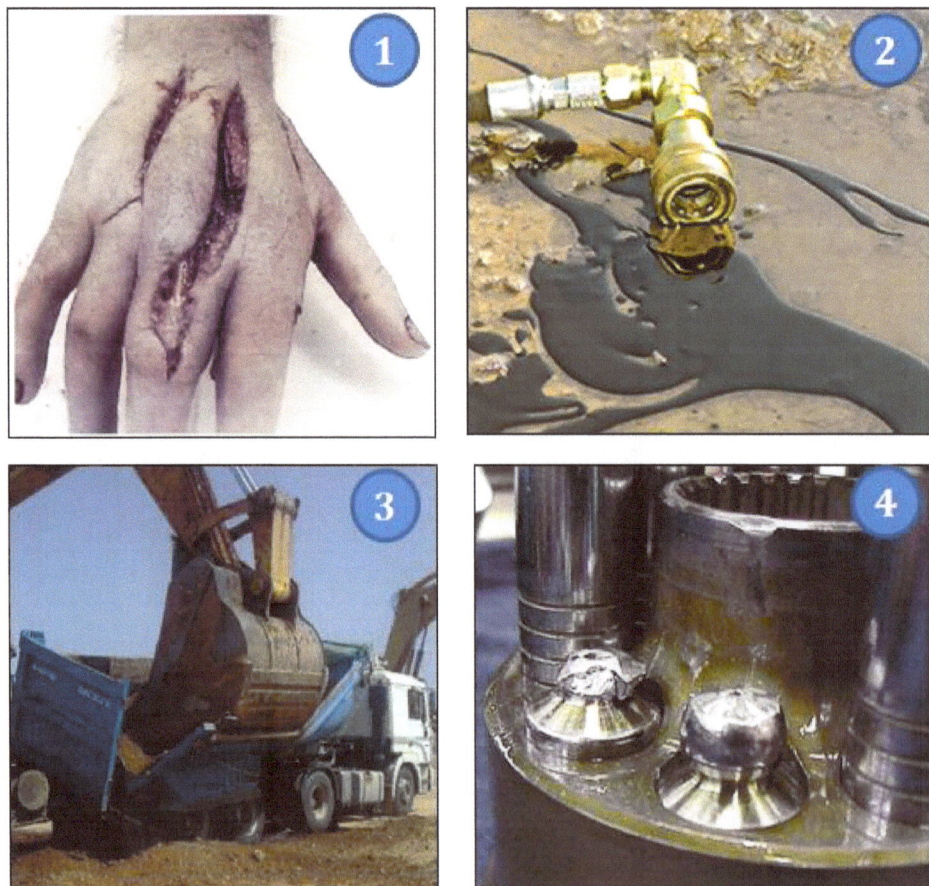

Fig. 1.5 – Why Hydraulics System Safety is Important?

1.3- Who is Responsible for Hydraulic Systems Safety?

It is a misconception that you are not responsible! You should not be one of those shown in the upper part of Fig 1.6. Safety is the responsibility of everyone including managers, system designers, supervisors, machine operators, and service crews. Regardless your position, you should react to emergencies based on pre-defined plans.

I'm not responsible   **I'm the only one responsible**

Based on my job duty, I should follow pre-defined best practices

Fig. 1.6 – Who is Responsible for the Hydraulic System Safety?

1.4- Where to Find General Industry Safety and Health Standards?

The *Occupational Safety and Health Administration* (*OSHA*) of the US Department of Labor describes and enforces safety standards at the industrial locations (***OSHA Standard #29 CFR 1910***). Publication # 2072 provides General Industry Guidelines for Applying Safety and Health Standards where hydraulic equipment is operated. These standards are categorized as follows:

❑ **Workplace Standards:**
  ▪ Safety of floors, entrance and exit areas, sanitation, and fire protection.

❑ **Machines and Equipment Standards:**
  ▪ Machine guards + inspection + maintenance techniques.
  ▪ Safety devices + mounting of equipment.
  ▪ Noise levels produced by operating equipment.

❑ **Materials Standards:**
  ▪ Toxic fumes, explosive dust particles, excessive atmospheric contamination.

❑ **Employee Standards:**
  ▪ Training, personnel protective equipment, medical and first-aid services.

❑ **Power Source Standards:**
  ▪ Power sources such as electric, hydraulic, pneumatic, and steam supply systems.

❑ **Process Standards:**
  ▪ Welding, spraying, abrasive blasting, and machining.

❑ **Administrative Regulations:**
  ▪ Displaying of OSHA posters, stating the rights and responsibilities of both the employer and employee.

1.5- What are the Sources of Best Practices for Hydraulic System Safety?

There are different sources of instructions for the safety of hydraulic systems. As shown in Fig. 1.7, machine manufacturers, industry standards, and related organizations are the main sources for such instructions. Additionally, there many accumulated experiences still in the minds of field experts that not been documented.

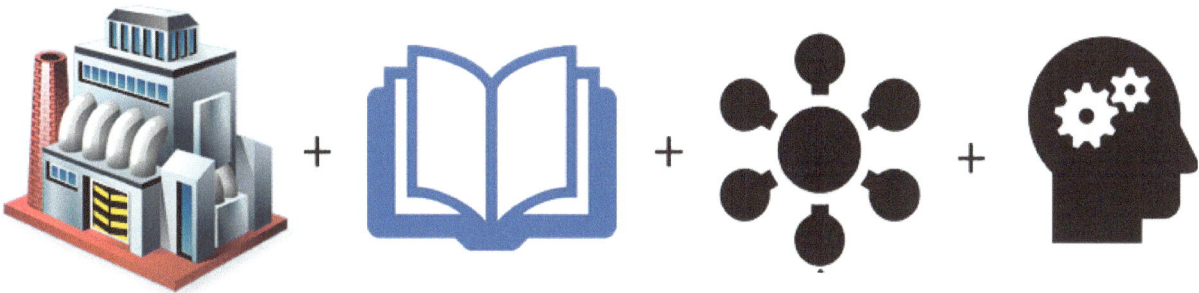

Fig. 1.7 – Sources of Best Practices for Hydraulic System Safety

1.6- When to Apply Hydraulic System Safety Best Practices?

As shown in Fig. 1.8, enforcing hydraulic system safety has many challenges.

1. Most of the involved personnel are unaware of the associated hazards.
2. Hazards from hydraulic systems range from minor injuries to loss of life.
3. A hydraulic system designer is not commissioning it, the commissioner is not operating it on the long term, and the operator is not maintaining or troubleshooting it. Therefore, the knowledge gab between these four groups of different people must be filled properly with predefined instructions, otherwise unsafe situations may arise.

1 Lack of understanding or not aware of the risks.

2 Hazards from Hydraulic systems

From
Minor Injuries

To
Loss of Life

3

| System Design | Starting Up | Operation | servicing |

Hydraulic Cylinder

Retract/Exten

Reservoir

Control valve

Filter Pump

Proper Instructions

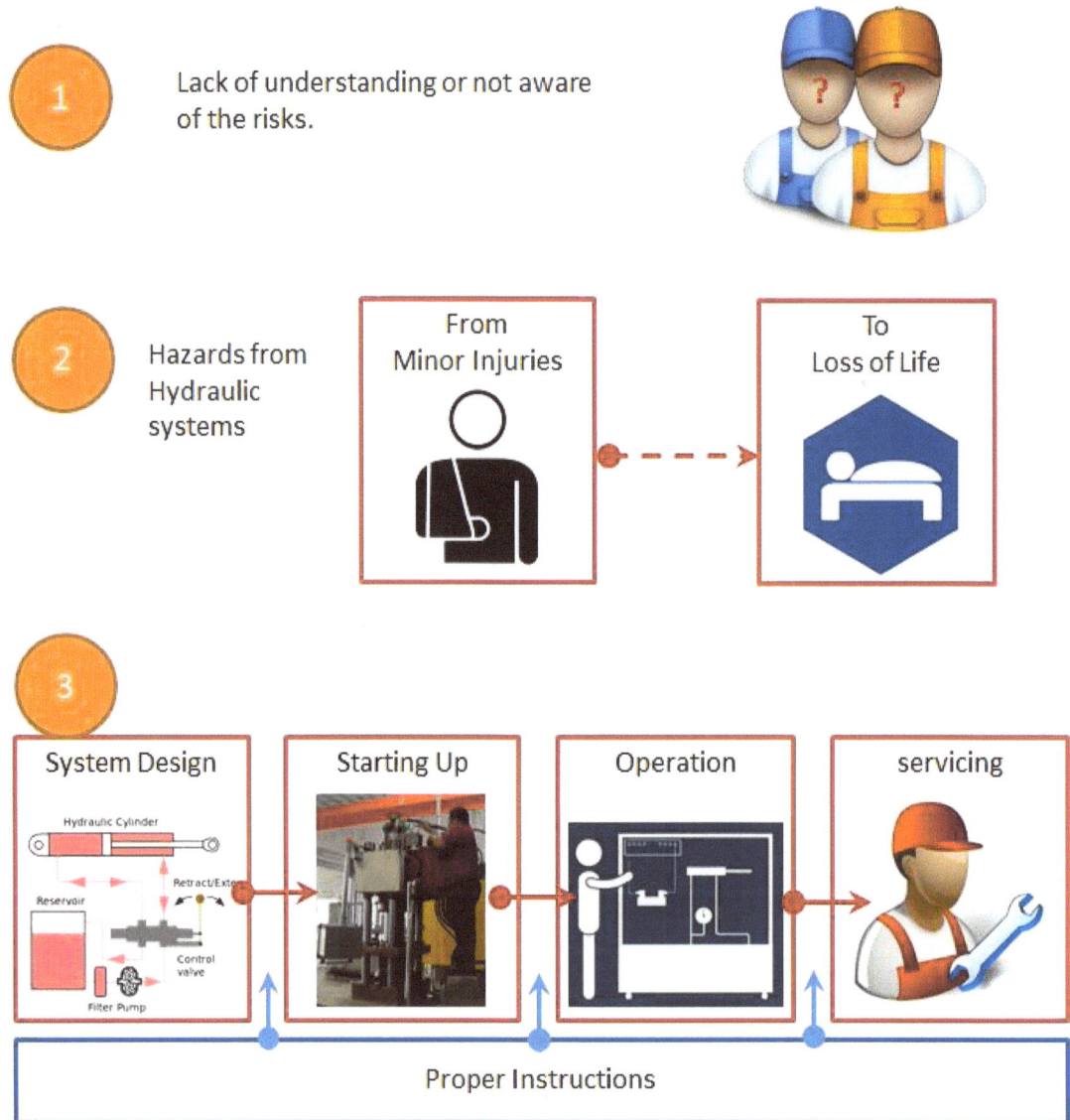

Fig. 1.8 – Challenges of Hydraulic System Safety

Miscommunication between system designer, commissioner, operator, and servicer can result in serious problems. As shown in Fig. 1.9, safety best practices must be applied during the following phases:

1. During system design.
2. During system commissioning (startup).
3. During system operation.
4. During system servicing (maintenance, troubleshooting, and testing).

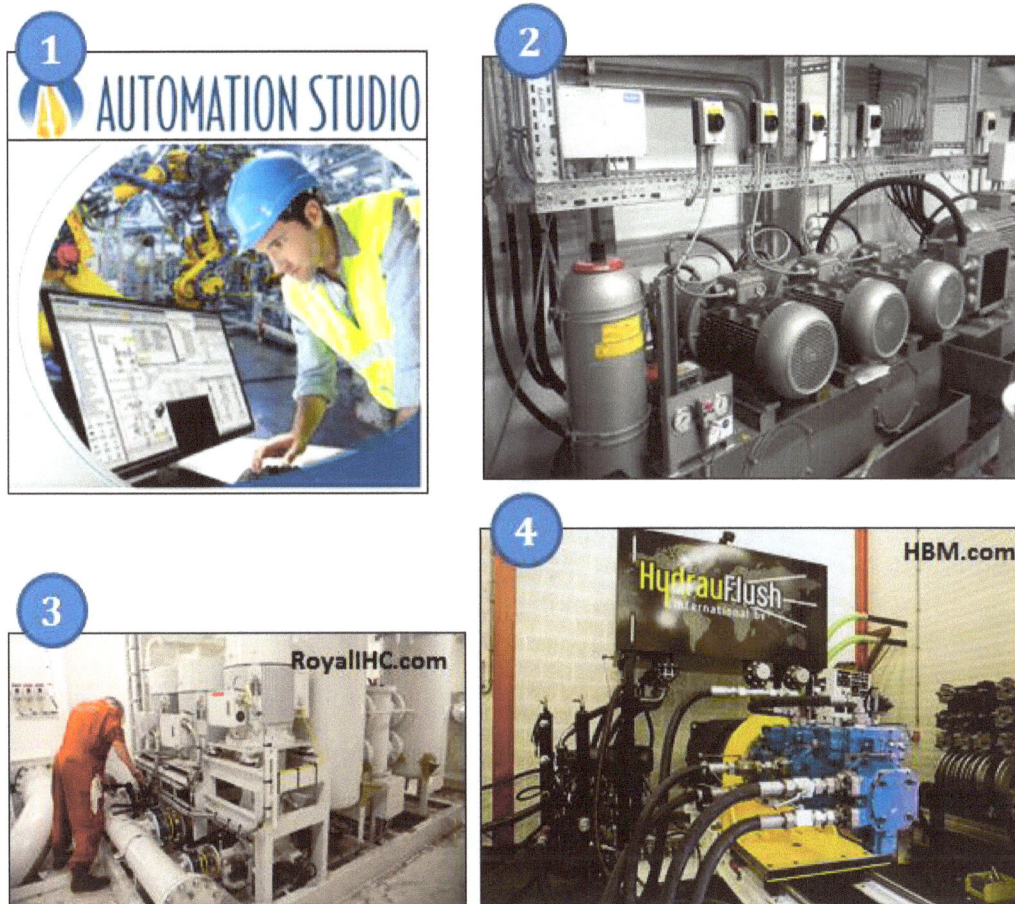

Fig. 1.9 – When to Apply Hydraulic System Safety Best Practices?

In this textbook, the author adopted a unique methodology of gathering these experiences in form of Best Practices lists. Each relevant chapter contains detailed interpretation of these lists with examples. This chapter will discuss the best practices for the safety of hydraulic systems in sequence as follows:

- BP-Safety-01: Design for Safe and Reliable Hydraulic Systems.
- BP-Safety-02: Safety of Hydraulic System Operators.
- BP-Safety-03: Safety of Hydraulic System Work Environment.
- BP-Safety-04: Safety of Hydraulic System Workspace.
- BP-Safety-05: Safe Startup of Hydraulic Systems.
- BP-Safety-06: Safe Operation of Hydraulic Systems.
- BP-Safety-07: Safe Servicing of Hydraulic Systems.
- BP-Safety-08: Safe usage of Hydraulic Powered Tools.

1.7- BP-Safety-01: Design for Safe and Reliable Hydraulic Systems

Designing safe hydraulic systems reduces of risks that could occur during machine operation. In addition to machine specifications predefined by the system manufacturer, the best practices list *BP-Safety-01* presents guidelines to be considered during hydraulic system design.

BP-Safety-01:
1. Review Manufacturers' Recommendations.
2. Work with Standards.
3. Limit Maximum Operating Pressure.
4. Limit Maximum Operating Temperature.
5. Provide Protection Against Surface Temperature.
6. Eliminate Risk of Fire and Explosion.
7. Adequately Size and Select Hydraulic Transmission Lines.
8. Adequately Size and Select Hydraulic Components.
9. Avoid Pump Cavitation.
10. Carefully Design Hydraulic Fluid Contamination Control Systems.
11. Minimize Noise and Vibrations of the System.
12. Apply Energy Saving Design Strategies for Hydraulic Systems.
13. Apply Fail-Safe Design Strategies for Hydraulic Systems.
14. Properly Design Condition Monitoring System.
15. Perform Dynamic Analysis Whenever Needed.

The following subsections provide detailed interpretation of the action items listed in the best practices list BP-Safety-01.

1.7.1- Review Manufacturers' Recommendations

A hydraulic system designer must review the technical specifications provided by the hydraulic components manufacturers. Usually components manufacturers provide data about the limits (minimum and maximum) of the operating conditions, e.g. pressure, temperature, pump speed, etc.

It is the responsibility of the system designer to make sure that the all selected components shall operate within manufacturer specifications. Otherwise, in the best-case scenario, system reliability reduced. Consequences of running unreliable hydraulic systems may be as worse as accidental machine breakage, which may be accompanied by uncontrolled loads and unsafe work environment. Providing examples is an endless process. The following two examples explore the idea of what could occur due to disrespect of manufacturer recommendations.

Example 1 (Figure 1.10) shows an example of operating conditions specified by a pump manufacturer. As shown in the figure, exceeding maximum viscosity may result in pump cavitation and overall system failure.

Variable Axial Piston Pump (A)A10VSO

Viscosity limits

The limiting values for viscosity are as follows:

v_{min} = 60 SUS (10 mm²/s)
short term (t ≤ 1 min)
at a max. permissible leakage oil temperature
of t_{max} = 195 °F (90 °C).

Please note that the max. fluid temperature of 195 °F (90 °C) is also not exceeded in certain areas (for instance bearing area). The temperature in the bearing area is approx. 7 °F (5 K) higher than the average leakage fluid temperature.

v_{max} = 7500 SUS (1600 mm²/s)
short term (t ≤ 1min)
on cold start
(p ≤ 435 psi/30 bar, n ≤ 1000 rpm, t_{min} = -13 °F/-25 °C)

**Fig. 1.10 – Operating Conditions Specified by a Pumps Manufacturer
(Courtesy of Bosch Rexroth)**

Example 2 (Figure 1.11) shows another example of operating conditions specified by a hose manufacturer. Exceeding the maximum working pressure reduces the hose service life and increases the possibility of blowout. Reducing the bend radius below minimum increases stresses on the reinforcement layer of the hose and cracks the outer protecting layer of the hose. Both conditions significantly affect the life of the hose.

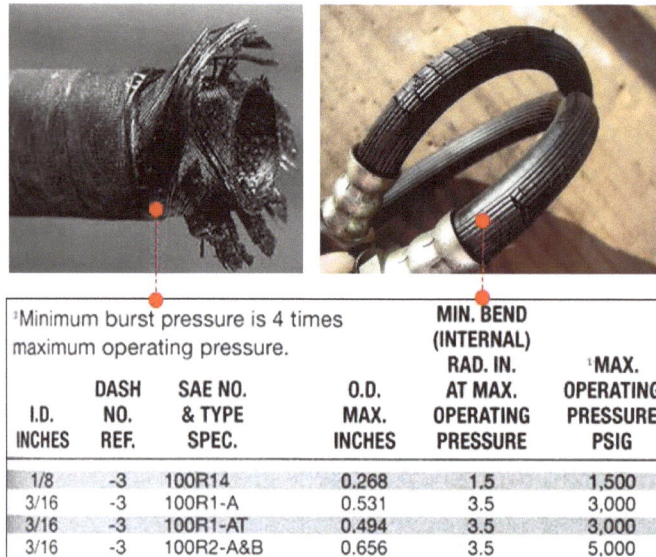

I.D. INCHES	DASH NO. REF.	SAE NO. & TYPE SPEC.	O.D. MAX. INCHES	MIN. BEND (INTERNAL) RAD. IN. AT MAX. OPERATING PRESSURE	MAX. OPERATING PRESSURE PSIG
1/8	-3	100R14	0.268	1.5	1,500
3/16	-3	100R1-A	0.531	3.5	3,000
3/16	-3	100R1-AT	0.494	3.5	3,000
3/16	-3	100R2-A&B	0.656	3.5	5,000

¹Minimum burst pressure is 4 times maximum operating pressure.

Fig. 1.11 – Operating Conditions Specified by Hose Manufacturer (Courtesy of Parker)

1.7.2- Work with Standards

Working on your own making inexperienced estimates and decisions may result in unsafe machines. It is always advisable to design hydraulic systems to conform to industry standards. If this is not obligatory, at least review general guidelines and rules of thumb provided by industry experts. As shown in Fig. 1.12, *Lightening Reference Handbook* is one good reference that contain significant fluid power related information. In case if contractually agreed to, design the hydraulic system in accordance with relevant standards.

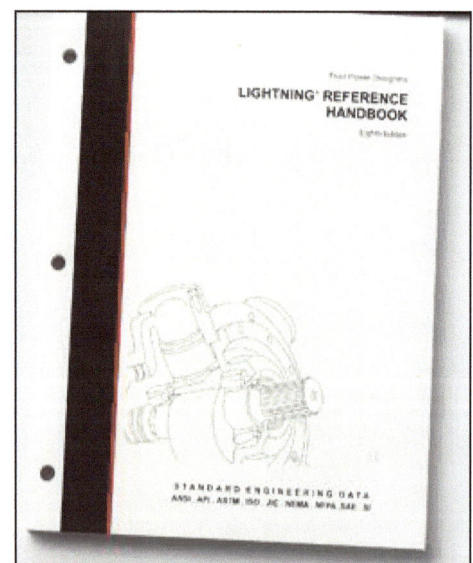

Fig. 1.12 – Lightening Reference Handbook

Communication Standards:

- *ISO Standard 1219* defines the graphic symbols and dimension codes for fluid power components.
- *ISO Standard 5598* defines terms and definitions, and other communication tools used in the fluid power industry.

Design Standards establish dimensions, tolerances and other physical characteristics of products. They ensure that fluid power products meet dimensional criteria that enable interfacing and interchangeability. New equipment must be designed to conform to:

- Hydraulics: *ISO Standard 4413 (NFPA T2.24.1)*.
- Pneumatics *ISO Standard 4414 (NFPA T2.24.2)*.
- Control Systems: The control systems shall be designed in accordance with the risk assessment. This requirement is met when *ISO Standard 13849-1* is used. System designers must comply with the actual wording of the standard.

Performance Standards provide a voluntary method of rating products. Pressure rating, particle counting methods used in contamination analysis, and methods of testing for strength and volume are typical performance standards. In the coming chapters of components maintenance, the relevant standard will be mentioned.

Table 1.1 provide different sources of standards that are used in fluid power. The table provides the weblink for each of the listed organization so that the readers can get more background about them. However, the following section presents a brief background about each source.

Source of Standards	Year Founded	Website
Society of Automotive Engineers (SAE).	1905	www.sae.org
Joint Industry Conference (JIC).	1950	-
National Fluid Power Association (NFPA).	1953	www.nfpa.com
International Fluid Power Society (FPS).	1960	www.ifps.org
American National Standards Institute (ANSI) B93	1961	www.ansi.org
International Standards Organization (ISO) TC131.	1970	http://www.iso.org

Table 1.1 - Standards that Contain Information about Fluid Power Systems

Society of Automotive Engineers (SAE): Dimensional interchange standards for hose, rigid tubing, hose and tube fittings, pump & motor mounting flanges and shaft configurations. Performance standards for hose and tube assemblies, pumps & motors, cylinders and valves.

Background: International technical society of professional individuals. Standards are developed by member committees comprised of experienced individuals in the field.

Joint Industry Conference (JIC) No longer an active organization: Originally developed some dimensional interchange standards for industrial tie rod cylinders, control valves and reservoirs. Most noted for the JIC 37 Degree flare fittings which ultimately became a SAE standard. Other JIC standards migrated into NPFA and/or ANSI standards.

Background: An association of some fluid power equipment manufacturers and the three big Detroit area automotive manufacturers.

National Fluid Power Association (NFPA): Dimensional interchange standards for industrial tie rod cylinders, valves and electrical connectors. Performance standards for most components including pumps & motors, valves, cylinders and filters. Most NFPA Standards have been submitted to ANSI B93 Committee which is responsible for approving them to be American National Standards as compared to just a trade association standard.

Background: A trade association of primarily component manufacturers, distributors and system integrators. Standards are developed by member committees comprised of experienced individuals in the appropriate areas. Currently, NFPA is not developing any new standards. When an existing standard is up for review, which is every 5 years, they can reaffirm it, withdraw it or send it up to ISO as a potential new work item.

International Fluid Power Society (IFFS): Provides educational and certification programs at various technical levels. Also provides educational materials, best practices and promotes safety of fluid power systems.

Background: International technical society of professional individuals. Certification programs are developed by member committees comprised of experienced individuals in the appropriate areas.

American National Standards Institute (ANSI): ANSI Committee B93 is responsible for all the country's fluid power standards. They review standards submitted by various US companies and trade associations to determine their applicability and relevance as a national standard. ANSI also submits its' standards to the ISO Secretariat to determine if they're applicable and relevant to be come and ISO Standards.

Background: National Committee of professional individuals with a broad base of knowledge and experience in the appropriate fields of fluid power technology

International Standards Organization (ISO): ISO Technical Committee TC131 is responsible for all global fluid power standards. Includes dimensional interchange standards for most components, performance and rating standards for most components, communication standards, system design guides and machinery safety standards. TC131 reviews standards submitted by member countries to determine their applicability at the international level.

Background: International Committee of professional individuals representing approximately 55 different countries who have a broad base knowledge and experience in the appropriate fields of fluid power technology.

1.7.3- Limit Maximum Operating Pressure

As best practice in hydraulic system design, all components of the system shall be protected against over-pressure and must work within the pressure limits specified by the manufacturers. Working pressure can increase unexpectedly due to many things such as: dynamic loads, pressure spikes, pressure intensification in differential areas cylinders, and thermal expansion of a confined volume of oil. Noncompliance with the maximum allowable pressure is a good reason for unsafe machine operation. The following examples show various ways of protecting a system against over pressure:

Example 1 (Fig. 1.13): The simplest and cost-effective solution is to use a main pressure relief valve to limit the pump outlet pressure.

Example 2 (Fig. 1.14): For the sake of energy efficiency of the system, a pressure compensated pumps maintains maximum pressure without flowing oil through a relief valve.

Example 3 (Fig. 1.15): A secondary relief valve protects other components in the system separately from the pump.

Example 4 (Fig. 1.16): A relief valve must protect pressure intensification at the rod side of a differential area cylinder during meter-out speed control.

Example 5 (Fig. 1.17): A special relief valve must protect pressure intensification due to *thermal expansion* of a confined oil volume. A screw-in, cartridge-style, hydraulic check valve for use as a blocking or load-holding device. The cartridge incorporates a low flow thermal relief valve intended to prevent cylinder damage resulting from temperature-induced pressure intensification. The valve allows flow from 1 to 2, while blocking oil flow in the opposite direction. If the pressure at 2 exceeds the thermal relief valve setting, a small amount of oil will be allowed to pass from 2 to 1, preventing cylinder damage from pressure intensification. Note: The relief valve feature is not intended for use in dynamic pressure limiting applications.

Example 6 (Fig. 1.18): An accumulator limits pressure spikes that may result during shock loads or valve shifting.

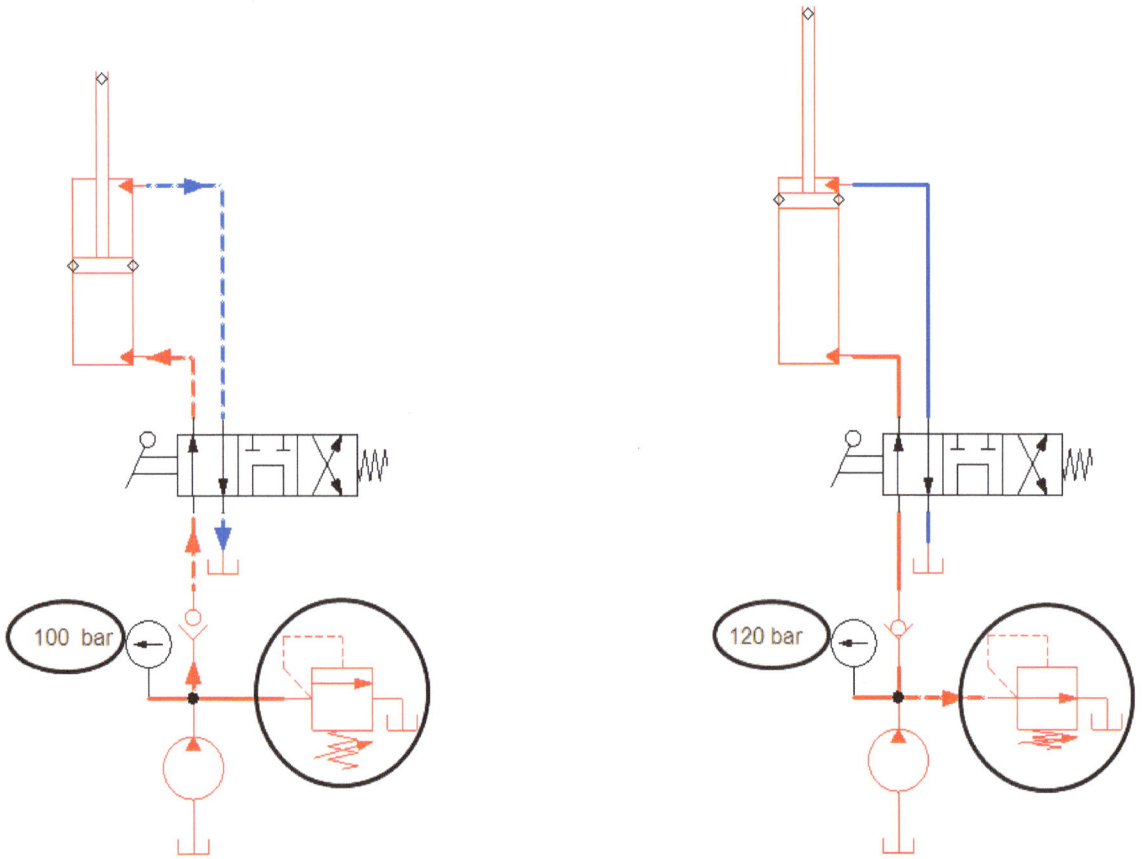

Fig. 1.13 – Limit Maximum Pressure at Pump Outlet

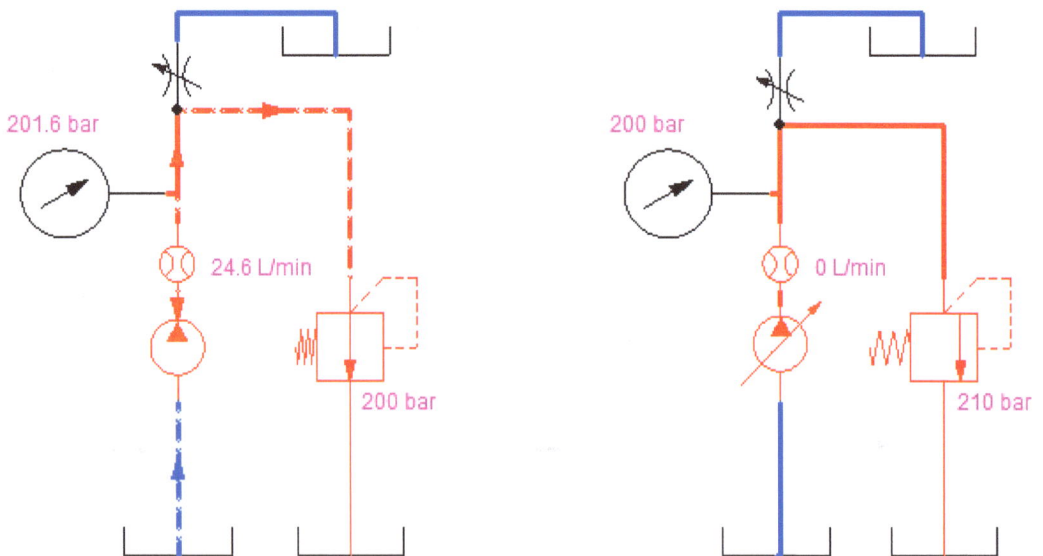

Fig. 1.14 – Pressure Compensated Pump Limits Maximum Pressure with Energy Saving

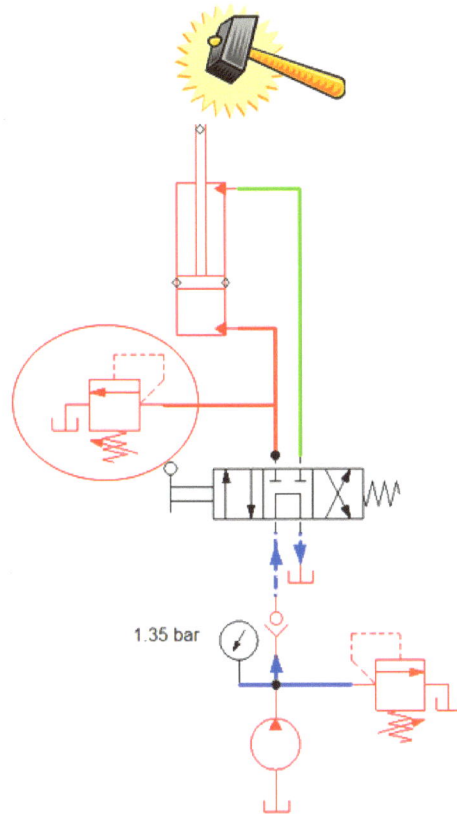

Fig. 1.15 – A Secondary Relief Valve Protects Other Components in the System

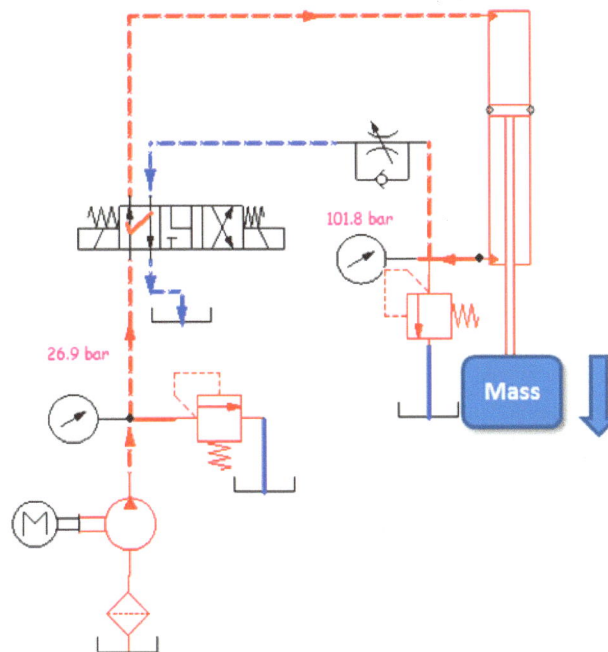

Fig. 1.16 – A Secondary Relief Valve Limits Pressure Intensification in a Differential Cylinder

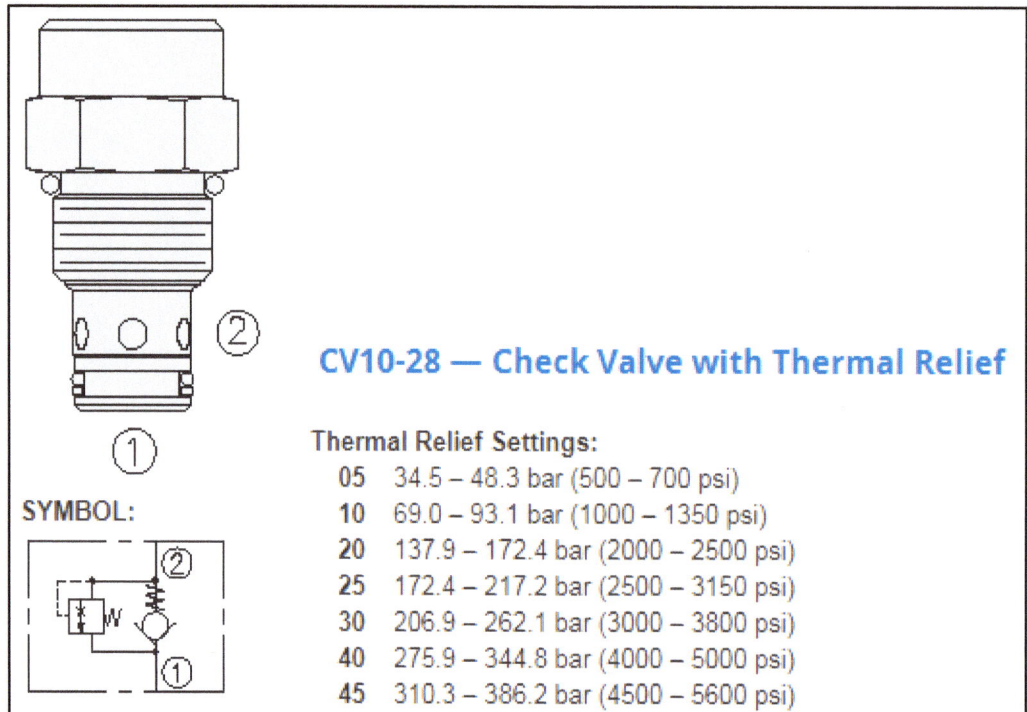

CV10-28 — Check Valve with Thermal Relief

Thermal Relief Settings:

05	34.5 – 48.3 bar	(500 – 700 psi)
10	69.0 – 93.1 bar	(1000 – 1350 psi)
20	137.9 – 172.4 bar	(2000 – 2500 psi)
25	172.4 – 217.2 bar	(2500 – 3150 psi)
30	206.9 – 262.1 bar	(3000 – 3800 psi)
40	275.9 – 344.8 bar	(4000 – 5000 psi)
45	310.3 – 386.2 bar	(4500 – 5600 psi)

SYMBOL:

**Fig. 1.17 – A Special Relief Valve Limits Pressure Increase due to Thermal Expansion
(Courtesy of Hydraforce)**

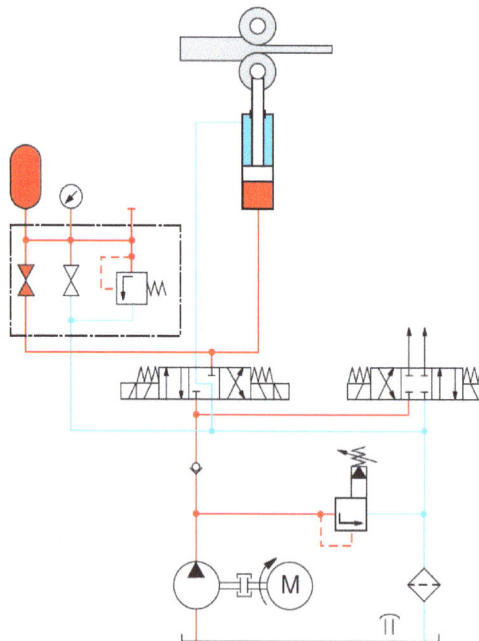

**Fig. 1.18 – An Accumulator Limits Pressure Spikes due to Shock Loads
(Courtesy of Bosch Rexroth)**

1.7.4- Limit Maximum Operating Temperature

In addition to contamination, overheating is one of the two strongest enemies to the hydraulic system reliability. Hydraulic fluid is the life blood of the hydraulic system and its physical properties are highly affected by working temperature. Therefore, the operating temperatures of a hydraulic system shall not exceed the limits beyond which the system operates unsafely. As shown in Fig. 1.19, for proper sizing of a heat exchanger, a system designer should preciously solve the heat balance equation for the hydraulic system.

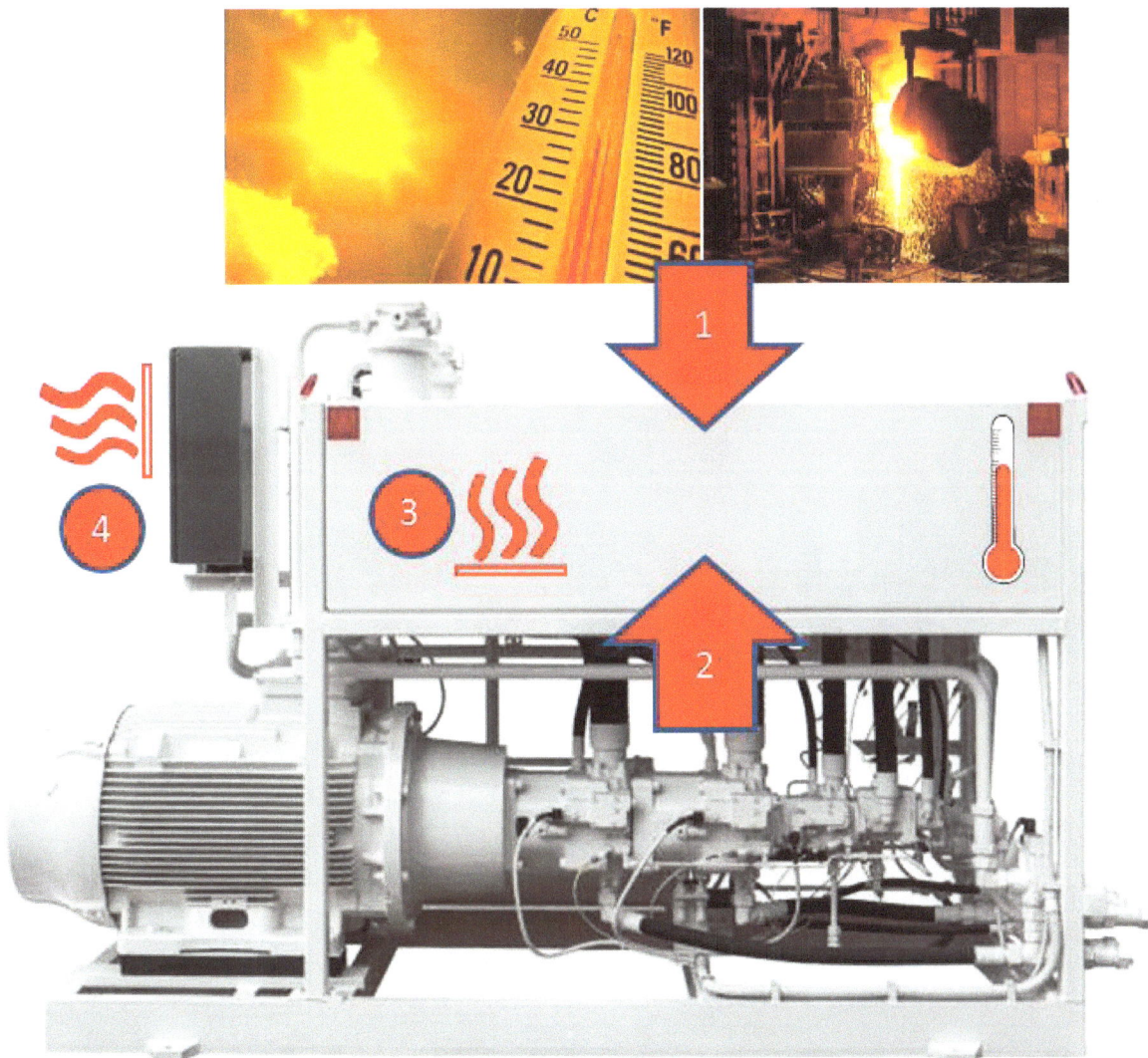

Heat Imported (1) + Heat Generated (2) – Heat dissipated (3)
= Size of Heat Exchange (4)

Fig. 1.19 – Solving Heat Balance Equation for Proper Sizing a Heat Exchanger

1.7.5- Provide Protection against Surface Temperature

Burns from Hot Surfaces: Figure 1.20 shows an example of hand burn as a result of accidental touching a hot surface of a hydraulic machine.

Burns from Conveyed Fluids: Contact with hot fluid, especially at temperatures over 60 °C [140 °F], can burn human skin.

Design Solutions: Hydraulic systems should be designed to protect personnel from *Surface Temperatures* that exceed touchable limits. Available solutions, as shown in Fig. 1.21, include, hose guards, pipe shields, or by placing the hot surface in an inaccessible location. When such protection is not possible, proper warnings shall be provided.

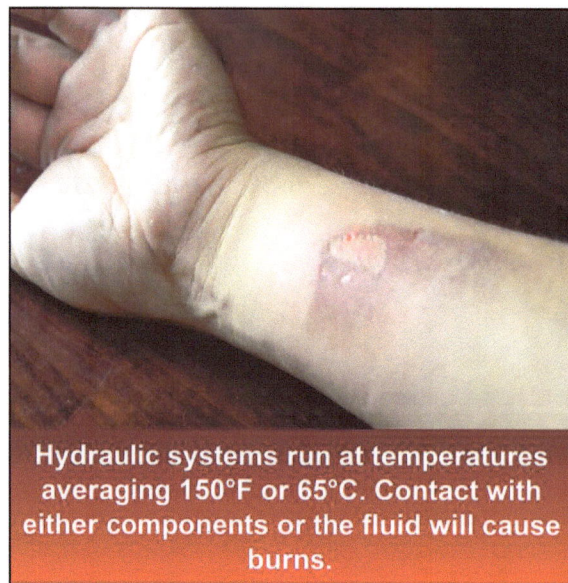

Hydraulic systems run at temperatures averaging 150°F or 65°C. Contact with either components or the fluid will cause burns.

**Fig. 1.20 – Hazard from Touching Hot Surfaces of a Hydraulic Machine
(Courtesy of the International Hydraulic Safety Authority)**

CAUTION HOT SURFACE DO NOT TOUCH

Fig. 1.21 – Protection against Surface Temperature

1.7.6- Eliminate Risk of Fire and Explosion

Fire and explosion of a hydraulic system could be due to number of sources. This section discusses sources and design solutions to eliminate risk of *fire and explosion*.

Source 1: Hydraulic Fluids:
- Mineral-based hydraulic fluids, including fire resistant ones are flammable by different degrees. Figure 1.22 shows an example of pressurized fluid spray escaping, then fluid flashing and explosion occurs upon contact with an ignition source.

Solutions:
- Use area guards, hose shields, and route fluid conductors to minimize the risk of combustion.

- Fire-resistant fluids are recommended for some applications that are associated with high heat sources such as die-casting, steel mills, mining, etc.

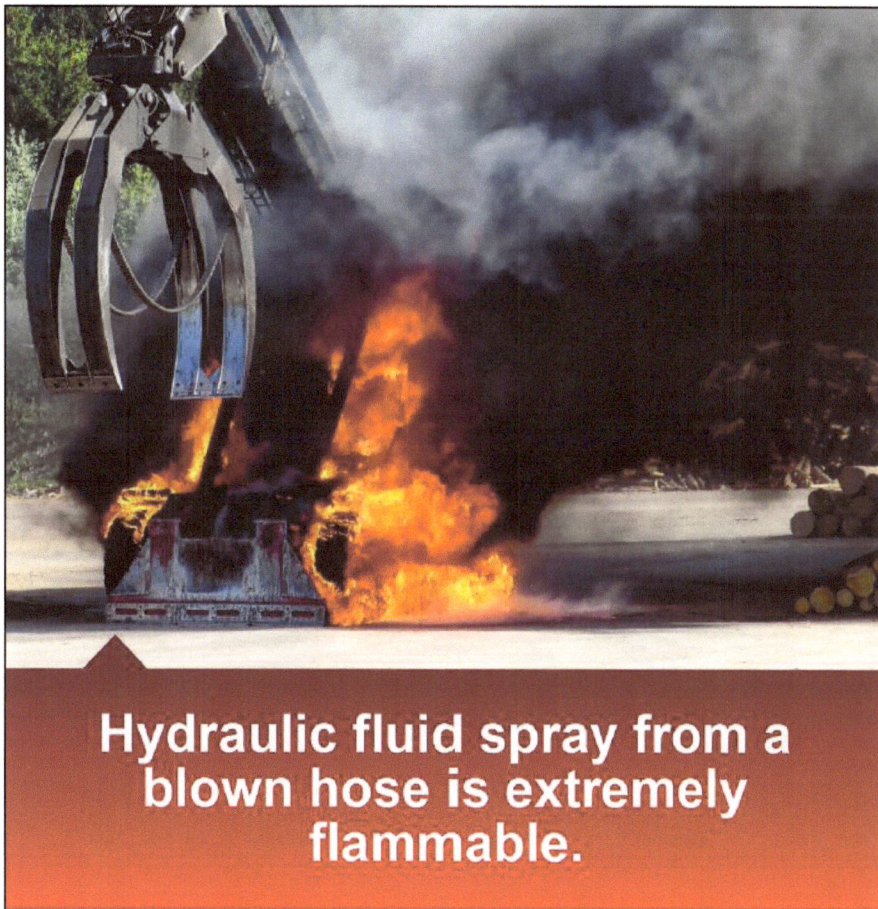

**Fig. 1.22 – Hazard from Burning Hydraulic Fluid
(Courtesy of the International Hydraulic Safety Authority)**

Source 2: Static Electric Discharge
- Fluid circulating through fluid conductors can generate static electric charge. Hose ends may be electrically conductive due to reinforcing layers
- Oil moving around inside the reservoir in mobile machines also creates static charges.
- Hydraulic systems located near high amperage electrical lines.
- When these *static charges* accumulate, it may create sparks that can ignite system fluids or gases in the surrounding atmosphere.

Solutions:
- Use *nonconductive hoses* (1) where there are potential electrical power lines and other hazards. Typically, the *non-conductive* hose is orange in color.
- Properly ground the metallic reservoir or isolate it by rubber basses (2).
- Use plastic or fiberglass type reservoirs (3) for mobile applications.

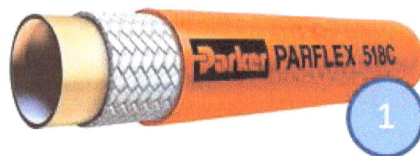

Non-Conductive high density fiber braid reinforced synthetic rubber hose for use in hydraulic systems where high voltage is present.

Fig. 1.23 – Solutions to avoid Electric Static Discharge

Source 3: potential sources of ignition

Solution:
Locate or shield potential sources of ignition, such as electric motors, hot surfaces and open flames away from hydraulic equipment.

1.7.7- Adequately Size and Select Hydraulic Transmission Lines

Improper sizing and selection of a hydraulic transmission line could be a good source of hazard. A system designer must be aware of:

- Different methods of sizing a hydraulic transmission line including rules of thumb, equations, charts, and software. The next volume of this textbook series provides these best practices.
- Best practices of selecting a type of the hydraulic line (pipe, tube, and hoses) depends on the line size, application, and pressure rating.
- Trusted sources of supplying hydraulic lines. Figure 1.24 (A) shows clearly the difference between standard and commercial hydraulic transmission lines.
- The **ISO Standard 3457** states that a hydraulic machine operator must be protected from any hose assembly located within 3 feet. One way to do that is to place the hose inside a protective sleeve. As shown in Fig. 1.24 (B), the Gates company patented a solution called *LifeGuard*. It is a sleeve surrounding the hose and does not affect its flexibility.
- Figure 1.24 (C) shows list of safety hazards as a result of improper selection of transmission lines and fittings.

International Hydraulic Safety Authority

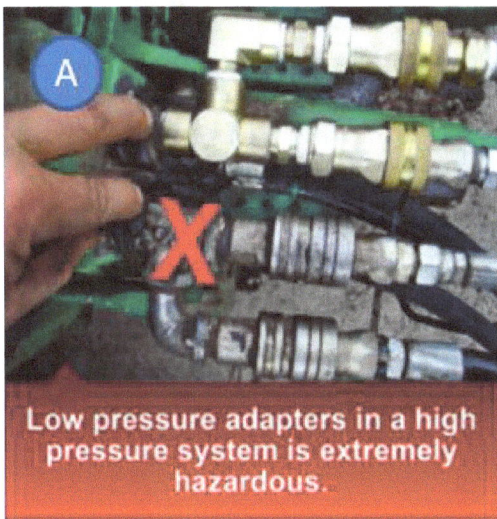

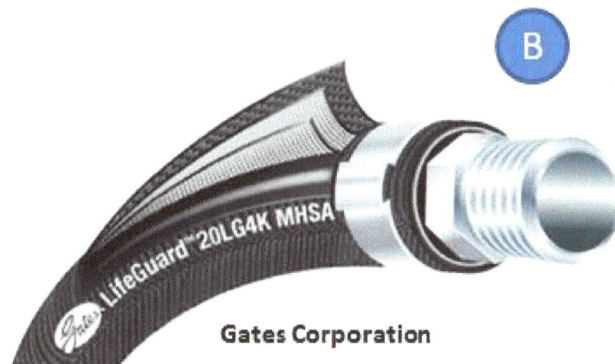

Low pressure adapters in a high pressure system is extremely hazardous.

Gates Corporation

Parker Safety Guide
⚠ **for Selecting and Using Hose, Tubing, Fittings and Related Accessories**
Parker Publication No. 4400-B.1
Revised: November, 2007

WARNING: Failure or improper selection or improper use of hose, tubing, fittings, assemblies or related accessories ("Products") can cause death, personal injury and property damage. Possible consequences of failure or improper selection or improper use of these Products include but are not limited to:

- Fittings thrown off at high speed.
- High velocity fluid discharge.
- Explosion or burning of the conveyed fluid.
- Electrocution from high voltage electric powerlines.
- Contact with suddenly moving or falling objects that are controlled by the conveyed fluid.
- Injections by high-pressure fluid discharge.

- Dangerously whipping Hose.
- Contact with conveyed fluids that may be hot, cold, toxic or otherwise injurious.
- Sparking or explosion caused by static electricity buildup or other sources of electricity.
- Sparking or explosion while spraying paint or flammable liquids.
- Injuries resulting from inhalation, ingestion or exposure to fluids.

Fig. 1.24- Transmission Lines Safety Considerations

1.7.8- Adequately Size and Select Hydraulic Components

Components such as valves, filters, heat exchangers, manifolds, transmission lines, etc. are sized based on the maximum expected flow across the component. System designers frequently make a wrong assumption. They assume "the maximum flow in the circuit is the pump flow!" Maximum flow in the circuit exceeds the pump flow in some cases, e.g.:

- During retracting a differential cylinder, such as in cycling a cylinder in machine tools.
- Flow surges after releasing high-pressure acting on a large volume of oil, such as in hydraulic loads.
- Flow surges due to overrunning loads, such as in earth moving machines.
- Flow surges from accumulators, such as in die-casting and injection molding machines.

While under-sizing valves will make them act like throttles creating excessive pressure drop regardless of the basic valve function, oversizing valves will reduce their controllability.

1.7.9- Avoid Pump Cavitation

As shown in Fig. 1.25, pump *cavitation* is a major cause of pump failure and consequently the rest of the system. A system designer must be aware of the best practices of a hydraulic system design, installation, and operation to eliminate possibilities of developing cavitation. These best practices will be discussed in the next volume of this textbook series and includes:

- Reservoir design and placement.
- Pump intake line sizing and installation.
- Protection instruments to monitor the vacuum conditions at the pump intake.
- Fluid selection.
- Pump driving speed.
- Maintaining recommended working temperature.

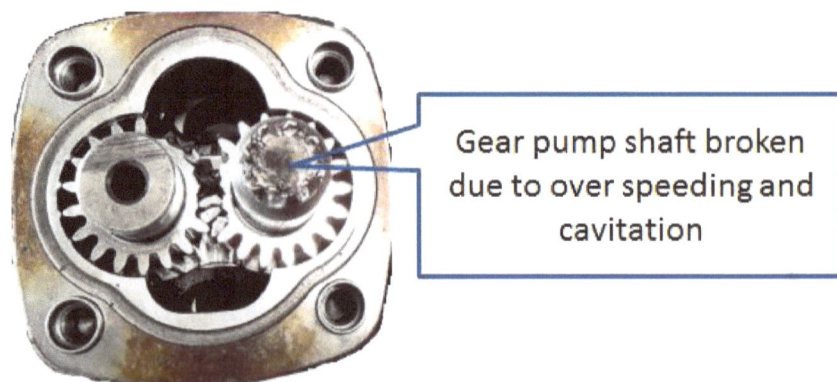

Gear pump shaft broken due to over speeding and cavitation

Fig. 1.25 – Hazards from Pump Cavitation

1.7.10- Carefully Design Hydraulic Fluid Contamination Control Systems

Knowing that 80% of the hydraulic components failure is due to contamination justifies the importance of proper design of the filtration system for the safe operation of a hydraulic system. Hazards from hydraulic fluid contamination are not limited to just wear in components, it could result in a disaster. An example of that is shown in Fig, 1.26, imagine what could happen as a result of blocking a of a servo valve nozzle in an airplane landing gear system! A hydraulic system designer must be aware of the best practices for controlling the contamination level of a hydraulic system. These best practices will be discussed in chapter 4 in this textbook.

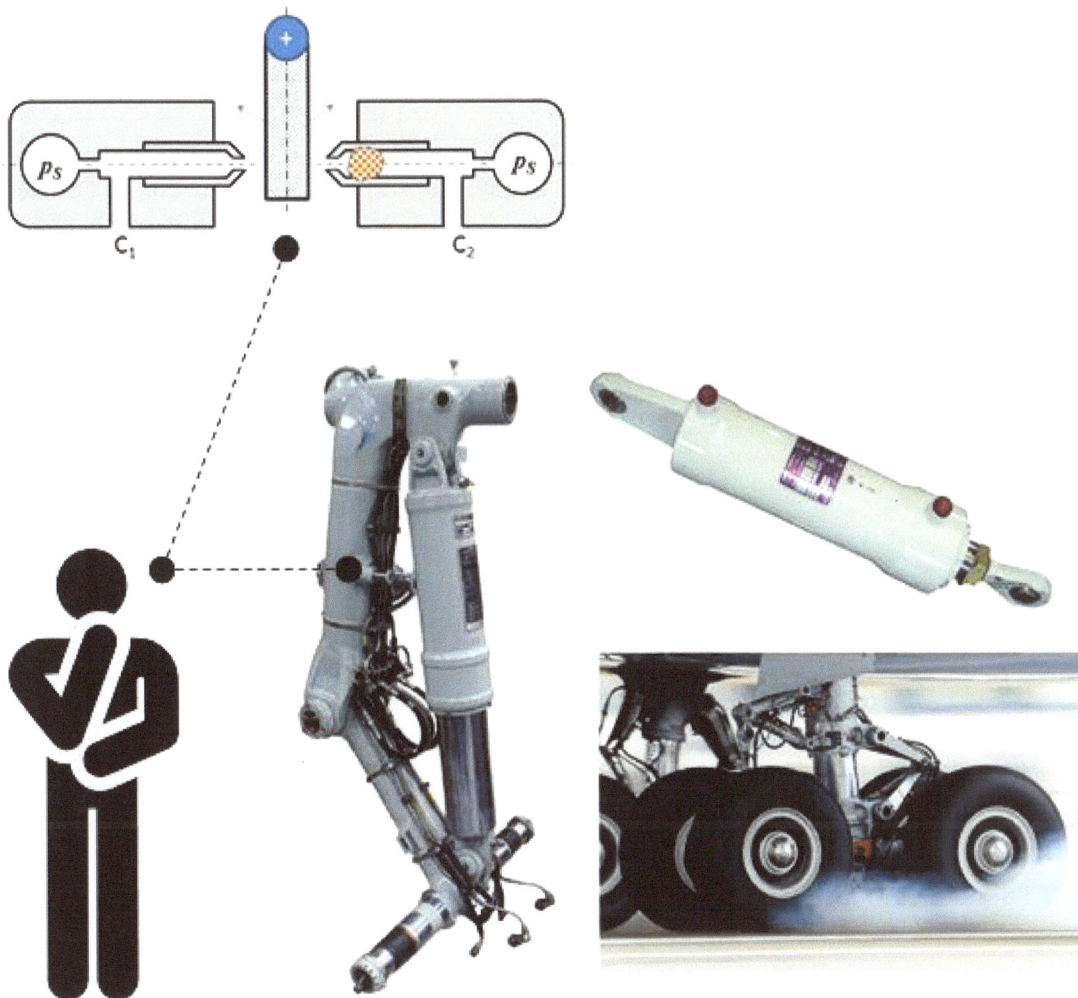

Fig. 1.26 – Hazards from Hydraulic Fluid Contamination

1.7.11- Minimize Noise and Vibration of the System

Hydraulic system vibration gradually loosens the fittings and subjects the system and transmission line to fatigue failure. Hydraulic system noise can also cause long-term personal disability. Structure-borne, fluid-borne and air-borne noise are due to many things such as pump-motor installation, pressure ripples from positive displacement pumps, layout of transmission lines, etc. In the design of hydraulic systems, these sources should be considered to minimize the risks caused by noise.

Best practices for minimizing the *Noise and Vibration* in a hydraulic system is <u>out of the scope of this textbook</u>. However, the following examples present a quick idea of how to minimize the hydraulic system noise and vibration.

Example 1: Using an Accumulator
Using an accumulator to minimize noise and vibration in a hydraulic system is classical and a very well-known solution. As shown in Fig. 1.27, an accumulator (1) can be located downstream of the pump to remove pressure ripple due to pulsating flow. The figure shows also using an accumulator (2) to remove pressure spikes in an automotive shock absorber system.

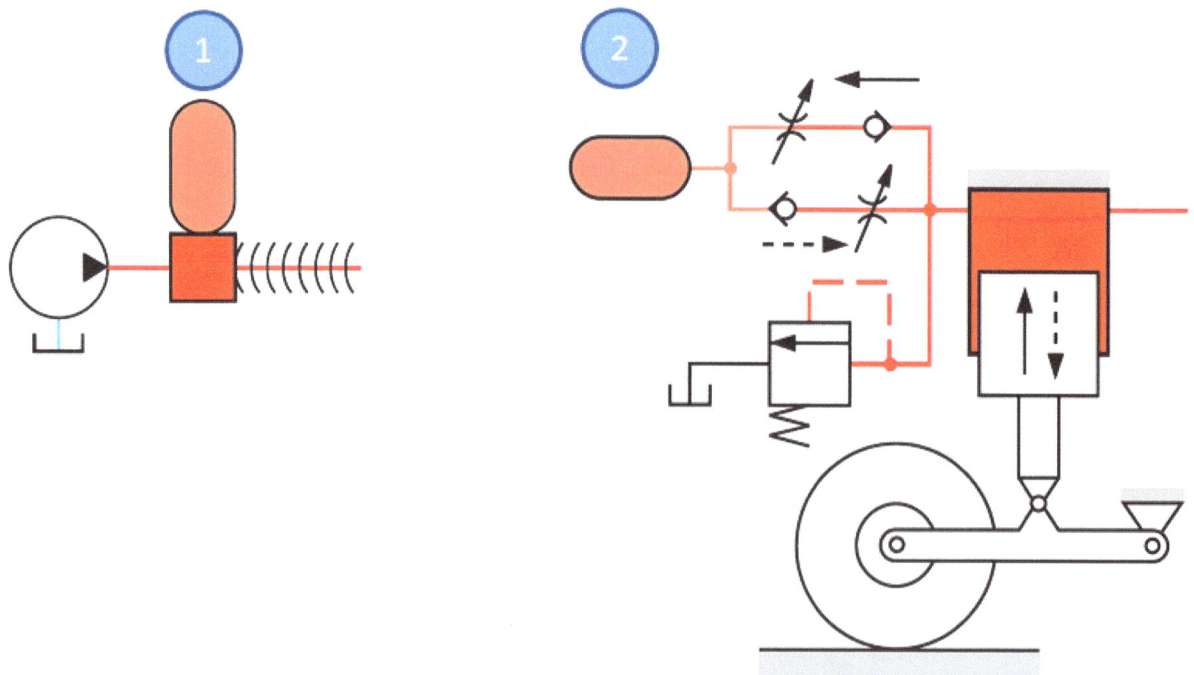

Fig. 1.27 – Using an Accumulator to Minimize Noise and Vibration (Courtesy of Bosch Rexroth)

Example 2: Suppress Pressure Pulsations in Hydraulic Lines

During changing the direction of motion of an actuator, pressure pulsations are sent through the lines causing undesirable noise and vibration. Figure 1.28 shows a cutaway in Parker's Pulse-Tone shock *suppressor*. It consists of an inner radial chamber with a series of 0.5-in. diameter holes, a compressed coil spring surrounding the inner chamber, an outer radial chamber dotted with 0.03in. diameter holes, and an elastomeric bladder around the outer chamber. A 0.25-in. gap separates the chambers. In operation, oil flows through the inner radial chamber, spring, and outer chamber. The 0.03-in. diameter holes maximize flow but keep the bladder from extruding through the outer chamber. Pulsations pass through the holes, strike and deflect the nitrogen-charged bladder. This deflection of the bladder reduces shock and noise.

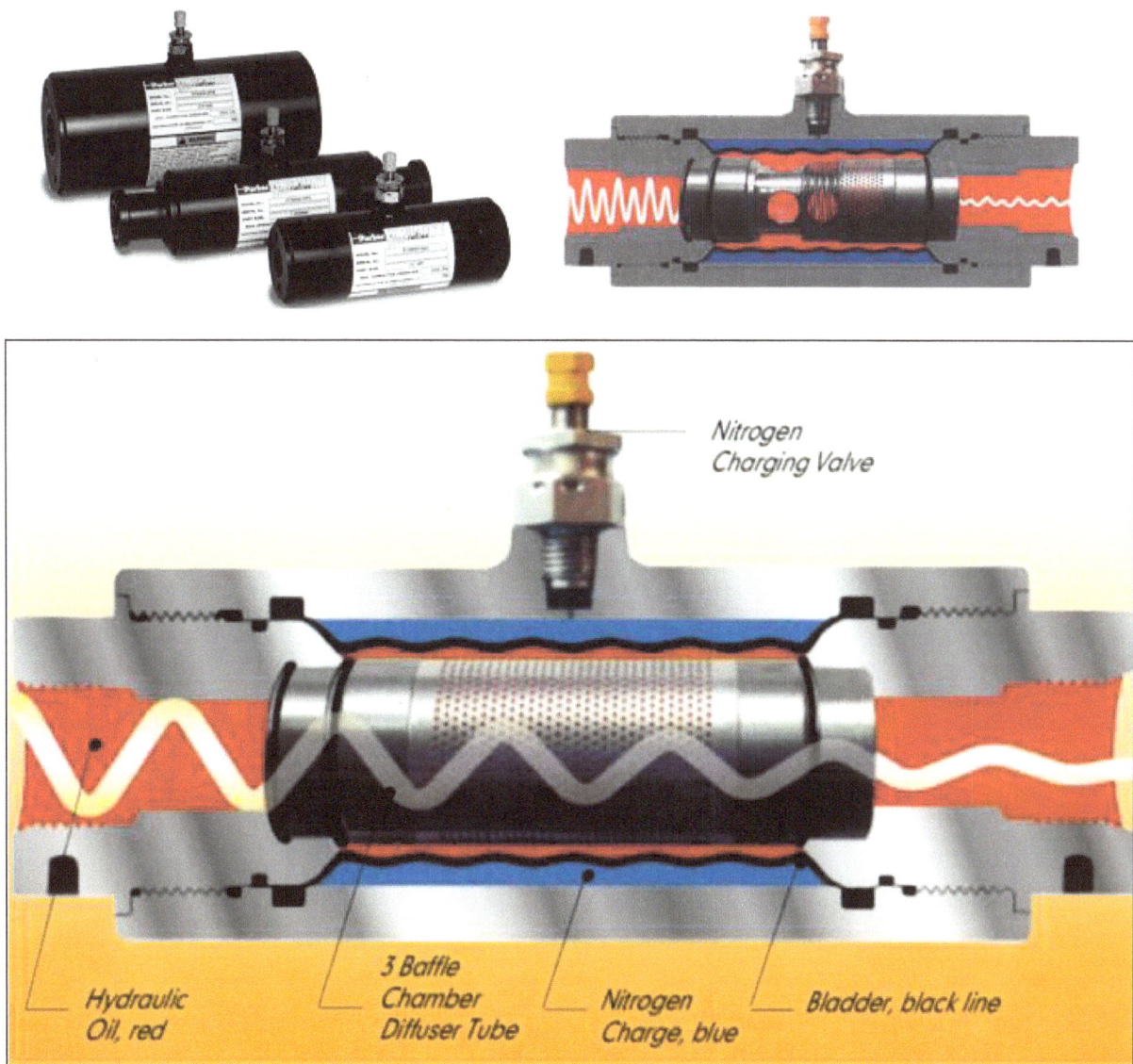

Nitrogen Charging Valve

Hydraulic Oil, red

3 Baffle Chamber Diffuser Tube

Nitrogen Charge, blue

Bladder, black line

Fig. 1.28 – Using a Shock Suppressor to Minimize Noise and Vibration (Courtesy of Parker)

1.7.12- Apply Energy Saving Design Strategies for Hydraulic Systems

Energy waste in a hydraulic system is the source that generates the most heat. Overheating is the second important source after contamination that causes hydraulic system failure. Best practices for "Energy Saving Design Strategies" are out of scope of this textbook. However, Fig. 1.29 shows some examples to explore simple solutions for energy saving in a hydraulic system such as: using a pressure compensated pump instead of a fixed pump (1), using over-center pump to control motion of an actuator rather than a servo or proportional valve (2), and using manifolds rather than plumbing (3).

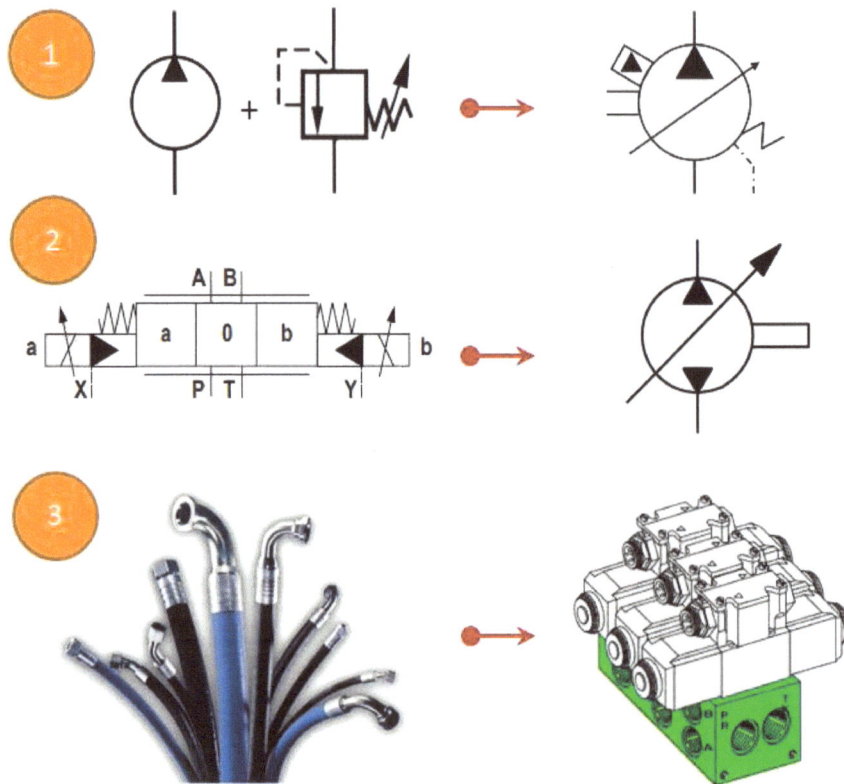

Fig. 1.29 – Examples of Energy Saving Solutions

1.7.13- Apply Fail-Safe Design Strategies for Hydraulic Systems

A system designer must consider, if a hydraulic system fails, how to make it fail safely considering all possible modes of failure. Best practices for "Fail-Safe Design Strategies" are out of the scope of this textbook. However, the following examples explore some ideas of fail-safe design concepts.

Example 1: Emergency Stop

As shown in Fig. 1.30, every hydraulic system must be equipped with a readily accessible *Emergency Stop*. Emergency stops must be energized independent of the rest of the machine.

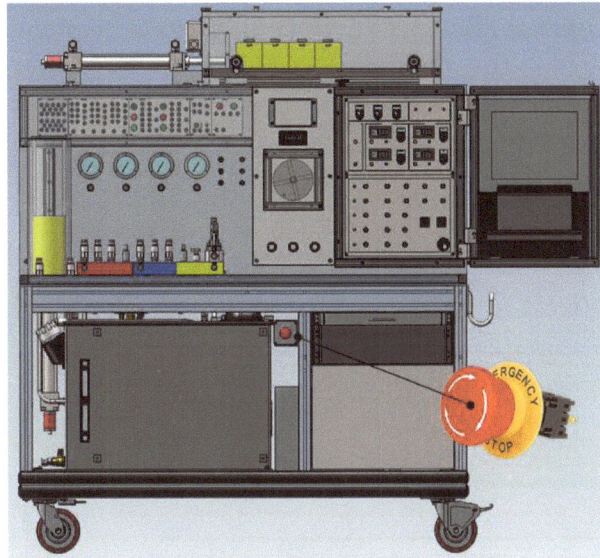

Fig. 1.30 – Emergency Stops

Example 2: Manual Override for Directional Valves

Solenoid-operated directional valves should be equipped with a *Manual Override* device. As shown in Fig. 1.31, the spool of the valve can still be shifted manually in case of power outage or loss of signal.

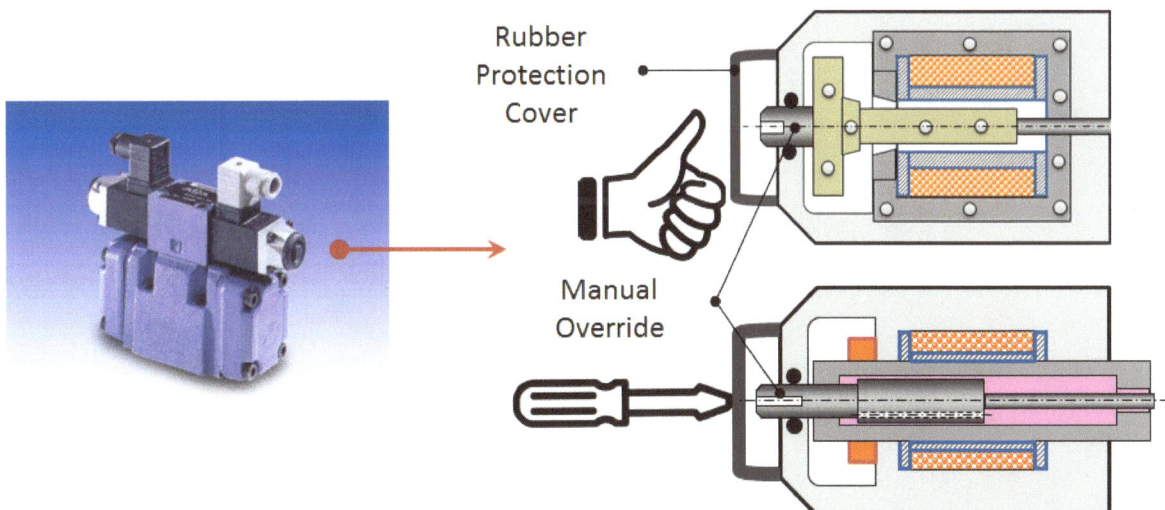

Fig. 1.31 – Manual Override for Solenoid-Operated Hydraulic Directional Valves

Example 3: Emergency Control

For additional safety, an *Emergency Control System* is used to drive the system towards a safe working condition (lock and stop, complete a cycle, or return). The emergency control could be a separate or part of the original system. The emergency control system shall be triggered automatically in case of emergency. The machine tool electro-Hydraulic control system, shown in Fig. 1.32, presents an example of that concept. Despite that the system is equipped with an emergency switch (S2), if the drilling force exceeds certain maximum limit, the drilling cylinder will automatically return and then the work piece will be unclamped. The pressure switch (PS) sets the maximum drilling force.

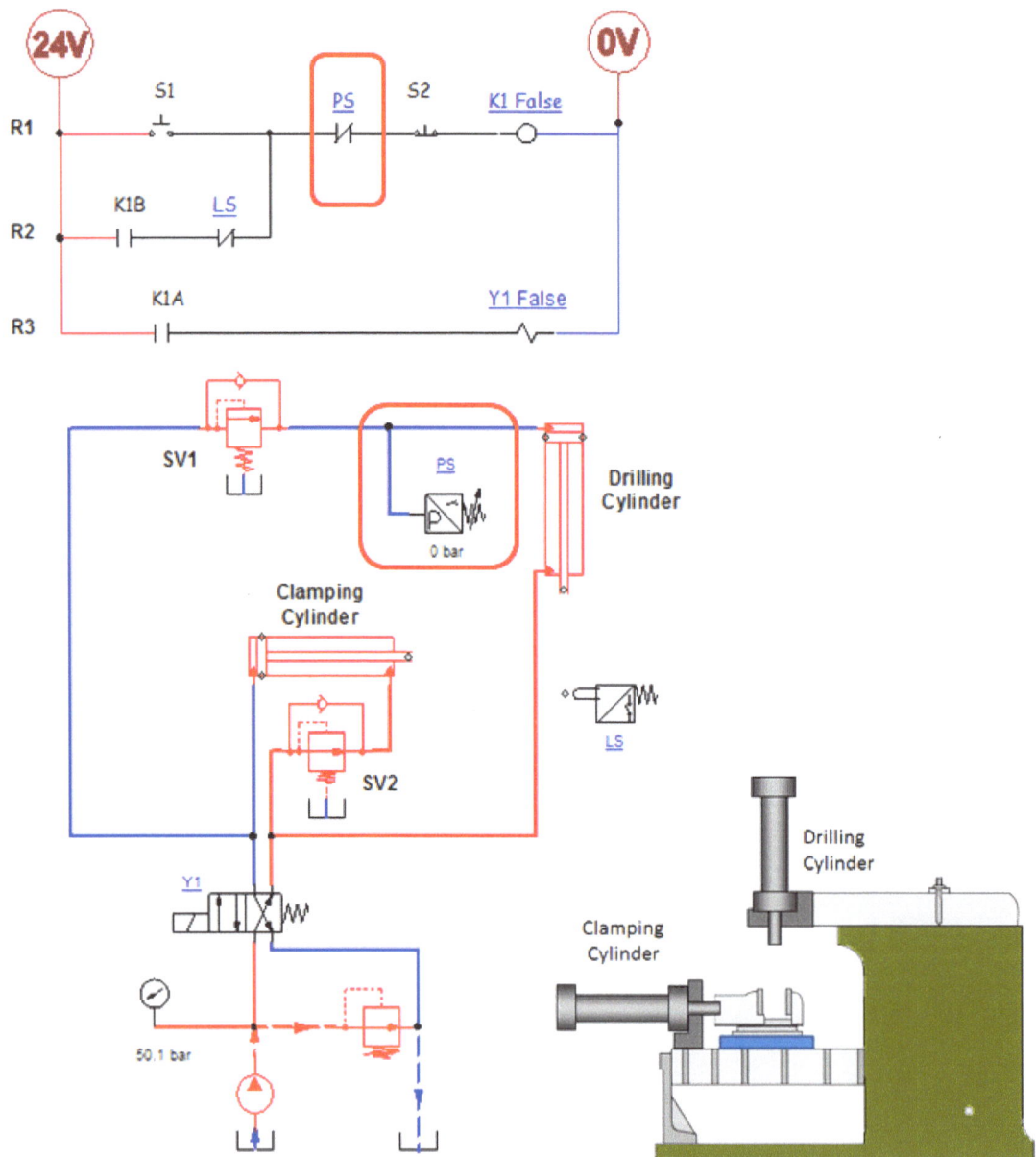

Fig. 1.32 – An Example of Emergency Control Concept in a Machine Tool

Example 4: Automatic Release of Stored Energy

Any form of *Stored Energy* in the system must be released in case of power outage or machine shutdown. The importance of that is to avoid unexpected movement of hydraulic actuators when the power returns. An example of this, is shown in Fig. 1.33, the normally-opened solenoid-operated 2/2 directional valve is used to assure automatic discharge of the accumulator in case of power outage or machine shutdown. If the accumulator is vented manually by a bleed valve, complete information should be given on or near the accumulator as well as on the circuit diagram. Review **OSHA Standard 3120** for proper labeling of stored energy equipment.

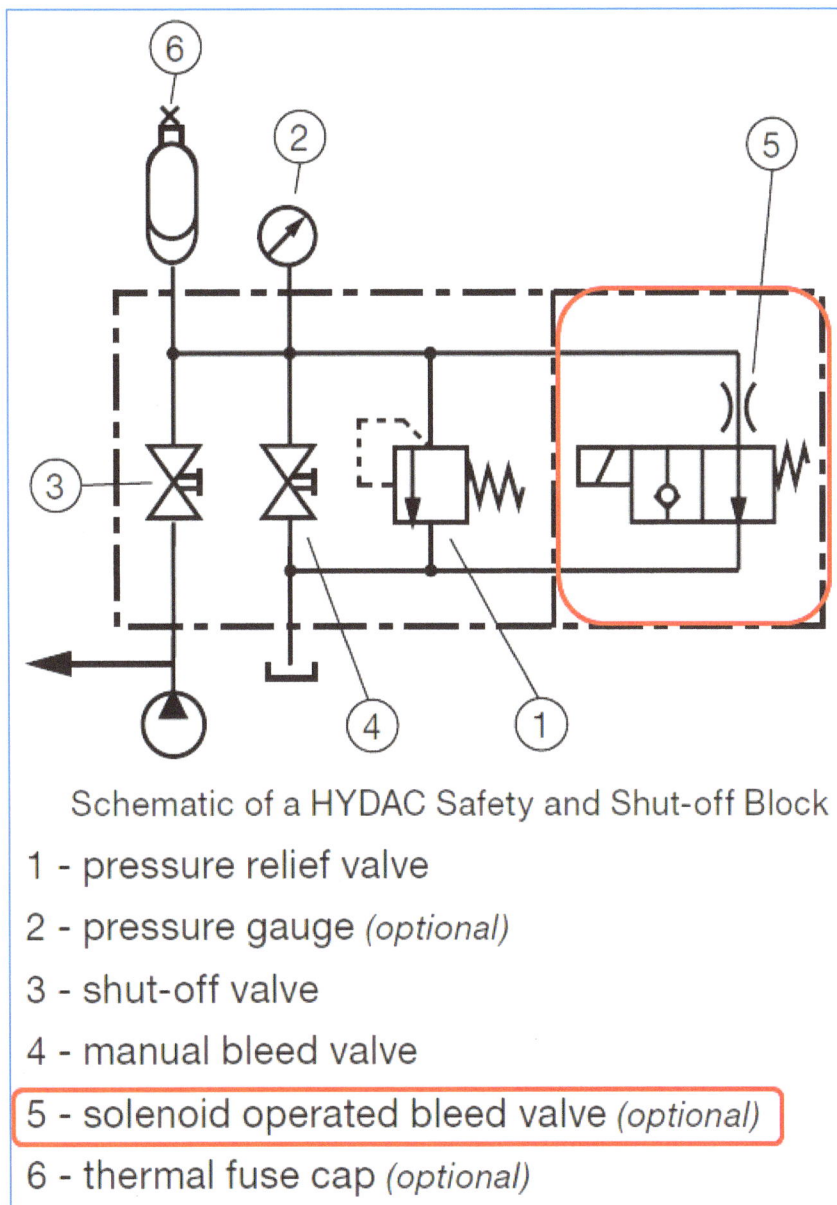

Schematic of a HYDAC Safety and Shut-off Block

1 - pressure relief valve

2 - pressure gauge *(optional)*

3 - shut-off valve

4 - manual bleed valve

5 - solenoid operated bleed valve *(optional)*

6 - thermal fuse cap *(optional)*

Fig. 1.33 – Automatic Release of Stored Energy

Example 5: Secure Overrunning Loads

Any form of overrunning load in the system must be secured in case of emergency. A hydraulic holding valve should be used to hold the load in case of pump failure. Various holding valves are used such as a counterbalance valve, throttle valve, or pilot operated check valve.

Figure 1.34 shows an example of using a pilot-operated check valve. The figure shows, regardless the type of holding valve the following design issues must be considered in order to safely secure an *overrunning load*:

- A float center DCV must be used in order to assure that the downstream side of the holding valve is completely vented. If any residual or back pressure is left at the downstream side of the valve, it could result in load creep.

- It is highly recommended to mount the valve directly on the actuator's body. If there is no way but to connect the valve to the actuator by hydraulic lines, DO NOT use flexible hoses. The hydraulic lines must be hard tubing made from appropriate material.

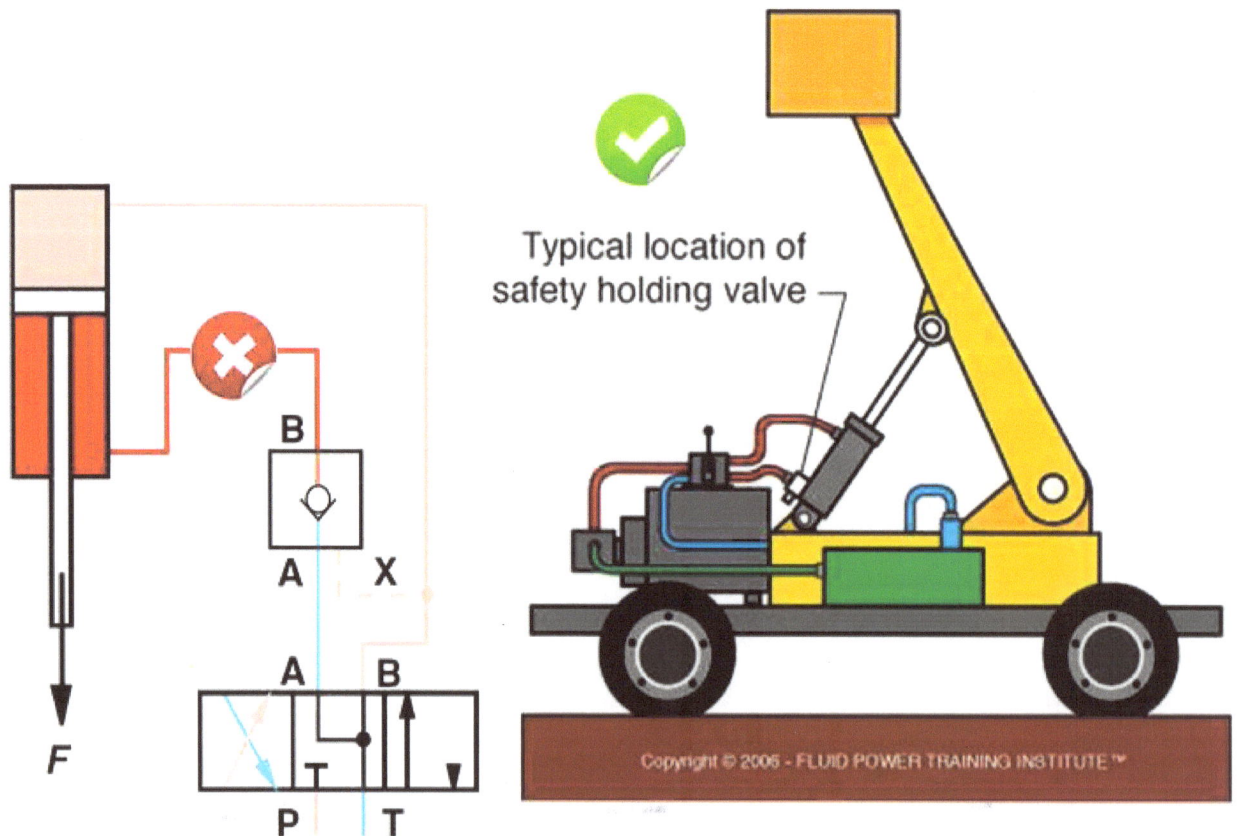

Fig. 1.34 – Securing Overrunning Load

1.7.14- Properly Design Condition Monitoring System

Hydraulic systems are non-transparent; It is not known what is going on inside the system unless system conditions are monitored using appropriate *measuring instruments*. Figure 1.35 shows various instruments to measure pressure, temperature, flow, oil level, linear position, linear speed, acceleration, RPM, vibration, sound level, etc. Gauges or sensors can measure each of the aforementioned working conditions. The next volume of this textbook series presents detailed explanation about the construction and operation of measuring instruments commonly used for condition monitoring of hydraulic systems.

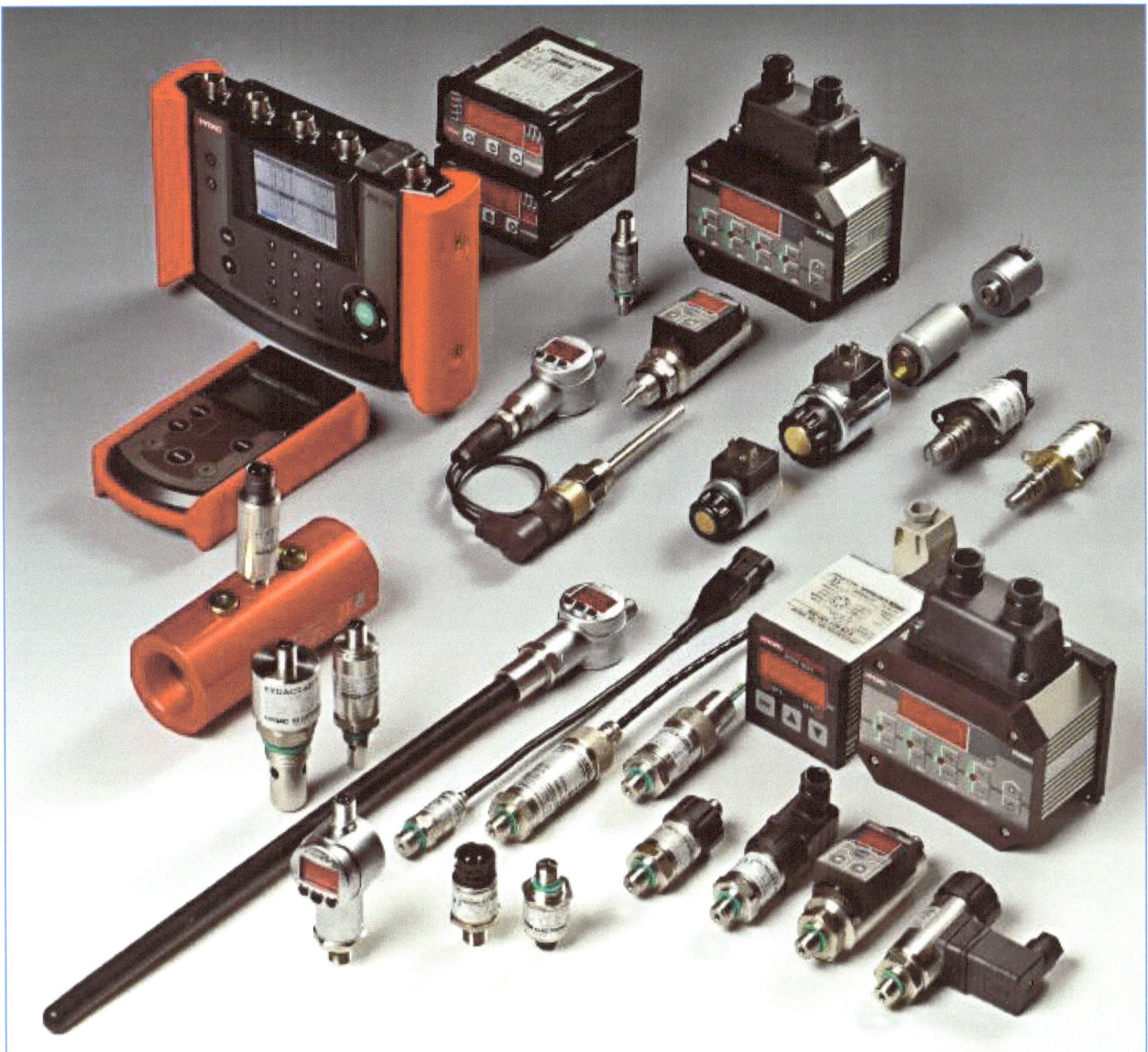

Fig. 1.35 – Hydraulic System Condition Monitoring Devices (Courtesy of Hydac)

1.7.15- Perform Dynamic Analysis Whenever Needed

Designing some hydraulic systems using time-invariant (*Steady State*) equations may not be sufficient to predict system response. Such systems should be designed using dynamic analysis, modeling and simulation to avoid unsafe and unexpected system responses. The following examples show the importance of *dynamic analysis* for some applications:

Example 1: Pressure Spikes in Highly Dynamic Systems

Figure 1.36 shows a hydraulic system that reciprocates a hydraulic cylinder while it drives a large inertial load. The steady state solutions can't see the pressure spikes created due to the inertial load every time the directional valve is shifted.

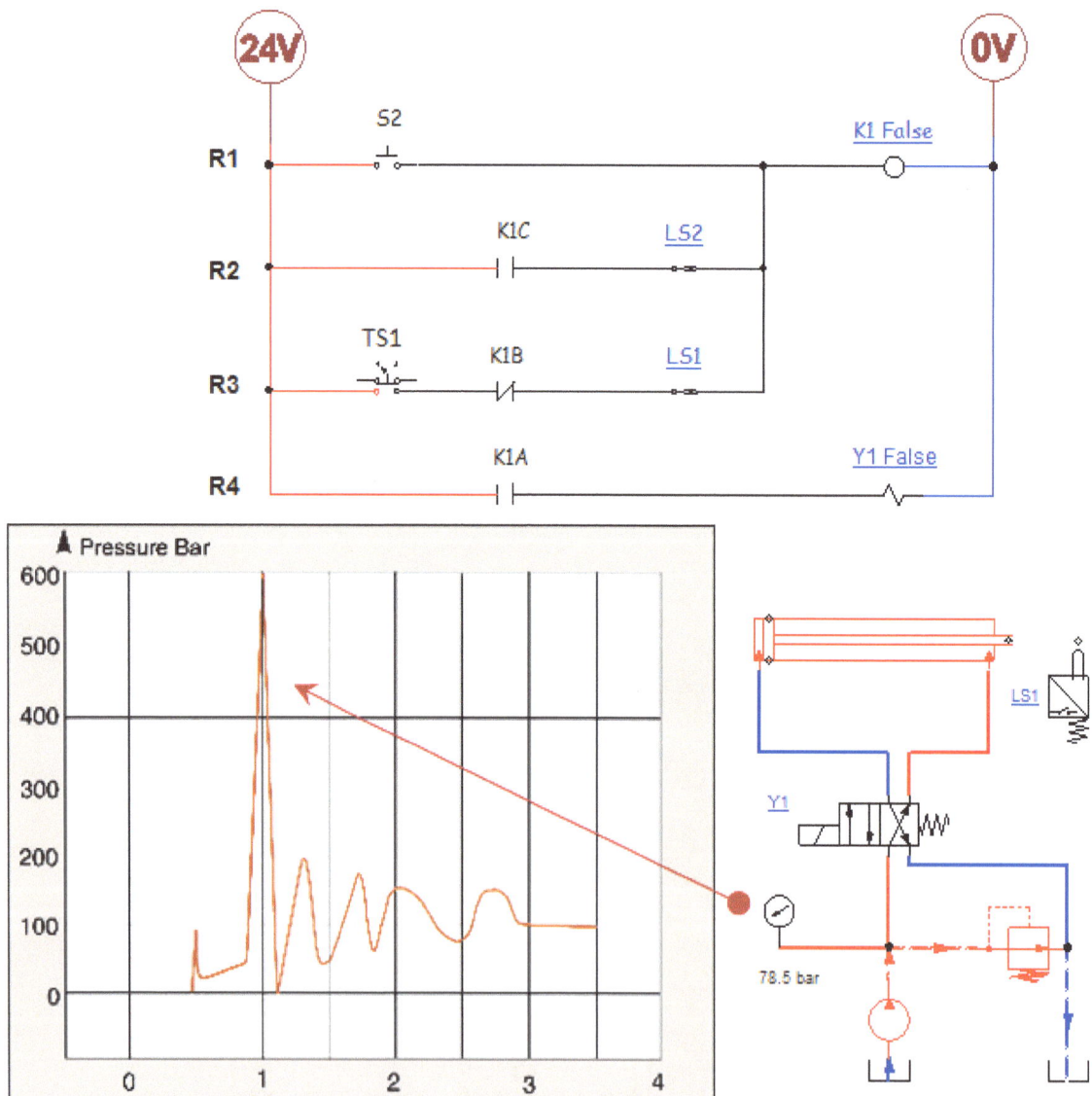

Fig. 1.36 – Pressure Spikes in a Cylinder Reciprocation System

Example 2: Effect of Oil Compressibility in High Pressure Applications

As shown in Fig. 1.37, in a hydraulic press, a large of oil volume is supplied to the ram under high pressure to perform the pressing function. If the system designer assumes that the oil is incompressible, its volume won't change. Such unfeasible assumption, the system designer made, may result in unsafe operation. However, hydraulic fluids are compressible. So, when the oil is depressurized to retract the ram, the oil will expand suddenly causing surge of flow that may cause unsafe operation and also damage the control valves.

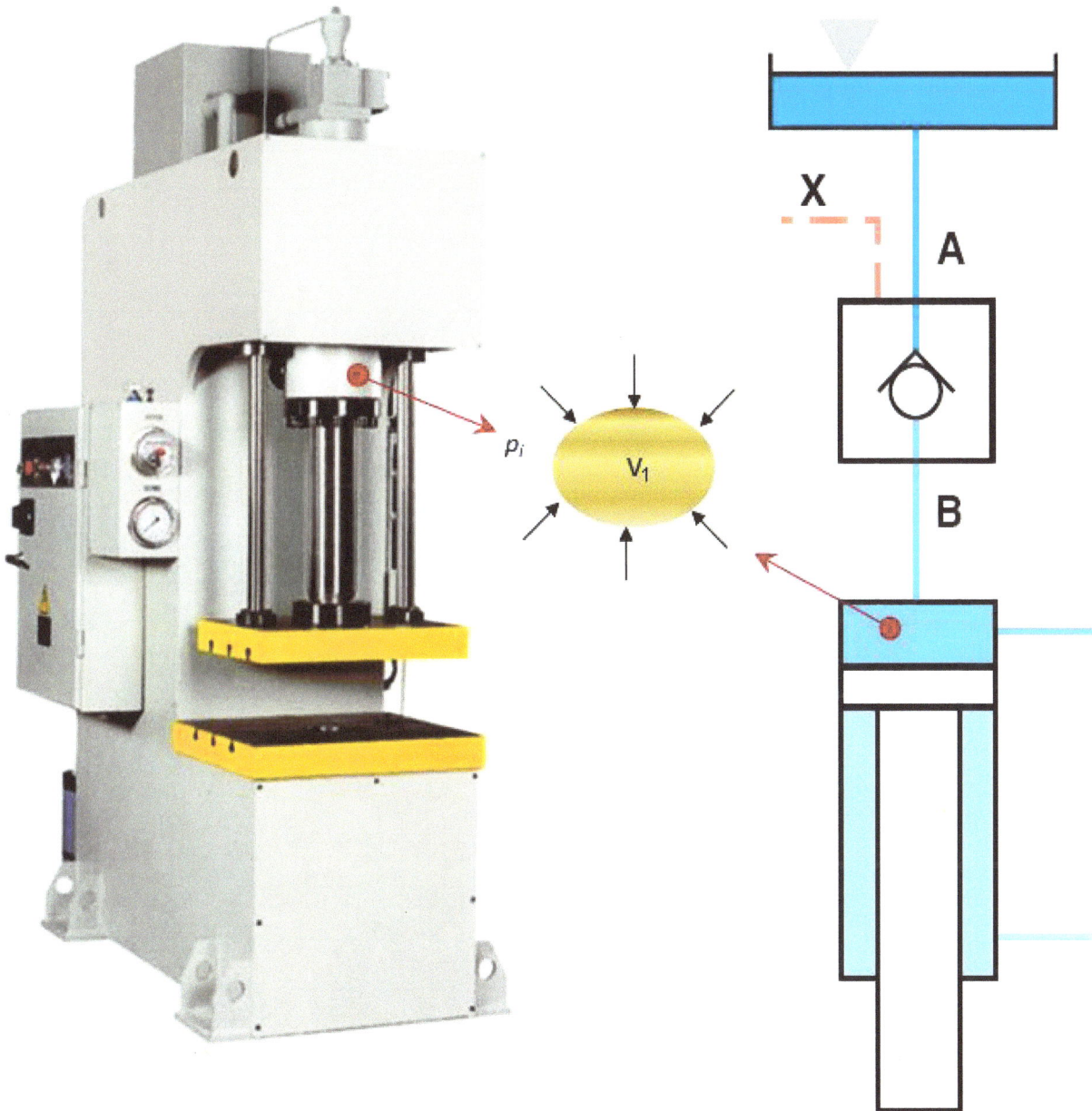

Fig. 1.37 – Effect of Oil Compressibility in High Pressure Applications

1.8- BP-Safety-02: Safety of Hydraulic System Operators

Designing for safe hydraulic systems is not the end of the story! Operating a hydraulic system must guarantee the safety of the operator, work environment, and workspace.

As the safety of the human comes first, the best practices list *BP-Safety-02* presents guidelines for the safety of hydraulic system operators and the required personal protective equipment.

PB-Safety-02:
1. Eye Protection.
2. Ear Protection.
3. Hand Protection.
4. Foot Protection.
5. Head Protection.
6. General Body Protection.

The following subsections provide detailed interpretation of the action items listed in BP-Safety-02.

1.8.1- Eye Protection

Oil leaks can occur without warning. Therefore, all hydraulic system operators must wear proper *Eye Protection* devices, particularly when disconnecting or connecting hydraulic lines. Figure 1.38 shows various types of eye protection devices such as plastic goggles and face shields. *Eye Washer* stations should be marked and made available at the work area.

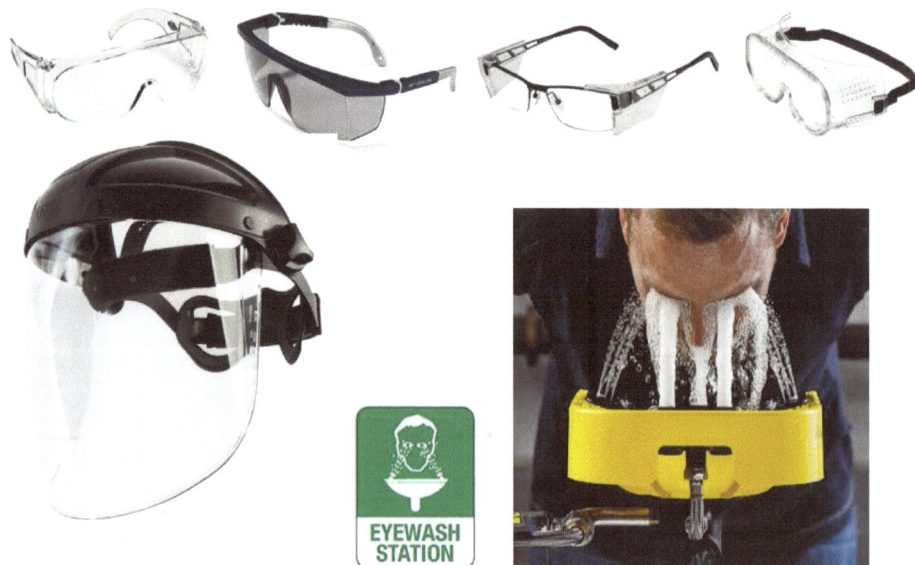

Fig. 1.38 – Eye Protection for Hydraulic System Operator

1.8.2- Ear Protection

Depends on the sound level, hydraulic system operators may be instructed to wear proper *Ear Protection* devices. Figure 1.39 shows various styles:

Earplugs are the most popular, inexpensive, and very commonly used by workers. They are made from rubber, plastic or foam.

Earmuffs are larger, more expensive, more effective, and more comfortable than earplugs.

Helmets are the most expensive form of hearing protection. They are usually used only in the most severe noise conditions or where a combination of a hardhat and hearing protection is required.

Earplugs   Earmuffs

Helmets

Fig. 1.39 – Ear Protection for Hydraulic System Operator

1.8.3- Hand Protection

Again, as shown in Fig. 1.40, oil leaks can occur without warning. Continual skin contact with hydraulic fluid is to be avoided. Careful skin cleaning of sticky fluid is required. Therefore, as shown in Fig. 1.41, all hydraulic system operators must wear special industrial *gloves* against oil injection, particularly when disconnecting hydraulic transmission lines.

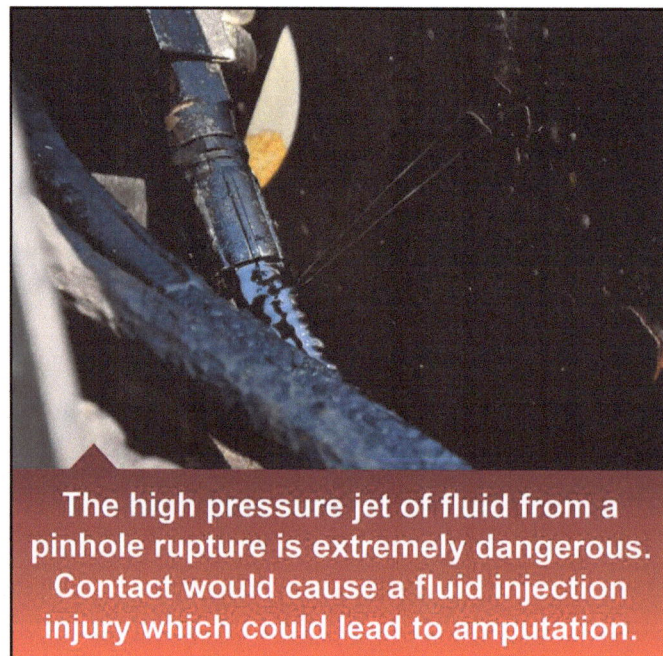

The high pressure jet of fluid from a pinhole rupture is extremely dangerous. Contact would cause a fluid injection injury which could lead to amputation.

**Fig. 1.40 – Jet of Fluid at High Pressure from a Pinhole
(Courtesy of the International Hydraulic Safety Authority)**

Fig. 1.41 – Industrial Gloves against Oil Injection

1.8.4- Foot Protection

Hydraulic components are heavy and can cause injuries if they fall accidently. Therefore, as shown in Fig. 1.42, hydraulic system operators must wear approved *safety shoes*. Open footwear is forbidden in the work area.

Fig. 1.42 – Industrial Safety Shoes

1.8.5- Head Protection

As shown in Fig. 1.43 *Safety helmets* (hard hats) are used to protect the head from injuries caused by the impact of falling or moving objects. Such hard hats are generally made from plastic that provide limited protection from heat and electrical shock. It can also be equipped with ear and eye protection.

Fig. 1.43 – Industrial Head Protection Equipment

1.8.6- General Body Protection

The following are general guidelines for body protection:

- Always remove rings, watches and jewelry before working with hydraulic systems.
- Do not wear ties or other loose pieces of clothing that can get between moving parts.
- Most hydraulic fluids are toxic. Therefore, food is forbidden in the work area.
- Hydraulic fluids can become HOT! Be aware of burn hazards.
- Hydraulic actuators can generate high forces and torques. DO NOT underestimate the load! Always remember that hydraulic power generates considerable forces.
- As shown in Fig. 1.44, avoid being in wrong position. Do NOT expose yourself to possible movement of a hydraulic actuator. DO NOT enter under hydraulically supported equipment unless it is mechanically locked in place.

**Fig. 1.44 – Avoid Being in Wrong Position
(Courtesy of Fluid Power Training Institute)**

Reported Case History:

Action: Figure 1.45 shows an assembly-line worker who was installing a hydraulic steering cylinder on a front-end loader.

Result: An accident claimed four of his fingers.

Reason: Improper air bleeding. Worker was not aware of crushing hazard. Cylinder moved erratic pinning his hand in-between the rod-eye and the frame anchor.

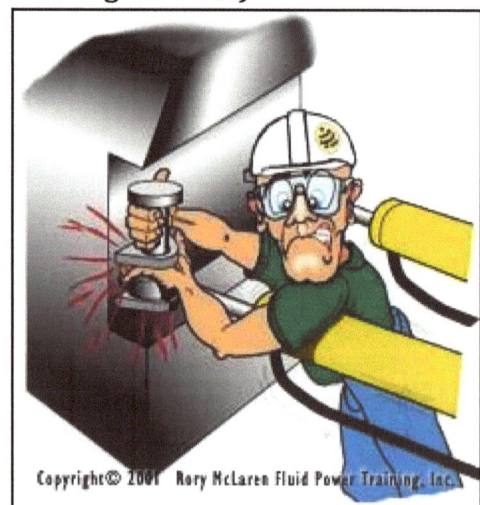

**Fig. 1.45 – Cylinder Crushing Operator's Hand
(Courtesy of Fluid Power Training Institute)**

1.9- BP-Safety-03: Safety of Hydraulic System Work Environment

Take care of your employees; they will take care of you. The effort you spent on improving the work environment increases production via increasing the comfort level of employees. Controlling the work environment is a vital step for preventing cumulative injuries. *Cumulative Injuries* are those injuries that occur from long-term exposure to unsafe environmental conditions. Work environment conditions must meet or exceed state and federal standards, including those of the Occupational Safety and Health Administration (OSHA).

The best practices list *BP-Safety-03* presents guidelines for the safety of a hydraulic system workspace

PB-Safety-03:
1. Air Quality.
2. Ambient Conditions.
3. Light Level.
4. Sound Level.

The following subsections provide detailed interpretation of the action items listed in BP-Safety-04.

1.9.1- Air Quality

Hydraulic system work environment should be adequately ventilated at all times. *Fumes* that are generated by some applications are good sources of health issues. Figure 1.46 shows fumes generated by the steel manufacturing process.

Fig. 1.46 – Fumes at Steel Works

1.9.2- Ambient Conditions

Ambient temperature and humidity are also important work conditions. Controlling such conditions, as much as possible, help reduce accidents by improving attention level during the duration of work.

1.9.3- Light Level

OSHA Standard 1926.56 stated that the minimum lighting requirements for a machine shop is 10 foot-candles (ftc). The unit (ftc) is the amount of *illumination* produced by a candle from 1-foot distance. OSHA requires that all emergency exit routes must be illuminated so that an employee can see the way to exit. This lighting must function even if there is a power failure. Each exit must also be marked with a sign stating "Exit."

1.9.4- Sound Level

Depending on the sound level and the duration of exposure per day, *noise* can cause cumulative damage to the hearing of individuals. Table 1.2 shows the permissible noise exposure level based on standards provided by **OSHA Standard (Act of 1970).** Sound level in work environment can be reduced by:
- Using sound absorbent materials on walls, ceilings, and partitions.
- Constructing sound-reducing enclosures around devices or machines that produce extreme noise.
- If the sound level cannot be reduced below the allowable maximum, ear protection must be provided in the workspace.

Hours/Day	Sound Level (db.)
8	90
6	92
4	95
3	97
2	100
1-1.5	102
1	105
0.5	110
0.25 or less	115

Table 1.2- Permissible Noise Exposure

1.10- BP-Safety-04: Safety of Hydraulic System Workspace

Safe operation of hydraulic systems begins before work starts. To ensure safety of hydraulic system operation, some safety regulations must be applied to the workspace. The best practices list *BP-Safety-04* presents guideline for the safety of a hydraulic systems workspace

PB-Safety-04:
1. General Cleanliness and Organization.
2. OSHA Floor Marking.
3. Medical First Aid.
4. Fire Fighting Equipment.
5. Secured Hazardous Areas.
6. Safety and Job Performance Posters.

The following subsections provide detailed interpretation of the action items listed in the best practices list BP-Safety-04.

1.10.1- General Cleanliness and Organization

General housekeeping has a great influence on the number of accidents and injuries in a hydraulic system workspace. As shown in Fig. 1.47, Workspace must be well organized and cleared from all unsafe conditions such as:

- Open flam or welding spots.
- Environmental temperature higher than (150 °F or 66 °C).
- Dangerous electrical connections.
- Trip hazards.
- Spilled liquids.
- Slippery floor and ladders.
- Standing water.

Concrete floors as foundations should be protected against fluids by being sealed or being painted with fluid-resistant paint.

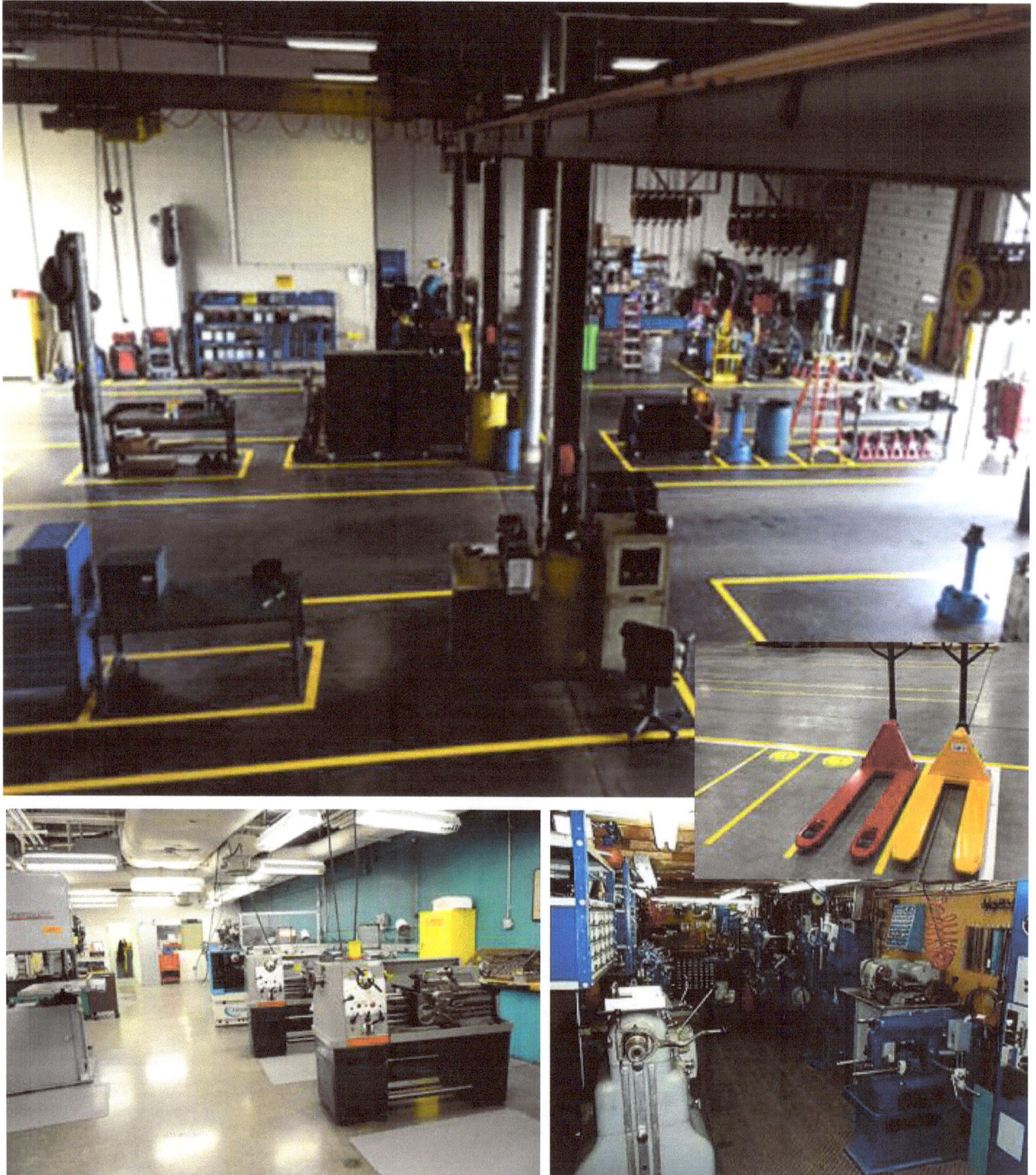

Fig. 1.47 – Clean and Organized Workspace

1.10.2- OSHA Floor Marking

As shown in Fig. 1.48, the floor of the workspace must be marked based on *OSHA Floor Marking Standard*. One of the top OSHA violations is "Walking/ Work Surface Violations" with an average fine of $1,632 per individual violation. These violations are issued when areas where employees walk, or workspace are not clearly marked to identify safe pathways or highlight dangers. *OSHA Standard 1910.22* dictates that all companies mark these areas to prevent accidents.

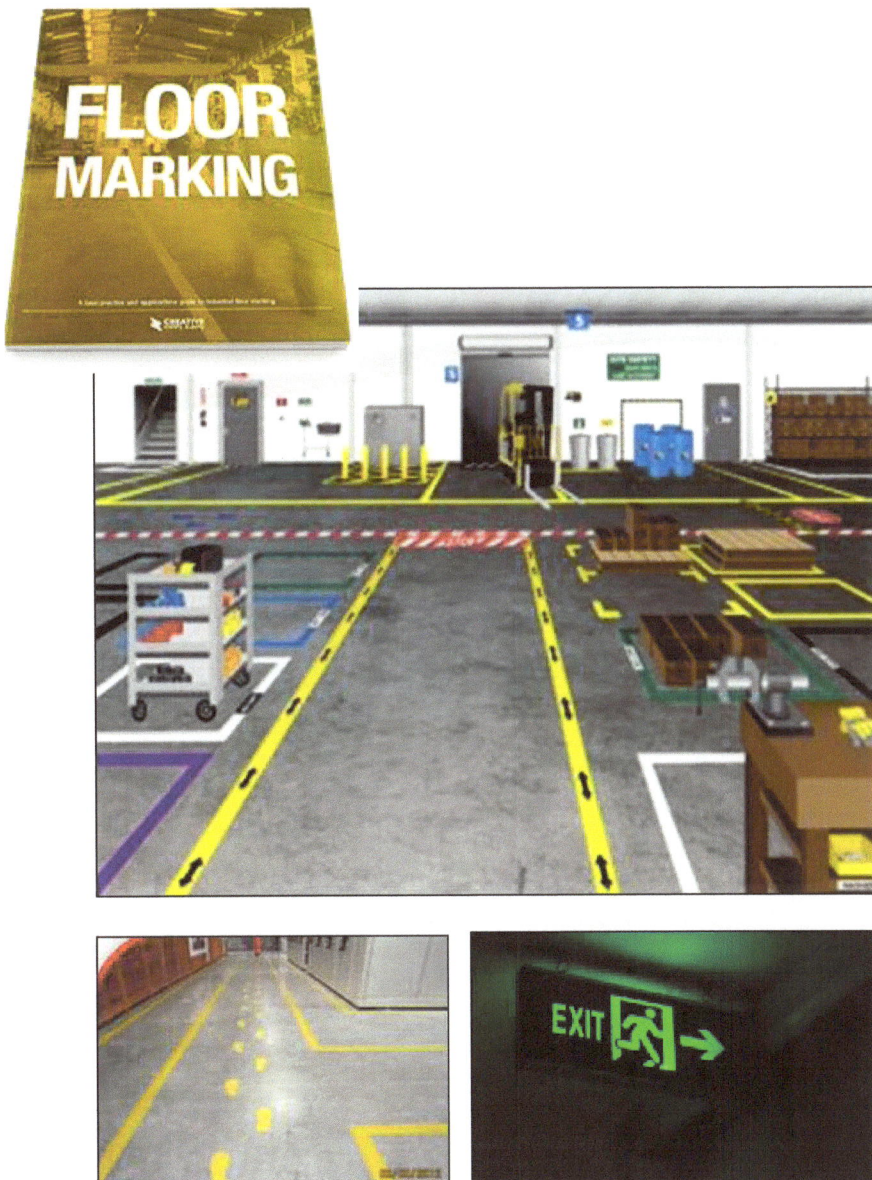

Fig. 1.48 –Workspace Floor Marking

1.10.3- Medical First Aid

Work injuries can happen even in an organized workspace. Rapid response to such work injuries is very important to minimize the consequences of the injuries. Industrial medical *First Aid Kits* should be available as well as appropriate employee training on use at the workspace. Figure 1.49 shows fixed and portable styles. They must be made available in predefined locations within the workspace.

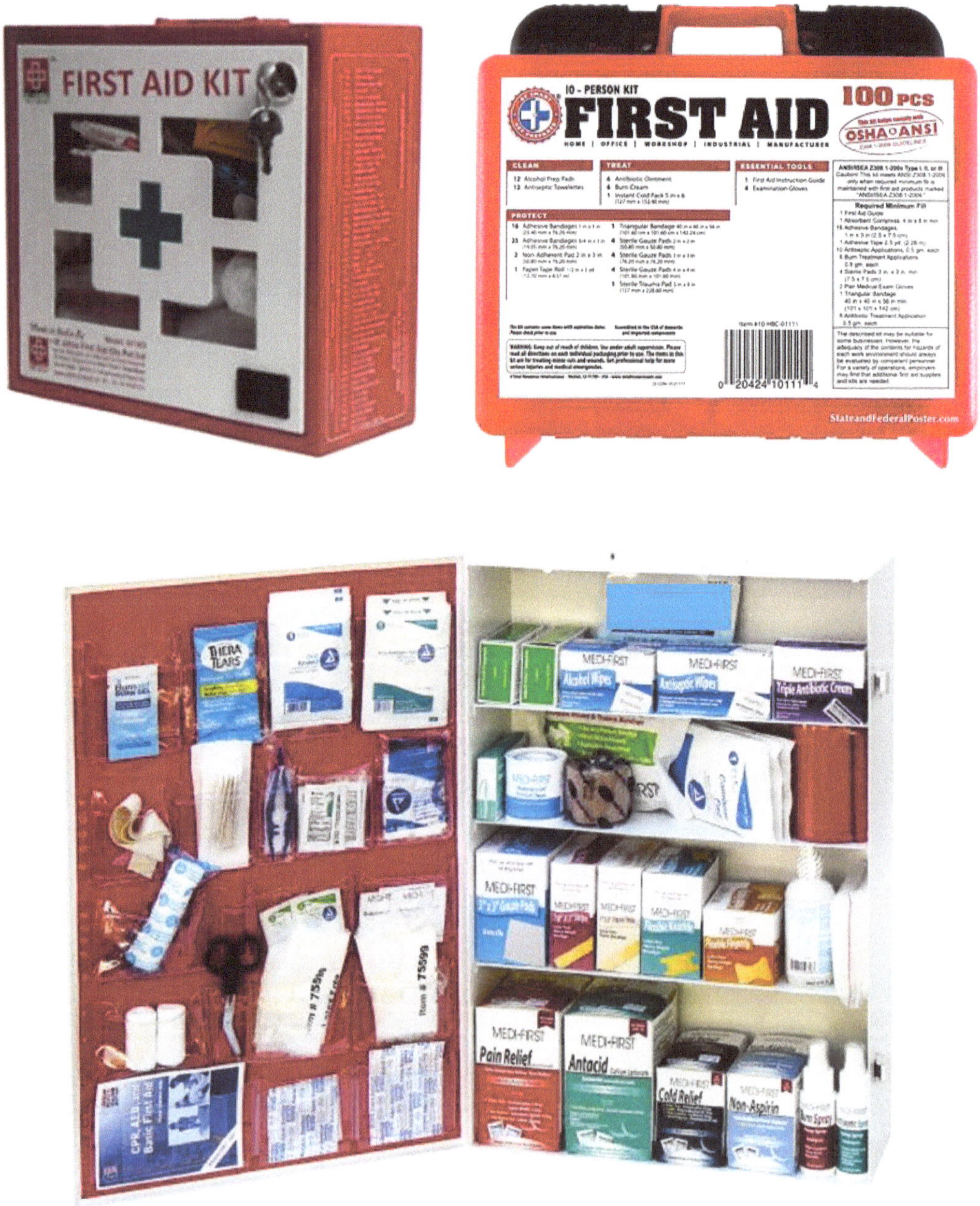

Fig. 1.49 –Industrial First Aid Kits in the Workspace

1.10.4- Fire Fighting Equipment

As shown in Fig. 1.50, following the recommendations of local fire officials, appropriate *Fire-Fighting Equipment* (1) must be available in the workspace and periodically inspected. *Fire blankets* (2) must be readily available and easily accessible. All *Flammable Liquids*, such as chemicals, solvents and fuels, must be stored in metal containers (3) in an area away from heat sources.

Fig. 1.50 – Fire Fighting Equipment in the Workspace

1.10.5- Secured Hazardous Areas

Some of hydraulic driven machines perform hazardous operations such as: Cutting, Pressing, Crushing, Shearing, and Flying Objects. As per **OSHA Standard 1910.307**, hazardous (classified) locations whose operation expose the employees to injuries shall be guarded. As shown in Fig. 1.51, work zones that shouldn't be accessible by non-authorized persons must be secured. Cordon off the area and post warning signs.

Fig. 1.51 – Secured Hazardous Area

1.10.6- Safety and Job Performance Posters

As shown in Fig. 1.52, it is highly recommended to post a number of safety posters within the workspace to remind employee about their safety. Job performance posters to increase the awareness of hydraulic machine operators about their job duties should also posted. Both types of posters should be demonstrative, convey the message to people of different backgrounds, and located in visible places.

Fig. 1.52- Informative Posters and Signs

1.11- BP-Safety-05: Safe Startup of Hydraulic Systems

After taking every action to ensure the safety of the operator, work environment, and the workspace, the machine is ready for commissioning. Many systems were found to work improperly after the first *startup* or during the testing period. Incorrect commissioning during startup can result in damaging hydraulic components. Usually the consequences of improper system commissioning are seen when it's too late and the component may need to be replaced!

Due to the interaction between the hydraulic equipment and the complete machine, the installation of the hydraulic equipment into the machinery will result in additional potential hazards. It is therefore essential for the manufacturer to prepare operating instructions for the complete machine. If there is no startup procedure provided by the machine manufacturer, the best practices list *BP-Safety-05* provides guidelines for safe startup of the hydraulic system.

BP-Safety-05:
1. **Visual Inspection:** Before installation, check the hydraulic power unit for visible transport damage e.g. cracks, missing lead seals, screws, protective covers.
2. **Emergency Stop**: Locate the emergency stop button.
3. **Local Safety Instructions:** review local safety instructions.
4. **Startup Instructions:** Review start up instructions if found.
5. **Communication:** Arrange for proper communication with people in charge (Fig. 1.53).
6. **Oil Level:** Check oil level in the reservoir and make sure it is above the minimum level.
7. **Priming:** Prime pumps, motors, and cylinders as per the manufacturer' instructions.
8. **Hydraulic Lines:** Make sure hydraulic conductors are properly tightened.
9. **Leakage Gutters:** Use leakage collection gutters or curbs around the equipment so that if external leakage occurs, it shall not be hazardous.
10. **Electrical Lines:** Make sure electrical connections are separated from hydraulic lines and are securely connected.
11. **Controls:** Make sure all controls and valves are in neutral.
12. **Pump Direction of Rotation:** Make sure that the pump direction of rotation is correct.
13. **Electric Motors:** If the prime mover is an electric motor, make sure it is wired to run in the correct direction. Start and stop to confirm that the motor runs in the right direction.
14. **Safe Position:** Take safe position to avoid unexpected actuators or machine movement. All unnecessary personnel must stay away from the system.
15. **Mechanical Safety Locks:** If the machine should move, remove all the safety interlocks.
16. **Easy Start:** Test the system for at least 10 minutes under conditions of no-load, low pressure, and low RPM (if possible).
17. **Variable Pumps:** For a variable pump or motor that receives external pilot pressure (Fig. 1.54), carefully and safely untighten the pilot line at the component to remove air.
18. **Closed Circuits:** On closed circuits (hydrostatic transmissions), monitor the charge pressure (Fig. 1.55). If the charge pressure specified by the manufacturer is not established within 20 to 30 seconds, shut down the prime mover and investigate the problem. Charge pressure is typically 110-360 PSI (8-25 bar). Losing charge pressure will result in pump cavitation.

19. **Easy Actuation:** Stroke cylinders slowly and run hydraulic motors at low speed until all air is removed from the components and the plumbing, and the actuators move smoothly.
20. **Inspection:** Inspect the system for leakage, unusual noise, vibration or unusual smells.
21. **Gradual Loading:** Increase the load gradually until the machine runs safely under full load and maximum pressure.
22. **Operating Temperature:** Observe the rate of temperature increase in the system.

**Fig. 1.53 – Safe Commissioning of Hydraulic System
(Courtesy of the International Hydraulic Safety Authority)**

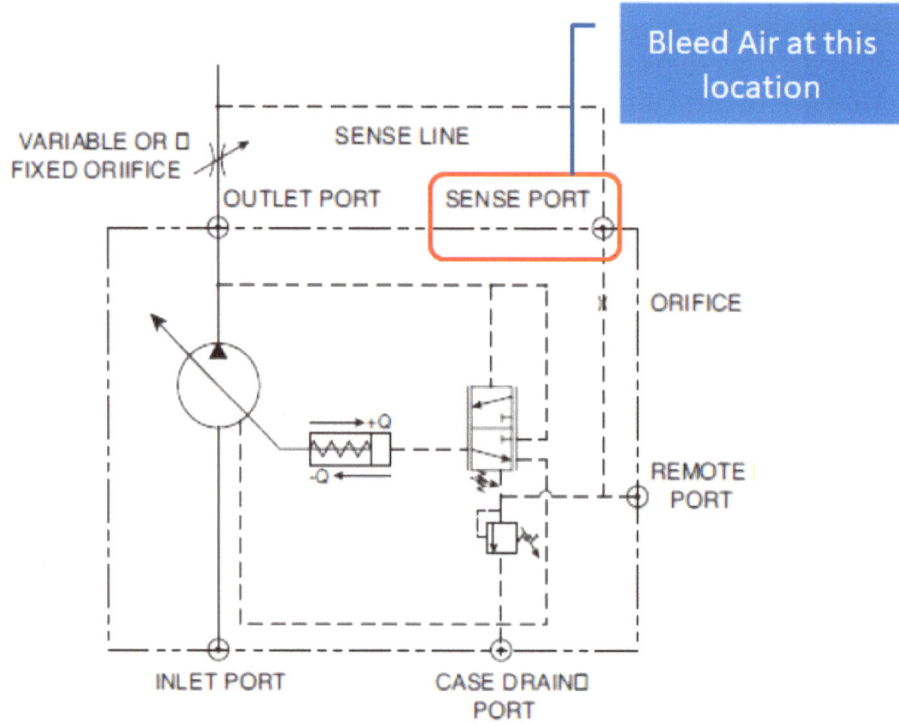

Fig. 1.54 – Pilot Pressure for a Load Sense Variable Pump (Courtesy of Parker)

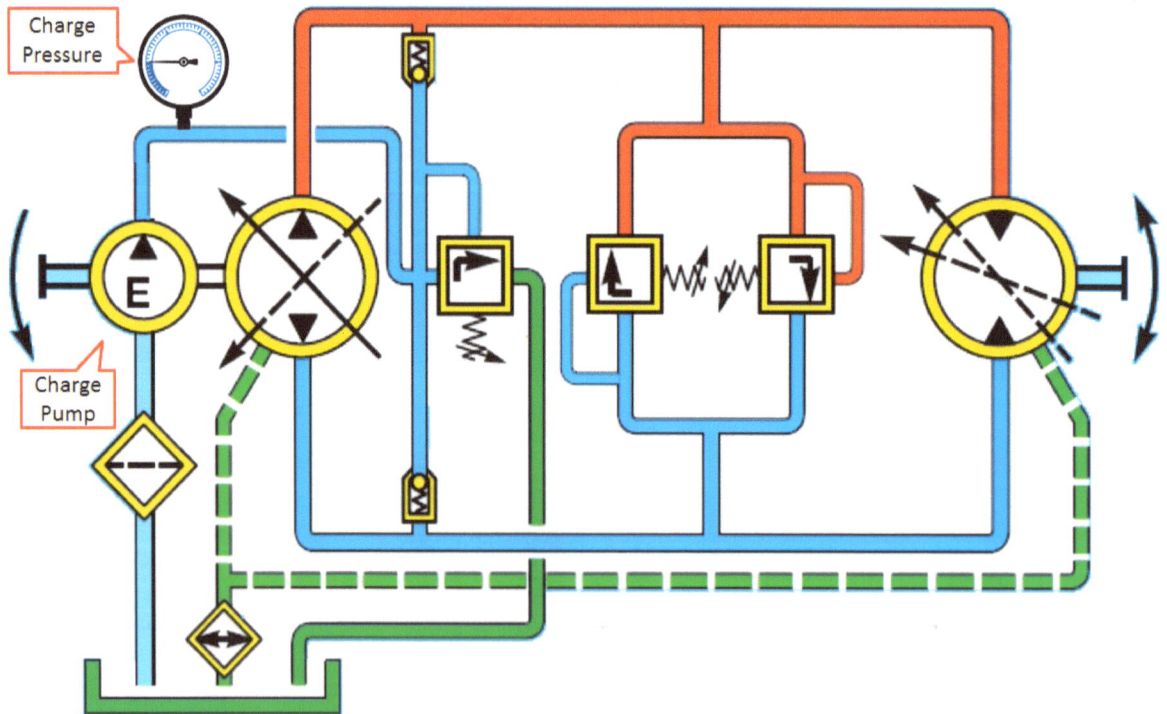

Fig. 1.55 – Charge Pressure in Hydrostatic Transmission (Courtesy of Bosch Rexroth)

1.12- BP-Safety-06: Safe Operation of Hydraulic Systems

After safe start up a hydraulic-driven machine, it must continue to operate properly and safely. The best practices list *BP-Safety-06* presents guidelines for safe operation of hydraulic systems.

PB-Safety-06:
1. Locate Emergency Shut Off.
2. DO NO Operate Leaking Machine.
3. Technical and Safety Training.
4. Continuous Condition Monitoring.
5. Continuous Contamination Control.
6. Periodic Inspection and Maintenance.
7. Reporting.

The following subsections provide detailed interpretation for BP-Safety-06.

1.12.1- Locate Emergency Shut Off

As shown in Fig. 1.56, the machine operator must be familiar with the locations of the machine emergency stop and the main circuit breaker for the machine if it is electrically driven.

Machine Emergency Stop   Main Emergency Shut Off

Fig. 1.56 - Emergency Shut Off

1.12.2- DO NOT Operate a Leaking Machine

Keep all hydraulically operated equipment and surrounding areas clean and free of fluid residue and combustible materials. As shown in Fig. 1.57, DO NOT operate a leaking hydraulic system. Serious injury may result.

Fig. 1.57 - Leaking Hydraulic System

1.12.3- Technical and Safety Training

Lack of understanding of "How it Works?" is a primary reason for unsafe operation of a hydraulic system. Therefore, as shown in Fig. 1.58, all hydraulic system operators should complete hands-on technical training. This technical training should include:
- Basic training about the construction and operation of hydraulic systems.
- Steps necessary to perform the job.
- Hazards associated with the job.

In addition to the technical training it is the responsibility of the employer to provide employees with *Safety Training* that provides the information needed to perform their jobs safely. This information should include:
- BP-Safety-03: Safety of Hydraulic System Operators.
- BP-Safety-05: Safety of Hydraulic System Workspace.
- Emergency Plan: predefined duties for each individual within the workspace in response to various emergency situations such as fire, explosion, injuries, etc.

It is the responsibility of the employee to learn seriously and follow the instructions strictly.

International Hydraulic Safety Authority (IHSA) (http://www.hsac.ca/) provides safety training focused on specific hydraulic systems applications.

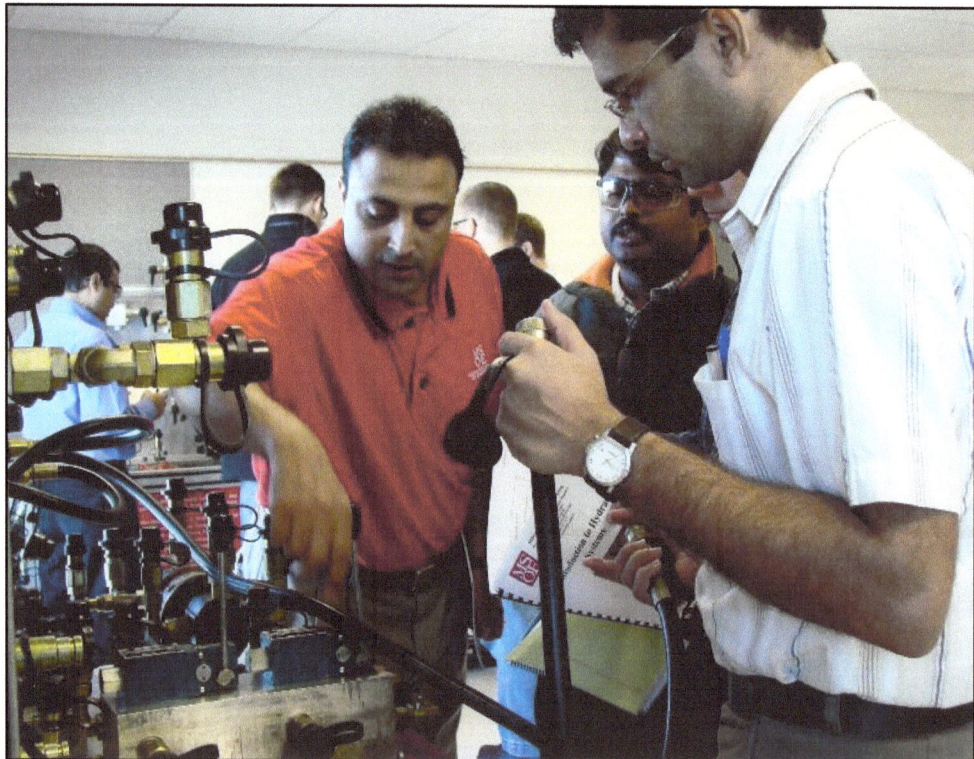

Fig. 1.58 - Hands-On Technical Training

1.12.4- Continuous Condition Monitoring

Monitoring the operating conditions, particularly temperature and pressure, of a hydraulic system is vital in avoiding unexpected accidents. Uncontrolled increase of these conditions could easily cause damage to the equipment and the operator. Other operating conditions are also required to be monitored such including, vibration, oil level, oil viscosity, air in the fluid, etc.

1.12.5- Continuous Contamination Control

The majority of hydraulic machines' failures are due to the contamination. That justifies the necessity for periodic cleaning of the surroundings and outer surfaces of the hydraulic components. During normal machine operation, every effort must be made to maintain the oil cleanliness level as specified by the machine manufacturer. Fluid sample analysis including contamination level and filter clogging status should be checked periodically based on machine manufacturers recommendation and/or past experience with the particular machine.

1.12.6- Periodic Inspection and Maintenance

In addition to professional scheduled inspections, daily visual inspection is always advisable. The following three symptoms reveal possible unsafe operational conditions at any time: leakage, unpleasant smell, and dark color of oil. Preventive and predictive maintenance are mandatory requirement for safe operation of hydraulic systems. Periodically check hoses for leakage, cracks, kinks or breaks.

1.12.7- Reporting

Report to the supervisor any hazardous conditions such as machine trips, leakage, unusal smell, vibration, increased temperature or pressure, oil color change, sustained foam in the oil, and low oil level.

1.13- BP-Safety-07: Safe Servicing of Hydraulic Systems

When maintaining and troubleshooting a hydraulic driven machine, some hydraulic components may be disassembled, cleaned, rebuilt, tested, and reassembled. The process of servicing a hydraulic system subject to some safety regulations. The best practices list *BP-Safety-07* presents guidelines for safe servicing of hydraulic systems.

PB-Safety-07:
1. Understand Job Tasks and Tools.
2. Understand Job Hazards.
3. Review Safety of the Operator and Workspace.
4. Follow Lockout Procedures.
5. Secure Overrunning Loads and Moving Parts.
6. Release Stored Energy.
7. Depressurize the System.
8. Wait Until the Machine Cools Down.
9. Prepare Service Location.
10. Prepare Service Spare Parts.
11. Prepare Service Utilities.
12. Be aware of Common Mistakes during Hydraulic System Maintenance.
13. Avoid Oil Spillage.
14. Careful Welding
15. Proper Cleaning and Painting.
16. Proper Storage and Transportation.
17. Be Predictive
18. Secure Wiping Hoses

The following subsections provide detailed interpretation of the action items listed in the best practices list BP-Safety-07.

1.13.1- Understand the Job Tasks and Tools

It is extremely important to understand how to perform your job safely. Therefore, as best practices, maintenance work force should:

- **Review History:** Review maintenance history of the machine.
- **Review Service Instructions:** Review available maintenance instructions and manuals.
- **Review Best Practices for Disassembling and Assembling:** Review predefined processes for assembling and disassembling all hydraulic components that are included in the maintenance plan. Next volume of this textbook series provides guidelines.
- **Review Circuit Diagram:** Review hydraulic circuit diagram and the associated electrical and control circuit diagrams.
- **Review Use of Proper Tools:** Some hydraulic components are heavy. Use caution while lifting these components. Otherwise, serious personal injury can occur.
- **Ask Questions:** If you are in doubt of anything, just stop and ask.

Example 1 (Fig. 1.59): Proper lifting of hydraulic components reduces the risk of injuries.

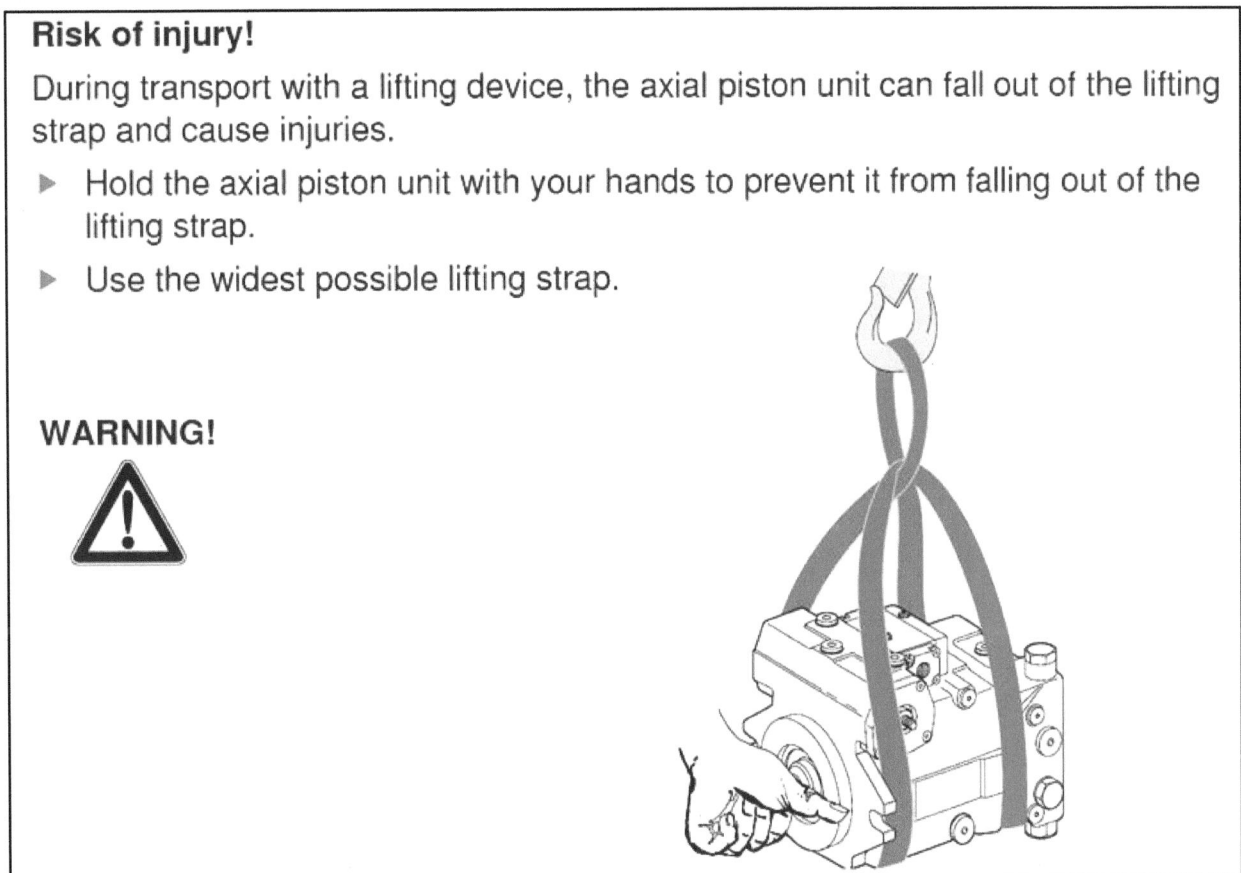

Risk of injury!

During transport with a lifting device, the axial piston unit can fall out of the lifting strap and cause injuries.

▶ Hold the axial piston unit with your hands to prevent it from falling out of the lifting strap.

▶ Use the widest possible lifting strap.

WARNING!

**Fig. 1.59 – Proper Transportation of Heavy Hydraulic Components
(Courtesy of Bosch Rexroth)**

Example 2 (Fig. 1.60): Use *Torque Wrench* to tighten bolts to the specified value.

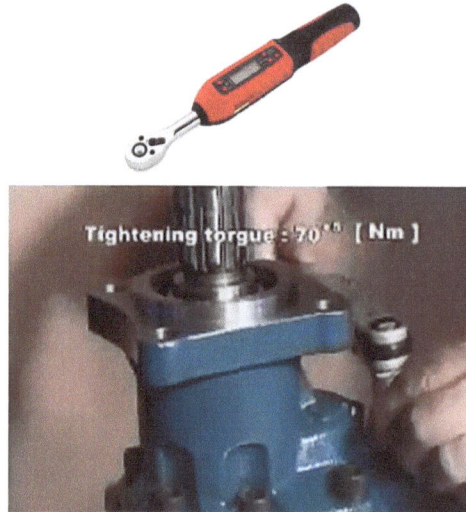

Fig. 1.60 – Using Torque Wrenches to Assemble Hydraulic Components

Example 3 (Fig. 1.61): Use the right support. Otherwise, load may fall.

**Fig. 1.61 – Use the Right Support
(Courtesy of the International Hydraulic Safety Authority)**

Example 4 (Fig. 1.62): Use the right hoisting device. Otherwise, unexpected load movement could occur.

Earth moving equipment is not a safe hoisting device. Earth moving equipment not engineered and equipped with safety devices is a hazard for anyone rigging to it. Cranes are equipped with safety devices designed for hoisting.

**Fig. 1.62 – Use the Right Hoisting Device
(Courtesy of the International Hydraulic Safety Authority)**

Reported Case History (Fig. 1.63):
Action: A maintenance technician installing a new cylinder on a production machine.

Result: When the valve shifted to test the cylinder, the cylinder did not move for few seconds, then accelerated and damaged the cylinder head. The technician suffered a non-injury accident.

Reason: Improper bleeding of air out of the cylinder. Pump flow pushes the air in the transmission lines ahead of the oil compressing it. When pressure overcame the initial resistance to move the cylinder rod, the compressed air expanded, causing the cylinder rod to accelerate at high velocity.

Prevention: Fill cylinder and plumbing with oil then bleed remaining air at low pressure points.

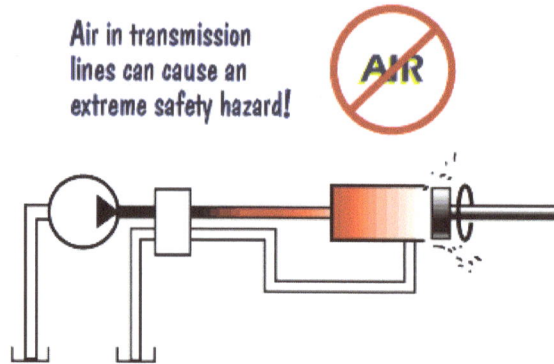

**Fig. 1.63 – Improper Cylinder Air Bleeding
(Courtesy of Fluid Power Training Institute)**

Reported Case History (Fig. 1.64):

Action: A maintenance technician, removed a hose from the cylinder to inspect piston seal leakage by supplying pressurized fluid from the rod side.

Result: No injuries, the fire caused $3.5 million in damages.

Reason: Piston seal failed. The damaged seal allowed oil to spray out through the open port at high velocity. The "atomized" oil ignited when it came into contact with the gas heater that was mounted above the machine.

Prevention: Connect open cylinder port directly to the reservoir and measure rod movement when pressurized.

FPSI™ © 2002

**Fig. 1.64 – Improper Cylinder Leakage Inspection
(Courtesy of Fluid Power Training Institute)**

Reported Case History (Fig. 1.65)

Action: A millwright is testing case drain leakage of a hydraulic motor on a conveyor system.

Result: The technician suffered an eye injury, minor burns, bruises, and abrasions as a result of an accident.

Reason: The technician improperly inspected the case drain leakage of the motor. He disconnected the case drain line from the port at the motor housing and held it in a receptacle. After running the system for few moments, without warning, the case drain oil surges and began to spray out in his face with such intensity that it jerked the hose violently.

Prevention: Install a low pressure drop flowmeter in the case drain line to measure flow and connect the case drain line to the reservoir as shown in the figure.

**Fig. 1.65 – Improper Motor Leakage Inspection
(Courtesy of Fluid Power Training Institute)**

1.13.2- Understand the Job Hazards

After understanding how to perform the job safely, it is extremely important to know about the hazards associated with the job. Knowing what can go wrong during servicing a hydraulic system increases the alertness of the individual and reduces the possibility of an accident. Therefore, it is the responsibility of the supervisor to:

- Explain the possible *Job Hazards*, how to avoid them, and how to respond if they occur.
- Define those parts, which may be moving during your servicing of the system.

General Job Hazards:
In addition to Job hazards for a specific job, the following set of examples shows general job hazards when servicing a hydraulic system:

Example 1: Hazards from Electrical Cords
As shown in Fig. 1.66, electrical cords are always a source of hazards. Frayed cords or damaged connectors are dangerous. They contribute to the possibility of electrical shock and fire. Therefore, as best practices make sure to replace all damaged or frayed *electrical cords* before servicing a hydraulic system.

Fig. 1.66 – Hazards from Electrical Cords

Example 2: Hazards from Control Panels

As shown in Fig. 1.67, control panels could be also a source of hazards. Unexpected loss of control due to rusty or corroded *control panel* could result in serious injuries, particularly if the control is safety-related. Therefore, as best practices make sure the control panels are in good shape before servicing a hydraulic system.

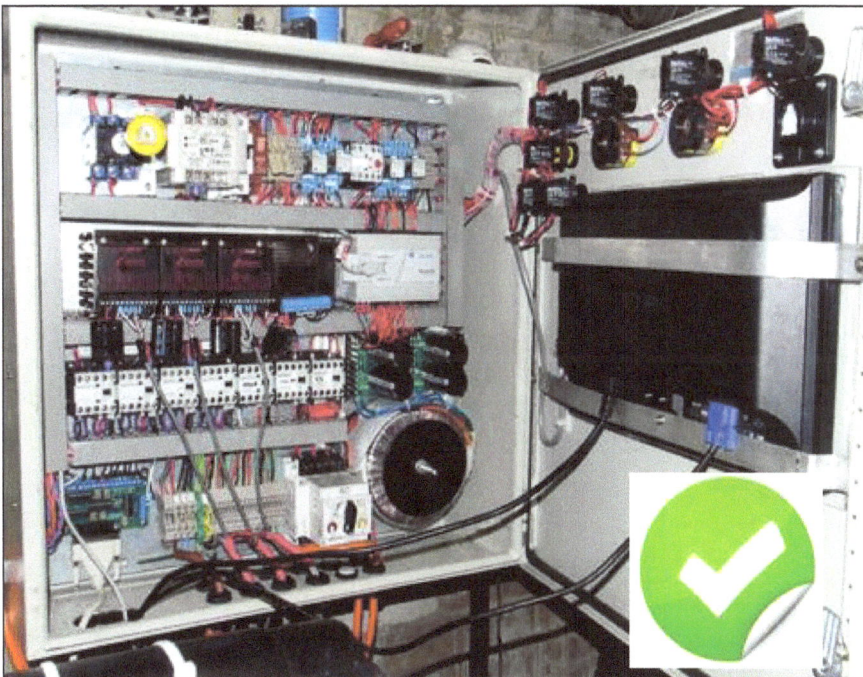

Fig. 1.67 – Hazards from Electrical Control Panels

Example 3: Hazards from Hand Pumps

As shown in Fig. 1.68, many of the workers are not aware that *hand pumps* can generate extreme high pressure. Improper use of such pumps could result in serious injuries.

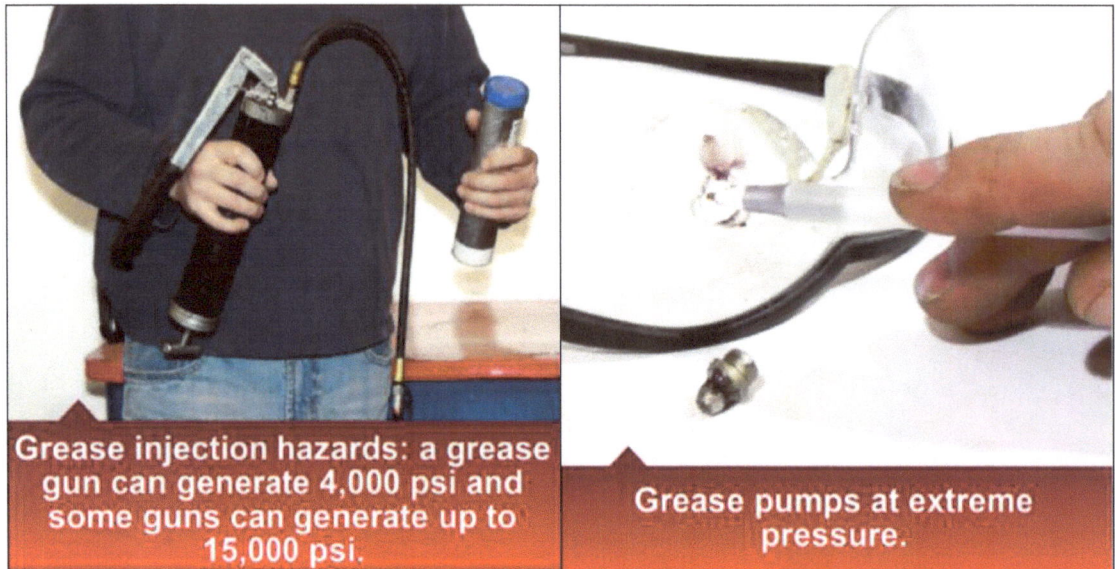

Grease injection hazards: a grease gun can generate 4,000 psi and some guns can generate up to 15,000 psi.

Grease pumps at extreme pressure.

Fig. 1.68 – Hand Pump Generates Extreme Pressure
(Courtesy of the International Hydraulic Safety Authority)

Example 4: Hazards from Unfastened Hoses

If a hydraulic hose failed from one end, it will whip around possibly striking an employee in the vicinity of the machine causing serious injuries. As shown in Fig. 1.69, hydraulic hoses must be secured in a way to avoid such accidents.

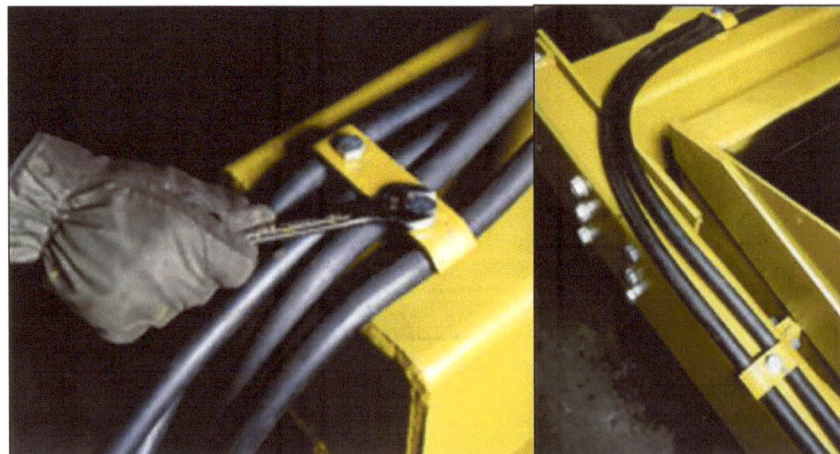

Fig. 1.69 – Hydraulic Hose Clamps Prevents Whipping if Hose Fails

1.13.3- Safety of the Operator and the Workspace

As shown in Fig. 1.70, DO NOT attempt to service a hydraulic-driven machine in an environment where safety regulations for the operator and the workspace are not established. Before servicing a hydraulic driven machine, review available safety regulations for servicing personnel and the workspace. If no regulations found, review the guidelines presented by the best practices lists BP-Safety-02/03/04/05/06.

Fig. 1.70 – Safety of Operator and Workspace

1.13.4- Follow Lockout Procedures

Before servicing a hydraulic-driven machine, follow the available *Lockout Procedure*. If no instructions available, follow the guidelines shown below:

Lockout Tags: As shown in Fig. 1.71, Use lockout tags whenever needed.

Turn off prime movers: Never perform maintenance on a running system unless the maintained component can be isolated from the system without stopping the machine. For example, in certain dual filter systems, one of the filters can be isolated and replaced.

Disconnect from main electric power: if portion of the machine is driven electrically, it is advisable to disconnect the machine from the main electrical power so that electro-hydraulic valves can't be actuated accidentally.

Disconnect from compressed air sources: if any of the valves are actuated by compressed air signals, it is also advisable to disconnect the compressed air source.

Fig. 1.71 – Lockout Tags

Reported Case History (Fig. 1.72)

Action: A millwright was standing on a stepladder tightening a leaking hydraulic connector in a steel hydraulic transmission line that was fixed to wall 12 feet above the floor. The hydraulic pump was running and driven by an electric motor.

Result: The connector unexpectedly failed. High-pressure hydraulic oil sprayed from the broken connector, striking him in the face and chest. He lost his grip on the ladder, and fell to the concrete floor below. He died as a result of the injuries he sustained from the fall.

Reason: Power unit left ON during maintenance.

Prevention: Lock out hydraulic power unit.

**Fig. 1.72 – Hazard of Maintaining a Running Hydraulic System
(Courtesy of Fluid Power Training Institute)**

1.13.5- Secure Overrunning Loads and Moving Parts

A hydraulic transmission line that supports an overrunning load is an extreme hazard when it fails. Additionally, a hydraulic-driven machine, may experience loss of power and/or control. The following set of bullets provide guidelines to avoid such hazards. Failure to comply with these can result in serious injury or death.

- Overrunning loads such as vertical and over-center loads must be properly lowered or secured in place during maintenance.
- Make sure moving parts are mechanically clamped or blocked to prevent motion.

Figure 1.73 shows the following examples:
- **Example 1:** A telescopic hydraulic cylinder of a hydraulic-driven elevator must have lowered to the last stage before servicing the hydraulic system.
- **Example 2:** A loader bucket must be lowered to the ground before servicing the loader.
- **Example 3:** A conveyor should be clamped before servicing the hydraulic drive system.

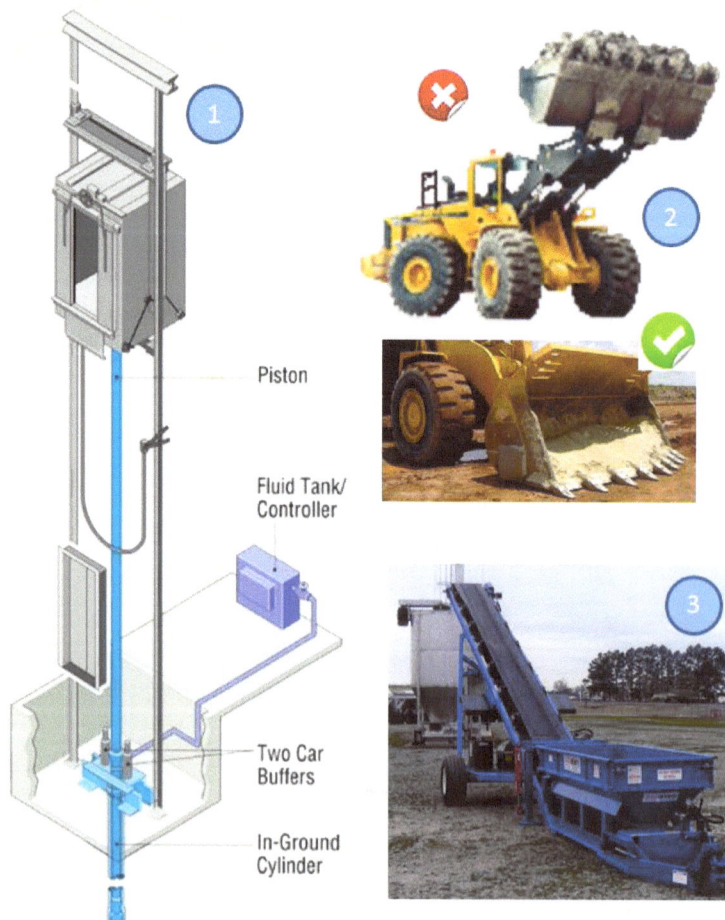

Piston

Fluid Tank/ Controller

Two Car Buffers

In-Ground Cylinder

Fig. 1.73 – Hazard of Maintaining a Running Hydraulic System

1.13.6- Release Stored Energy

A hydraulic-driven machine may be turned off while it contains mechanical, or hydraulic energy stored in some spots. As shown in Fig. 1.74, stored energy is a hidden enemy that can be unleashed at any time. Figure 1.75 shows that an accidental or unprofessional releasing of such energy may result in movement of actuators resulting in serious injuries.

Fig. 1.74 – Hazard from Stored Energy
(Courtesy of the International Hydraulic Safety Authority)

Fig. 1.75 – An Example of Unprofessional Release of Stored Energy
(Courtesy of Fluid Power Training Institute)

Therefore, as best practices, workers must learn how to control the release of electrical, mechanical, and other types of hazardous energy safely as per OSHA regulations.

Example 1: One example of mechanical stored energy is a compressed spring in a single-acting spring-return cylinder. Such cylinder must return to its initial position where the return spring is relaxed. Otherwise, it may accidentally have retracted by unintentional valve opening even when the machine is off.

Example 2: One example of fluid stored energy is oil stored in an accumulator under gas pressure. An accumulator may contain a significant amount of energy stored from last time the machine was in operation. Uncontrolled release of this stored energy may result in accidental movement of actuators and possible injuries. Therefore, accumulators must be drained to tank, then isolated by a shutoff valve before servicing the machine.

Example 3: Discharging the gas from an accumulator must be based on predefined instructions. A supervisor must explain to the workforce the process of safely releasing the gas from the accumulator before performing the maintenance.

Reported Case History (Fig. 1.76)

Action: disassembling gas-charged accumulator.

Result: A maintenance helper lost his hand as a result of an accident he suffered.

Reason: Energy stored in the accumulator released in a very short time not in accordance with OSHA regulations.

Prevention: Properly discharge gas prior to disassembling the accumulator.

Fig. 1.76 – Hazard from Unprofessional Gas Discharge from an Accumulator
(Courtesy of Fluid Power Training Institute)

1.13.7- Depressurize the System

After lowering all suspended loads and releasing all stored energies, there may still be some residual pressure present inside some parts of the system. Therefore, as shown in Fig. 1.77, the system must be completely depressurized prior to removing any components from the system. Figure 1.78 shows that, as best practices, depressurize the system by moving the handles of manually-operated directional valves several times in both directions. Use *manual overrides* to shift the spool of solenoid-operated directional valves to release any trapped pressure.

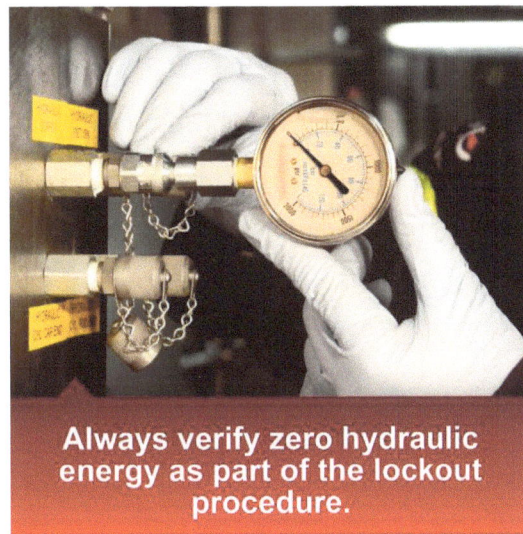

**Fig. 1.77 – Depressurize the System before Service
(Courtesy of the International Hydraulic Safety Authority)**

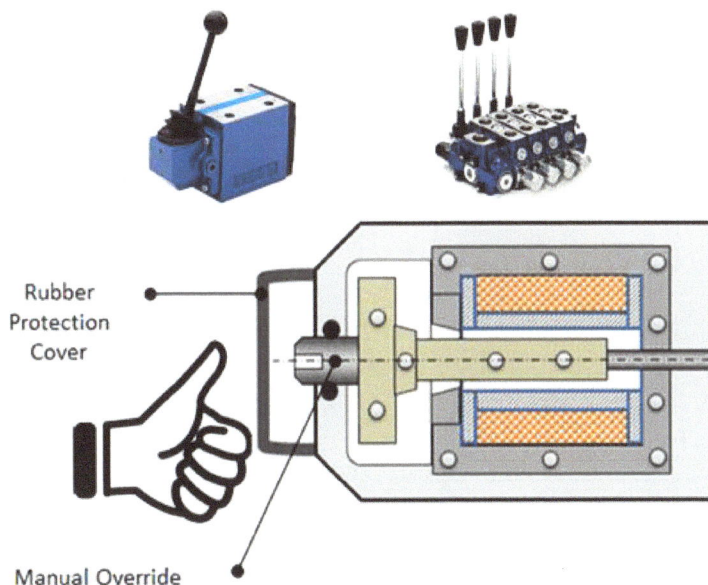

Rubber
Protection
Cover

Manual Override

Fig. 1.78 – Move Directional Valves to Release Trapped Pressure

1.13.8- Wait Until the Machine Cools Down

Wait until the system cools down to room temperature before disassembling it. Disassembling a hydraulic component while it is hot may result in permanent damage and serious injuries.

Example: if the directional valve shown in Fig. 1.79 is assembled while it is still hot, the tight clearance between the valve spool and sleeve will be changed. Even a slight change in the original dimensions due to thermal stresses may result in changing the static and dynamic characteristics of the valve.

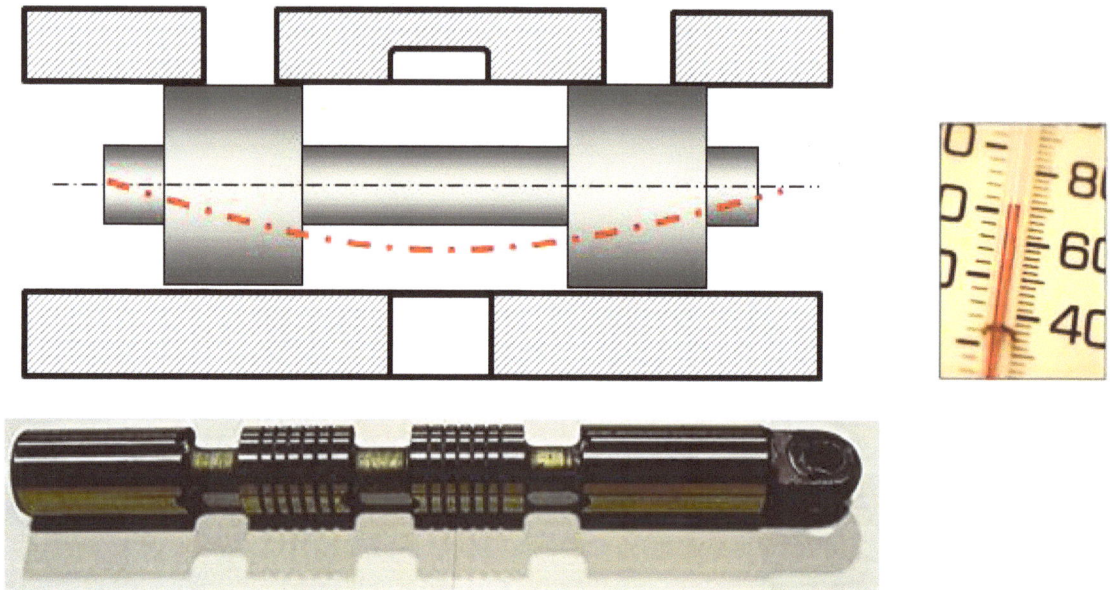

Fig. 1.79 – Dimensional Change when Dissembling a Hot Component

1.13.9- Prepare Service Location

Major requirements of the hydraulic service area are to be clean and dry to eliminate the possibility of contamination to the disassembled components space. Also having an organized workplace reduces the risk of injuries. Therefore, as best practices:

- Make the service area well organized, not a messy place.
- Cordon off the area per OSHA regulations.
- Post warning signs in the area.

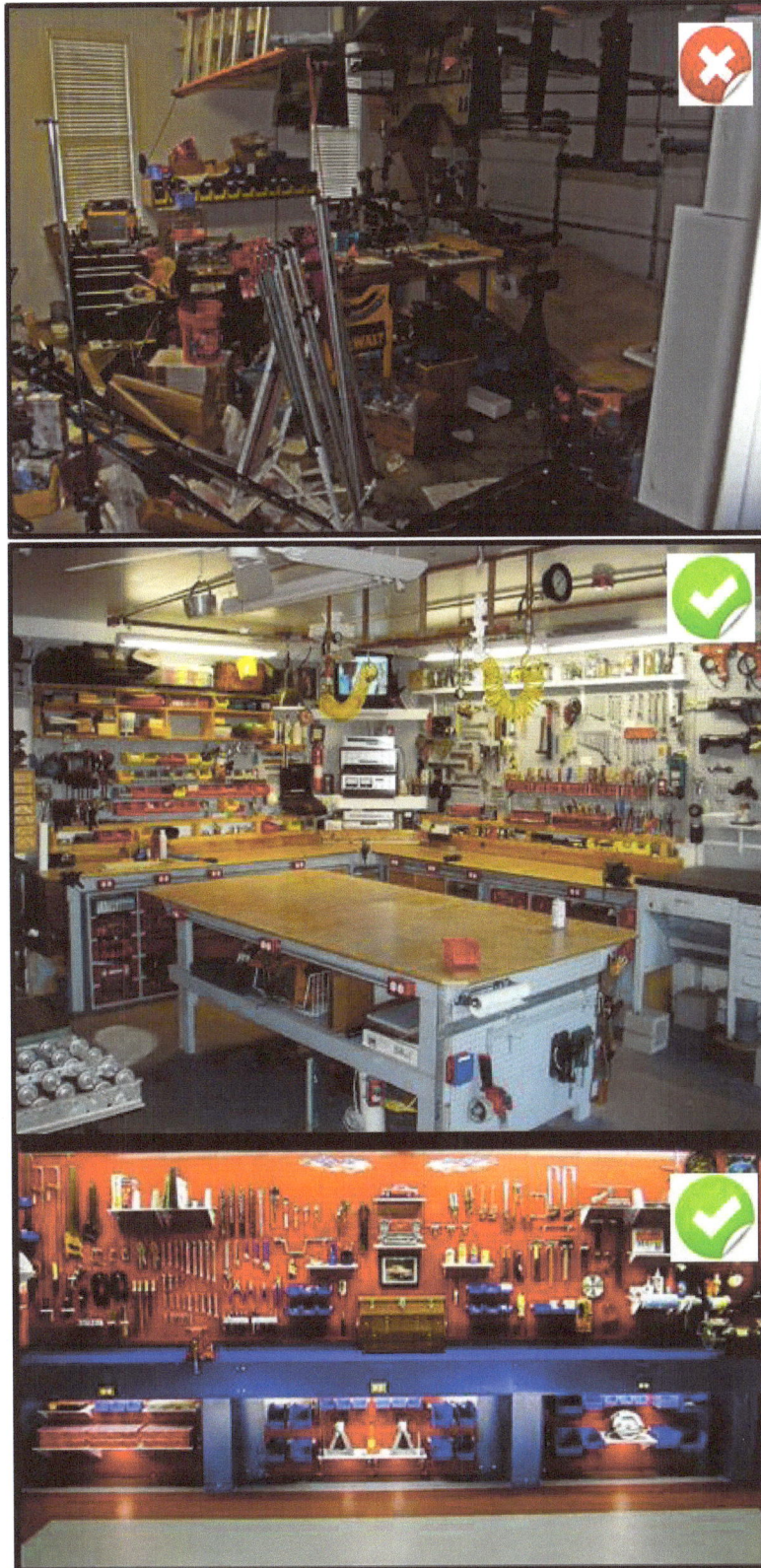

Fig. 1.80 – Organized Service Location Helps to avoid Risks of Injuries and Contamination

1.13.10- Prepare Service Spare Parts

Hydraulic spare parts are vital element in safe servicing of hydraulic systems. Therefore as best practices, as shown in Fig. 1.81:

- Order the repair kits and spare parts as per the manufacturer's instructions.
- Make sure the part numbers match the manufacturer's instructions.
- Source the genuine spare parts only from certified or trusted sources.
- Using non-original spare parts may result in voiding the warranty.
- Check if the spare part is reusable or non-reusable. Some parts are used only once. These elements perform their jobs by being deformed. Examples of that are rubber O-Rings and metallic cutting (sealing) rings.

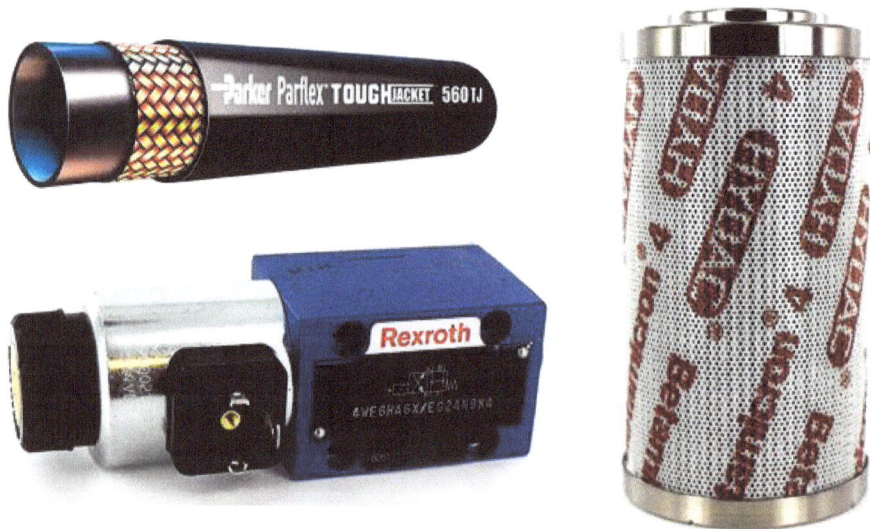

Genuine Spare parts

Cutting (Sealing) Metallic Rings

Rubber O-Rings

Fig. 1.81 – Importance of using Genuine Spare Parts in Servicing Hydraulic Systems

1.13.11- Prepare Service Utilities

Maintenance utilities should be made available. Examples of these utilities are shown in Fig.1.82 as follows:

- Lint-Free *Cleaning Towels* (1) for wiping the bench and the components.
- Source of clean compressed air (2).
- Protective *Plastic Covers* (3) to seal the ports of hydraulic components during storage and transportation.
- Washing Fluids (4). If no particular solvent is specified, *Kerosene* is used to clean hydraulic components because it is lubricant and combatable with most seals. DO NOT use *gasoline*. Federal, state, and local codes may require the use of Eco-Friendly solvents baths.
- *Ultrasonic* (5) for cleaning washable filters.

Fig. 1.82 – Examples of Maintenance Utilities

1.13.12- Be aware of Common Mistakes during Hydraulic System Maintenance

Avoid common mistakes during system maintenance. Such common mistakes will be reported in the next volume of this textbook series.

1.13.13- Avoid Oil Spillage

Sources of spilled oil:
- Leaking hydraulic systems.
- Disassembling hydraulic components may contain some oil that may spilled during components transportation. Spilled oil may result in risk of injuries.

Result of spilled oil:
- As shown in Fig. 1.83, spilled oil may result in *slippage* and falls.
- Spilled oil may cause a fire.
- Oil spillage wastes money.

Best Practices to avoid risks from spilled oil:
- Prepare for collecting oil from the disassembled components.
- Spilled and other dirty oil should be stored in a container labeled "Oil Waste".
- Dispose of used oil and system filters as required by federal, state, or local regulations.
- Keep your hands dry. Wet hands are slippery and transport contamination to components.
- When the service work is completed, thoroughly clean any spilled oil from the equipment.

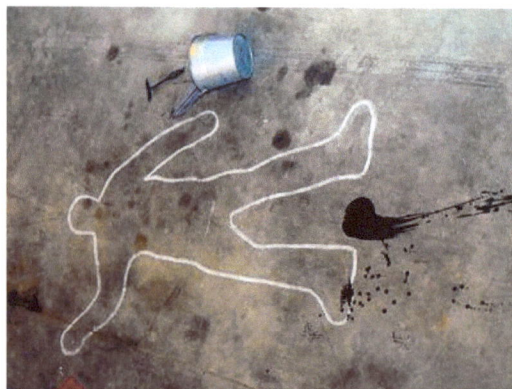

Hydraulic Fluid Spills... don't let this be the result!

Ensure immediate availabilty of spill kits suited for the volumes contained in your hydraulic equipment.

Fig. 1.83 – Risks of Oil Spillage
(Courtesy of the International Hydraulic Safety Authority)

1.13.14- Careful Welding

Do not cut or weld in any area where hydraulic fluids are used until the area is free of all oil deposits and the system is shut down and depressurized.

1.13.15- Proper Cleaning and Painting

During external cleaning and painting of machinery, sensitive materials and parts of hydraulic components must be protected against incompatible liquids. Examples of parts that should be protected, as shown in Fig. 1.84, piston rod and pin eye bearing surfaces.

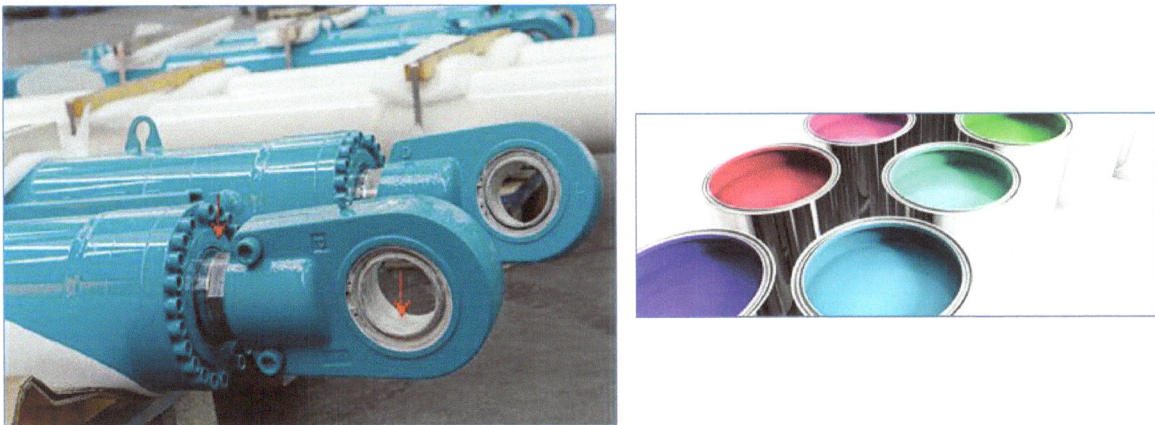

Fig. 1.84 – Protect Sensitive Surfaces during Painting

1.13.16- Proper Storage and Transportation

Improper packaging, transportation, and storage may result in damaging hydraulic components. The next volume of this textbook series provides best practices to cover this subject for the various hydraulic components. Generally speaking as best practices:

Packaging: All parts of the hydraulic system shall be packaged for transportation in a manner that preserves their identification and protects them from damage, distortion, contamination and corrosion.

Transportation: Follow manufacturer's instructions. The following are just two examples:
- Accumulators should be discharged from both gas and oil before shipping.
- Hydraulic power units should not be filled with oil during shipping.

Storage: Follow manufacturer's instructions for each hydraulic component.

1.13.17- Be Predictive

Be predictive means that, especially when a service man works with partners, he should expect a result as follows:
- Do not make a change just to see what will happen.
- Every time an adjustment is made predict what should happen.
- Every time a valve is shifted, predict what will happen, particularly actuators movement.

1.13.18- Secure Wiping Hoses

If a pressurized hose/tube assembly blows apart, the fittings can be thrown off at high speed, and the loose hose can flail or whip with great force. Where this risk exists, consider the use of guards and hose restraints to protect against injury. Do not temporarily drape a return line hose into the reservoir.

1.14-BP-Safety-08: Oil Injection Avoidance and Treatment

How Oil Injection Happens?
Fine streams of escaping pressurized fluid can penetrate the skin and thus enter the human body causing, as shown in Fig. 1.85, serious injuries, loss of organs, or even death. This is known as *Oil Injection*.

Challenges of Oil Injections:
- According to the "Occupational Injuries Handbook," oil can penetrate the skin at pressures as low as 100 PSI (7bar).
- The symptoms of oil injection sometimes don't appear until the injected area becomes in critical condition.
- Fluid injected into the skin must be surgically removed within hours or gangrene may occur resulting in amputation of the affected area.
- Some medical care providers are unfamiliar with fluid injection injuries.

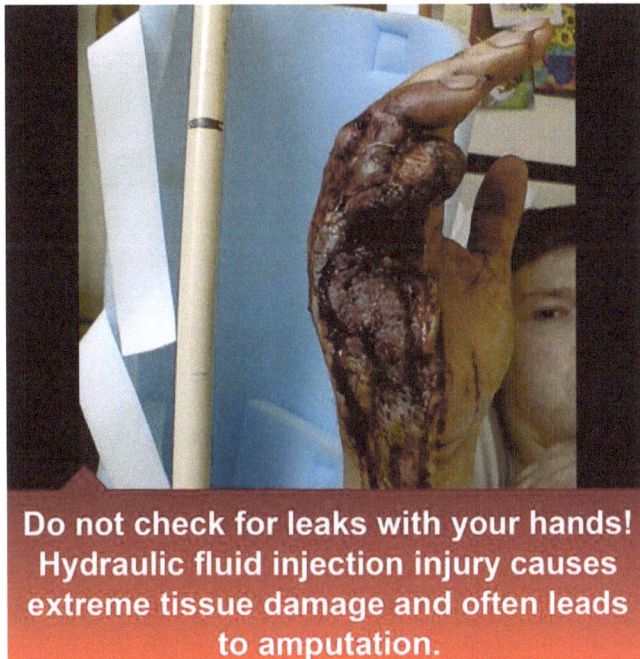

**Do not check for leaks with your hands!
Hydraulic fluid injection injury causes
extreme tissue damage and often leads
to amputation.**

**Fig. 1.85 – Risks of Oil Injection
(Courtesy of the International Hydraulic Safety Authority)**

The best practices list *BP-Safety-08* provides guidelines for situation of oil injection.

Best practices to avoid oil injection:
- Make the people aware of oil injection risks.
- Never use your hand or fingers to search for leaks.
- Wear protective gloves when servicing hydraulic transmission lines.

Best practices if fluid injection occur:
- Report the case immediately to your supervisor.
- DO NOT allow the injured person to drive himself to the medical facility.
- DO NOT give food or drink to the injured person.
- DO NOT treat injections as a simple cut!
- DO NOT delay treatment.
- See a medical specialist doctor immediately.
- Prepare the case information as shown in the safety card provided by the *International Fluid Power Society* (IFPS), shown in Fig. 1.86

Fig. 1.86 – Safety Focus Card
(Courtesy of the International Fluid Power Society)

Reported Case History (Fig. 1.84):

Action: A technician used his hand to check leakage from a hose spraying oil.

Result: 2 fingers lost.

Reason: High-pressure oil injection not treated immediately

For more information about the subject matter, the Fluid Power Training Institute. released a demonstrative video shown in Fig. 1.87. The video can be ordered from www.fluidpowersafety.com.

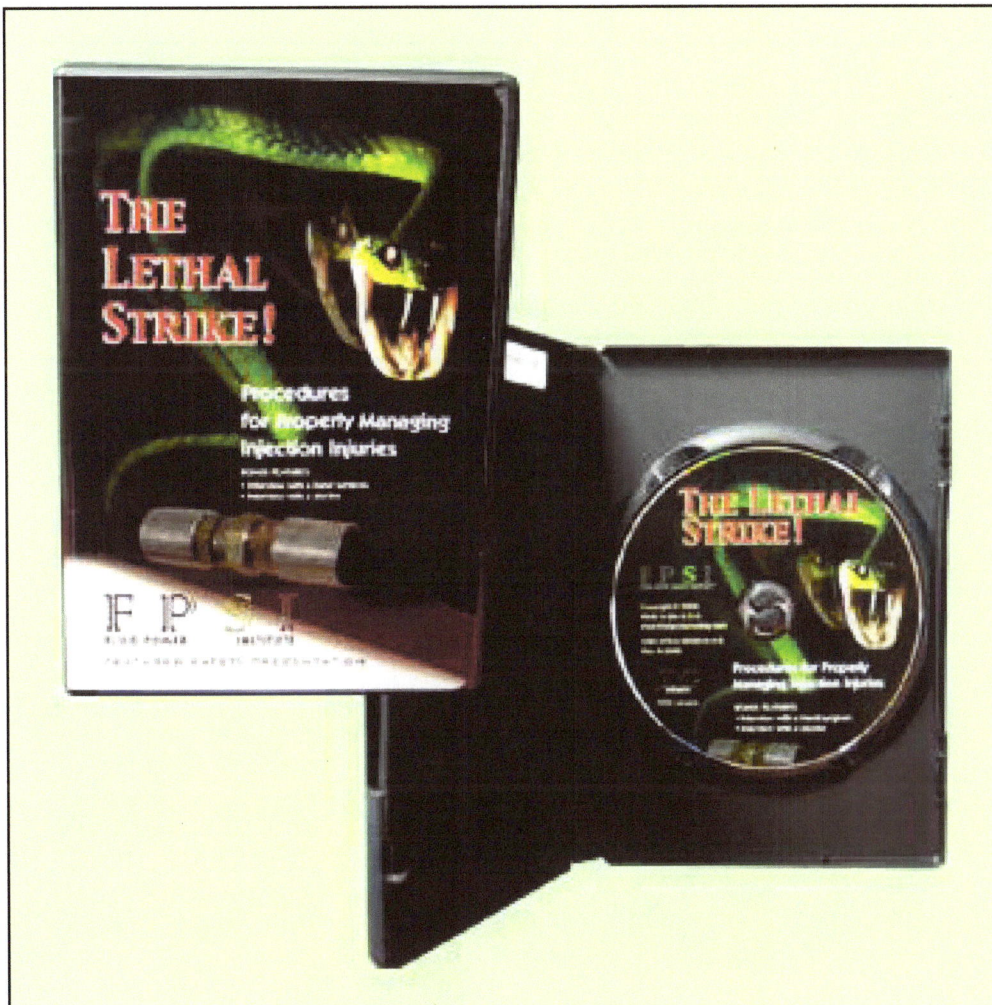

**Fig. 1.87 – Oil Injection Training Video
(Courtesy of Fluid Power Training Institute)**

1.15- BP-Safety-09: Safe usage of Hydraulic Powered Tools

Hydraulic powered hand tools are widely used in various applications such as civil engineering, mining, oil and gas, power generation, shipbuilding, etc. Fig.1.88 shows various types of hydraulic *powered tools* such as lifting cylinders, bolting tools, pullers, work holding tools, and custom tools.

Improper use of hydraulic powered tools under high pressure may result in serious injuries. The following sections present number of best practices for safe operation of hydraulic powered tools. Part of the context and the pictures have been extracted from a training video provided with permission from ENERPAC.

The best practices list *BP-Safety-09* presents guideline for safe operation of hydraulic powered tools.

PB-Safety-09:
- ❏ BP-Safety-09-Powered Tools-A: Safe Operation of Cylinders.
- ❏ BP-Safety-09-Powered Tools-B: Safe Operation of Hoses.
- ❏ BP-Safety-09-Powered Tools-C: Safe Operation of Pumps.

The following subsections provide detailed interpretation of the action items listed in BP-Safety-09.

Fig. 1.88 – Hydraulic Powered Tools (Courtesy of ENERPAC)

1.15.1- BP-Safety-9-Powered Tools-A: Safe Operation of Cylinders

Cylinders in hydraulic powered tools vary widely in terms of sizes and strokes. Make sure to follow the best practices shown below to ensure the safe operation of cylinders in hydraulic powered tools. the best practices list *BP-Safety-09-Powered Tools-A* provides guidelines for safe operation of cylinder in hydraulic powered tools.

BP-Safety-09-Powered Tools-A:
- **Cylinder Inspection** (Fig. 1.89): Inspect the cylinder of the tool for signs of wear, corrosion or damage.
- **Modified Cylinders** (Fig. 1.90): Never use cylinders that have been modified.
- **Air Bleeding** (Fig. 1.91): Bleed air from the cylinder before first use as per the instruction manual.
- **Saddle Usage** (Fig. 1.92): Properly use a *saddle* to ensure even load distribution on the cylinder plunger. Without a saddle, the cylinder rod may be permanently damaged.
- **Saddle Placement** (Fig. 1.93): The entire saddle must be in full contact with the load. Placing only portion of the saddle under the load puts regular stress on the plunger, which can damage equipment.
- **Cylinder Placement** (Fig. 1.94): Make sure the surface under the cylinder is flat as possible and use the cylinder support bases to offer more stability.
- **Side Loading** (Fig. 1.95): Avoid side loading. If the cylinder rod is bent, stop pumping, retract the load, and determine what changes are required to complete the task safely.
- **Maximum Load** (Fig. 1.96): Do not exceed 80% of the manufacturer stroke and load levels.
- **Pump Handle** (Fig. 1.97): Remove the handle when it is not being used to avoid creating a tripping hazard.
- **Body Protection** (Fig. 1.98): Never put your body under the load without making sure that the load in mechanically secured.

Fig. 1.89 - Cylinder Inspection (Courtesy of ENERPAC)

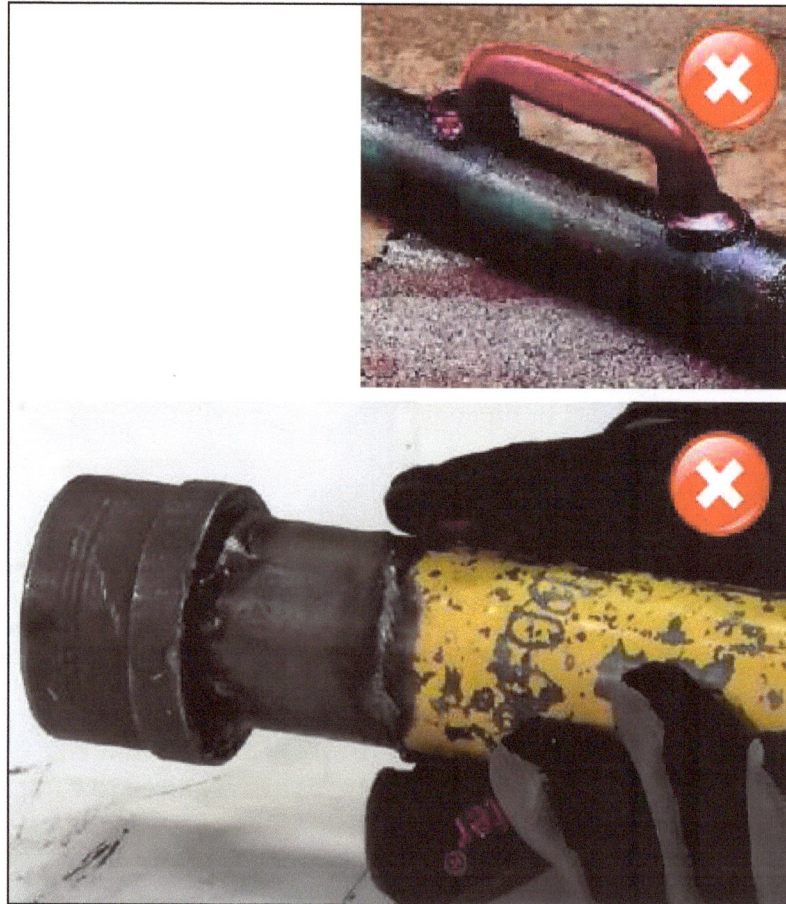

Fig. 1.90 - Modified Cylinders (Courtesy of ENERPAC)

Fig. 1.91 - Cylinder Air Bleeding (Courtesy of ENERPAC)

Fig. 1.92 - Importance of using a Saddle (Courtesy of ENERPAC)

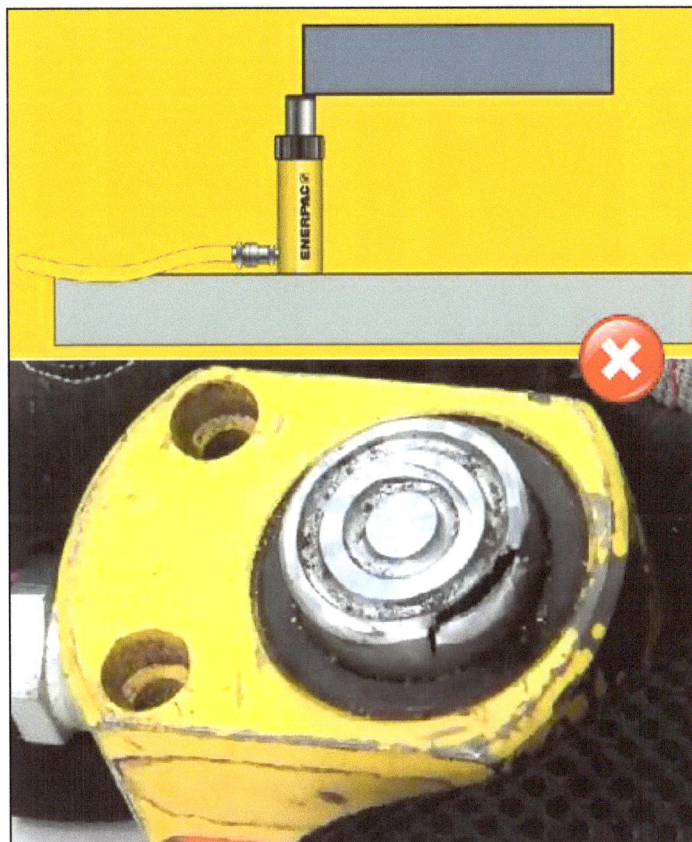

Fig. 1.93 - Hazard from Improper Saddle Placement (Courtesy of ENERPAC)

Fig. 1.94 - Proper Cylinder Placement (Courtesy of ENERPAC)

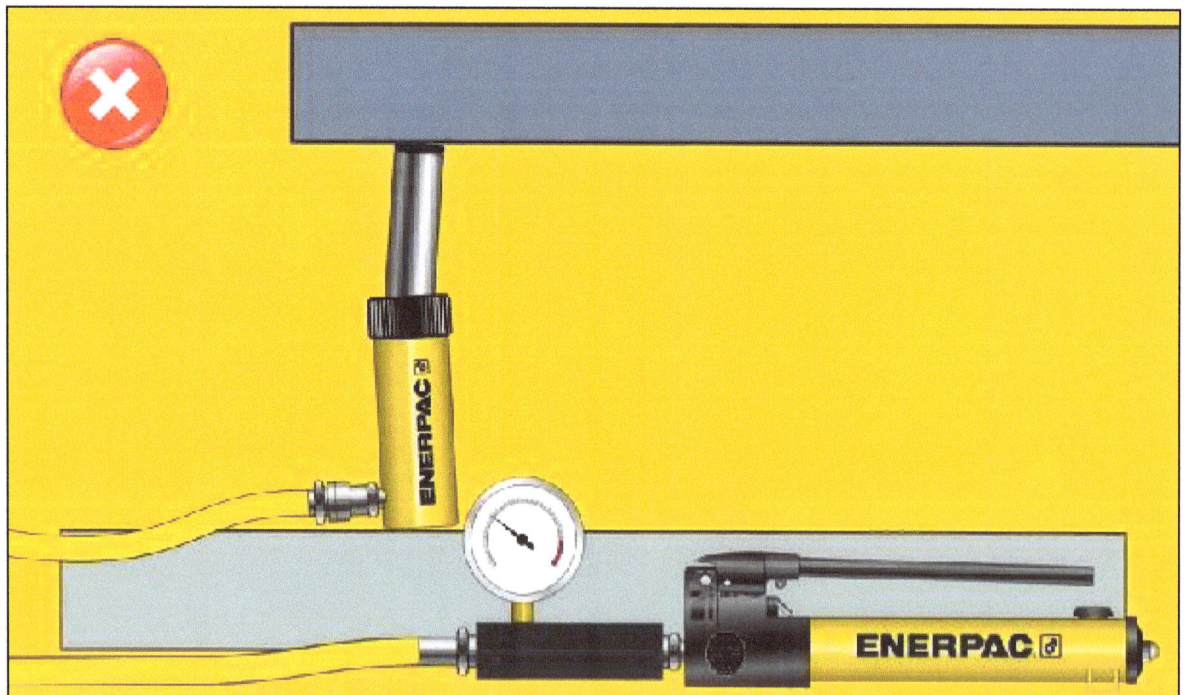

Fig. 1.95 - Effect of Side Loading (Courtesy of ENERPAC)

Fig. 1.96 - Rule for Maximum Load (Courtesy of ENERPAC)

Fig. 1.97 - Remove Handle when if not needed (Courtesy of ENERPAC)

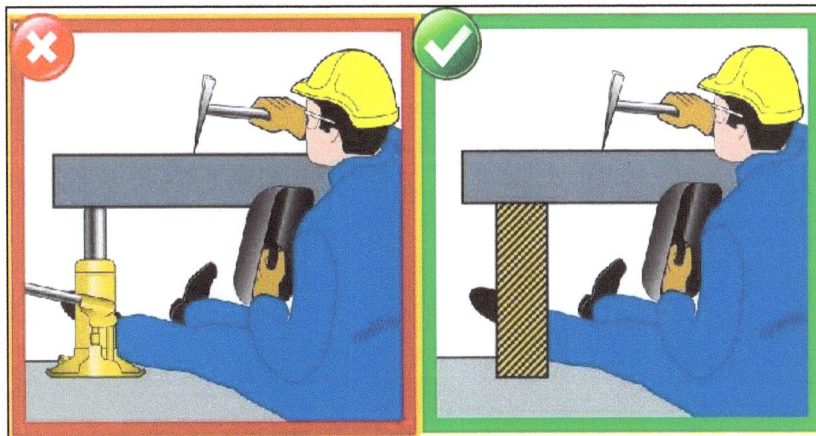

Fig. 1.98 - Body Protection (Courtesy of ENERPAC)

1.15.2- BP-Safety-09-Powered Tools-B: Safe Operation of Hoses

Hoses in hydraulic powered tools are the greatest source of unsafe incidents in hydraulic powered tools. Make sure to follow the best practices shown below to ensure the safe operation of hoses in hydraulic powered tools. The best practices list *BP-Safety-09-Powered Tools-B* provides guidelines for the safety of hoses in hydraulic powered tools.

BP-Safety-09-Powered Tools-B:
- **Hose Inspection** (Fig. 1.99): Before each use, inspect the hoses for cuts, cracks, abrasion, or any signs of damage. It is recommended to use hoses beyond 6 years from the manufacturing date even if they appear to be in good condition.
- **Tool Handling** (Fig. 1.100): Never lift, carry, or drag equipment from the hoses.
- **Care of Hoses** (Fig. 1.101): Avoid dropping objects on hoses or driving equipment over a hose.
- **Hose Placement** (Fig. 1.102): Keep hoses away from areas enclosed by the lifted load.
- **Minimum Bend Radius** (Fig. 1.103): Avoid short bends or kinking hoses. DO NOT exceed the minimum bend radius specified by the manufacturer.
- **Hose Rupture** (Fig. 1.104): If a hydraulic hose ruptures, immediately release system pressure and retract the load.
- **Hose Detachment** (Fig. 1.105): Never detach a hose or a coupling while the system is under pressure.
- **Inspect Hose Couplings** (Fig. 1.106): Before attaching couplings, check both parts to make sure there are no signs of defects or damage.
- **Dust Caps** (Fig. 1.107): Use dust caps when the couplings are not used.
- **Improper Hose Coupling** (Fig. 1.108): Never use low pressure couplings or fittings that are not compatible with high-pressure operation.

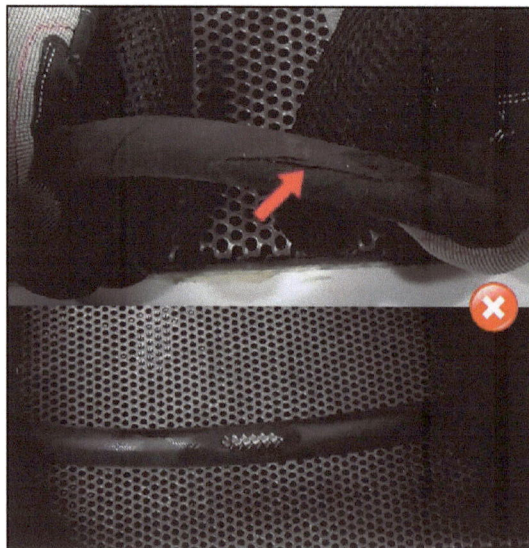

Fig. 1.99 - Hose Inspection (Courtesy of ENERPAC)

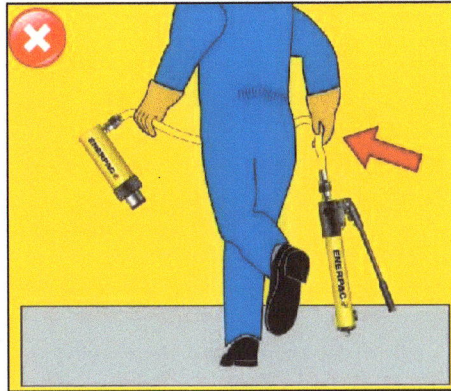

Fig. 1.100 - Improper Tool Handling (Courtesy of ENERPAC)

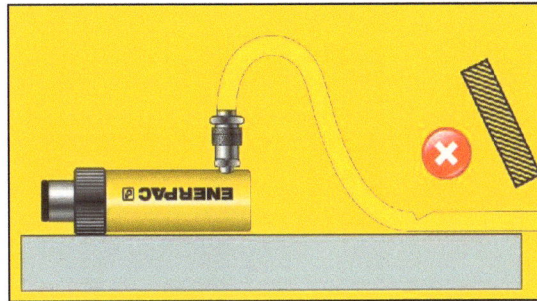

Fig. 1.101 - Improper Care of Hoses (Courtesy of ENERPAC)

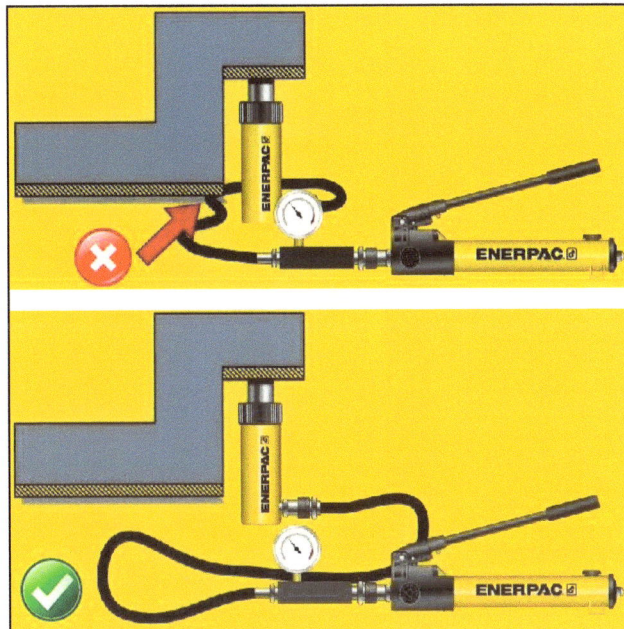

Fig. 1.102 - Hose Placement (Courtesy of ENERPAC)

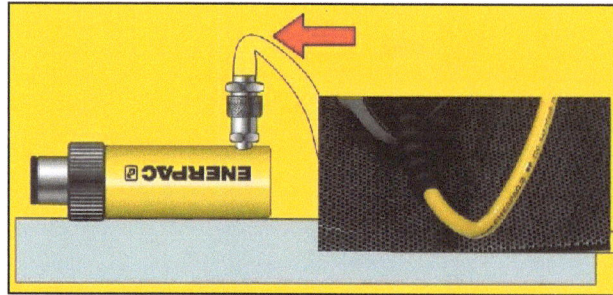

Fig. 1.103 - DO NOT Exceed Minimum Bend Radius of the Hose (Courtesy of ENERPAC)

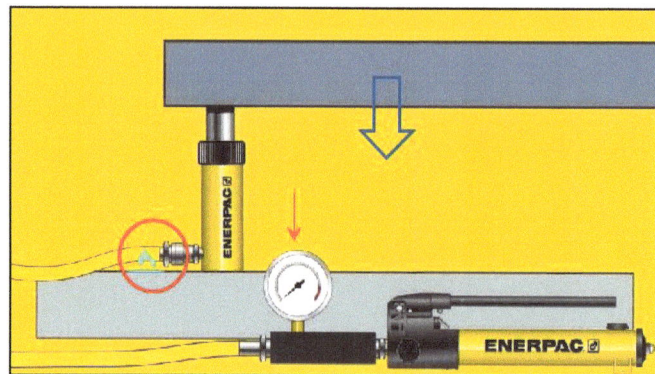

Fig. 1.104 - What to do if a Hose Ruptures (Courtesy of ENERPAC)

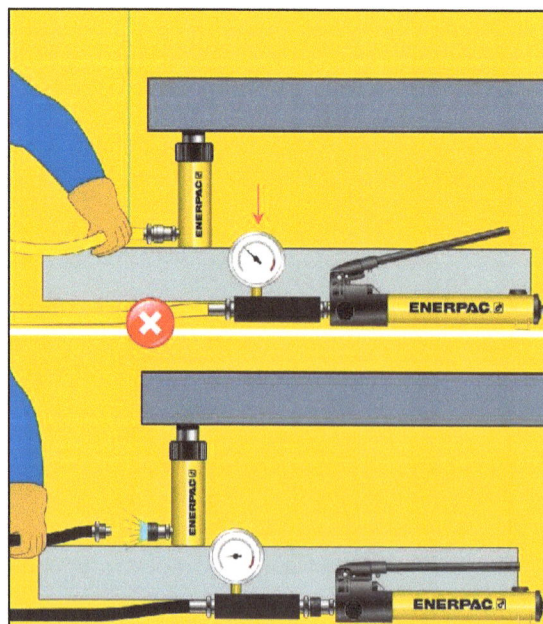

Fig. 1.105 - Improper Hose Detaching (Courtesy of ENERPAC)

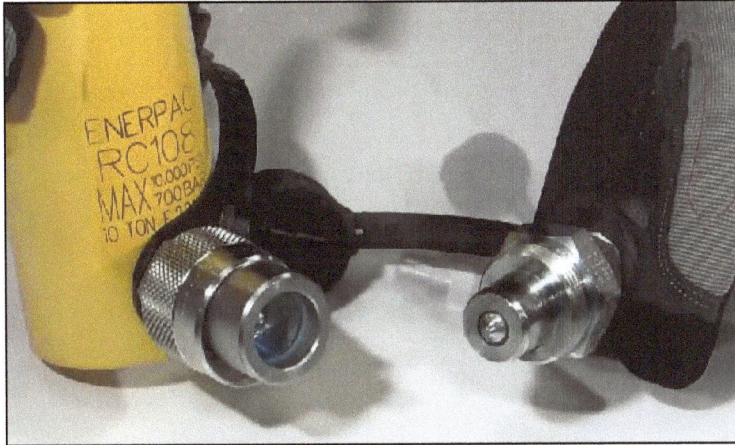

Fig. 1.106 - Inspect Coupling before Connecting (Courtesy of ENERPAC)

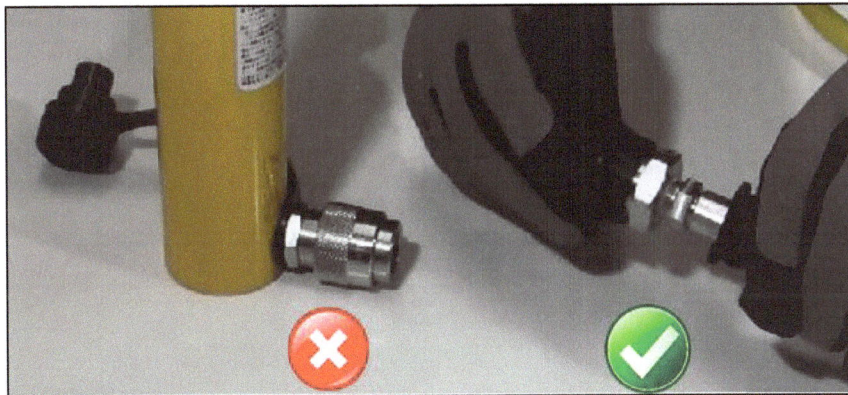

Fig. 1.107 - Use Protective Caps When Couplings are not Connected (Courtesy of ENERPAC)

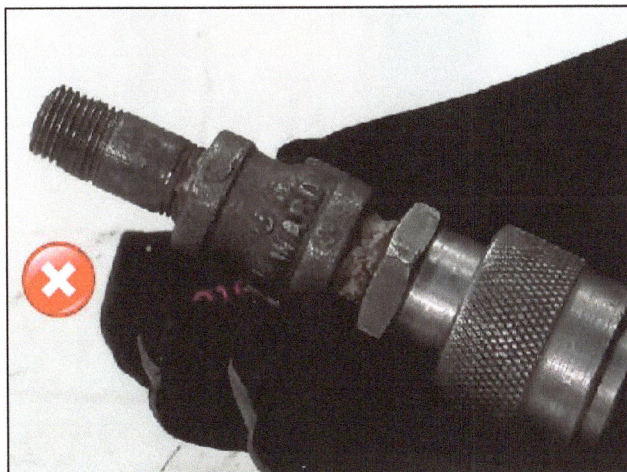

**Fig. 1.108 - Use Fittings and Couplings Rated for System Operative Pressure
(Courtesy of ENERPAC)**

1.15.3- BP-Safety-09-Powered Tools-C: Safe Operation of Pumps

Pumps in hydraulic powered tools can be hand-operated, electrical powered, battery powered, or air powered. Make sure to follow the best practices shown below to ensure the safe operation of pumps in hydraulic powered tools. The best practices list *BP-Safety-09-Powered Tools-C* provides guidelines for safe operation of pumps in hydraulic powered tools.

BP-Safety-09-Powered Tools-C:
- **Pump Priming:**
  - (Fig. 1.109): If the pump is connected to a cylinder, make sure the cylinder is fully retracted before adding oil.
  - (Fig. 1.110): When using a hand pump, fill it only to recommended level. Too much oil may limit performance and may result in a leak.
  - (Fig. 1.111): Close the release knob moderately. Too much force will damage the valve.
- **Pump Handle** (Fig. 1.1121): Never use an extension on the handle for more leverage. Hand pumps are designed to work efficiently with the built-in handle.
- **Pressure Setting** (Fig. 1.113): When force setting is critical, always use a calibrated pressure gauge. Never override the factory setting of the relief valve and do not exceed 80% of the cylinder stroke length.
- **Electrical Connection** (Fig. 1.114): Make sure that the electrical cord is not damaged, and the plug matches the electrical receptacle voltage and the current requirements.
- **Pump Battery** (Fig. 1.115): For cordless electric pumps, follow the manufacturer's guide for battery, charger, and pump operation.
- **Pump Start Up** (Fig. 1.116): Regardless of the type of the powered pump, never start the powered pump for first time under load. Always make sure that the directional control valve is in neutral position and it matches the application.

Fig. 1.109 - Retract Attached Cylinder before Filling the Pump (Courtesy of ENERPAC)

Fig. 1.110 - Fill the Pump to Recommended Level (Courtesy of ENERPAC)

Fig. 1.111 - Close the Release Knob Moderately (Courtesy of ENERPAC)

Fig. 1.112 - DO NOT Use Extensions on Pump Handles (Courtesy of ENERPAC)

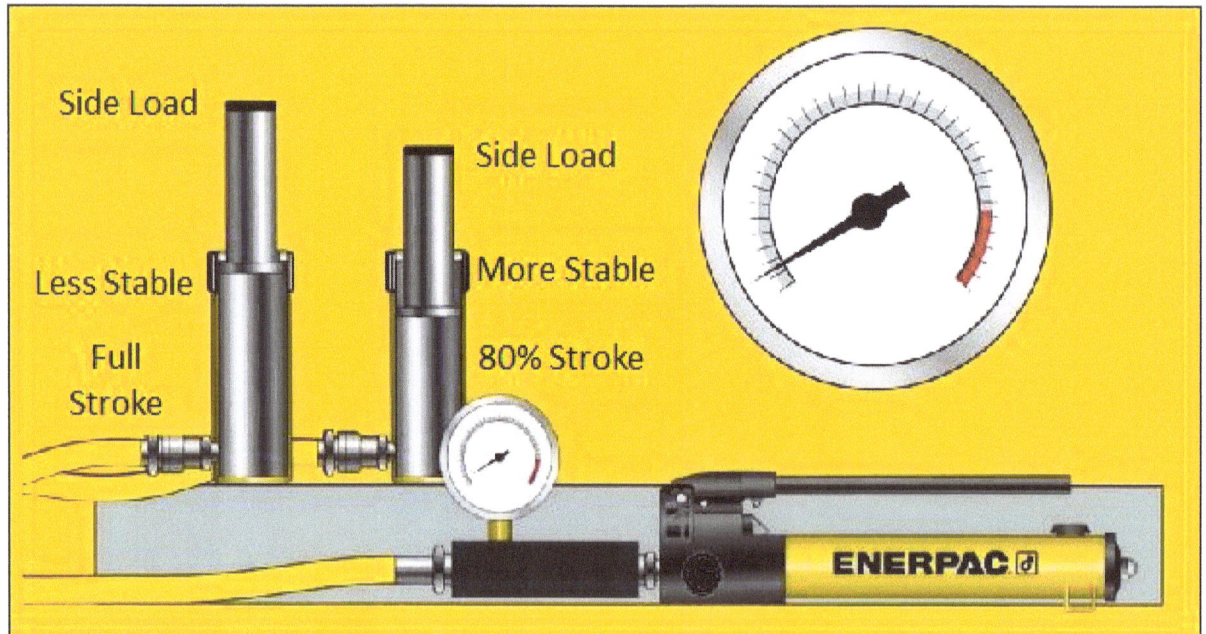

Fig. 1.113 - DO NOT Exceed 80% of the Cylinder Stroke Length (Courtesy of ENERPAC)

Fig. 1.114 - Proper Electrical Connection (Courtesy of ENERPAC)

Fig. 1.115- Battery-Operated Pump (Courtesy of ENERPAC)

Fig. 1.116 - Never Start a Powered-Pump Under Load (Courtesy of ENERPAC)

1.16- BP-Safety-10: Safe Storage and Transportation of Hydraulic Systems

If not lifted appropriately, the hydraulic power unit may lose its stability and thus be knocked over, fall or move in an uncontrolled way causing danger to life. If there is no startup/transportation are provided by the machine manufacturer, the best practices list *BP-Safety-10* provides guidelines for safe startup of the hydraulic system.

BP-Safety-10:
- **Transportation of Hydraulic Power Unit:**
- **Personnel Involvement:** Ensure that no unauthorized persons are within the hazard zone.
- **Oil:** Hydraulic power units shouldn't be transported with oil in or accumulators charged.
- **Packaging:** Review federal/state packaging instructions and regulations for shipping.
- **Weight & CG:** Check the weight and location of the center of gravity.
- **Foundation:** Place the product on a suitable foundation/ground that insures its stability.
- **Attachments Points:** Only the intended locations and attachment points should be used for securing or lifting the hydraulic power unit. Hydraulic power units must never be attached to or lifted at the mounted components (piping, hoses, manifolds, electric motors, accumulators, etc.).

- **Storage:**
- **Ambient Conditions:** Review the ambient conditions available in the product specific documentation.
- **Package Removal:** The packaging should not be removed until directly before assembling the unit. If the package has to be opened e.g. for inspection purposes, you should reseal the packaging to the condition in which it was supplied.
- **Stability:** Provide for sufficient stability before removing any packing/transit materials or fixtures.

Chapter 2

Basic Concepts of Hydraulic System Maintenance

Objectives

This chapter covers basic rules of hydraulic system maintenance and skill set required for service workers. Impact of maintenance on system reliability and various maintenance techniques are presented. Common mistakes and reasons to void warranty are discussed. Best practices of maintaining a specific hydraulic component will be presented in the relevant chapter including guidelines for selection, replacement, installation, storage, maintenance scheduling, and standard testing.

Brief Contents

2.1-Introduction
2.2-Hydraulic Systems Maintenance Programs
2.3-Factors that affect the Frequency of Maintenance
2.4-Hydraulic Systems Maintenance Record Keeping
2.5-Skill Set for Maintenance and Troubleshooting Team
2.6-Setting Hydraulic Systems Stages of Services
2.7-Common Mistakes during Hydraulic System Maintenance
2.8-Reasons to Void Warranty During Hydraulic System Maintenance
2.9-Standard Tests for Hydraulic Components

Chapter 2: Basic Concepts of Hydraulic System Maintenance

2.1- Introduction

2.1.1- Impact of Maintenance on Hydraulic Systems Reliability

Hydraulic systems are designed to be:
- Safe.
- Dry.
- Efficient.
- Quit.
- Reliable.

Therefore, lack of maintenance results in unsafe, leaking, inefficient, noisy, and less productive hydraulic systems.

2.1.2- Major Causes of System Failures

It is well known statistically that major causes of hydraulic system failure are:
- Overheating.
- Fluid contamination.

Defending a hydraulic system against these two enemies improves system reliability.

2.1.3- Factors that Reduce Service Life of a Hydraulic System

Service life of hydraulic systems are drastically reduced due to one or more of the following:
- Hydraulic fluid physical, chemical, and contamination instability.
- Poor components maintenance.
- Poor component quality.
- Abuse or improper use of the components or systems.
- Ignoring manufacturers' operational and maintenance recommendations.

2.2- Hydraulic Systems Maintenance Programs

Maintenance of hydraulic systems should be planned as follows:
- Done on predetermined bases.
- Done by qualified workers.
- Based on instructions from the system/component manufacturer.

As shown in Fig. 2.1, various maintenance methodologies are applicable for all industry sectors including hydraulic systems:

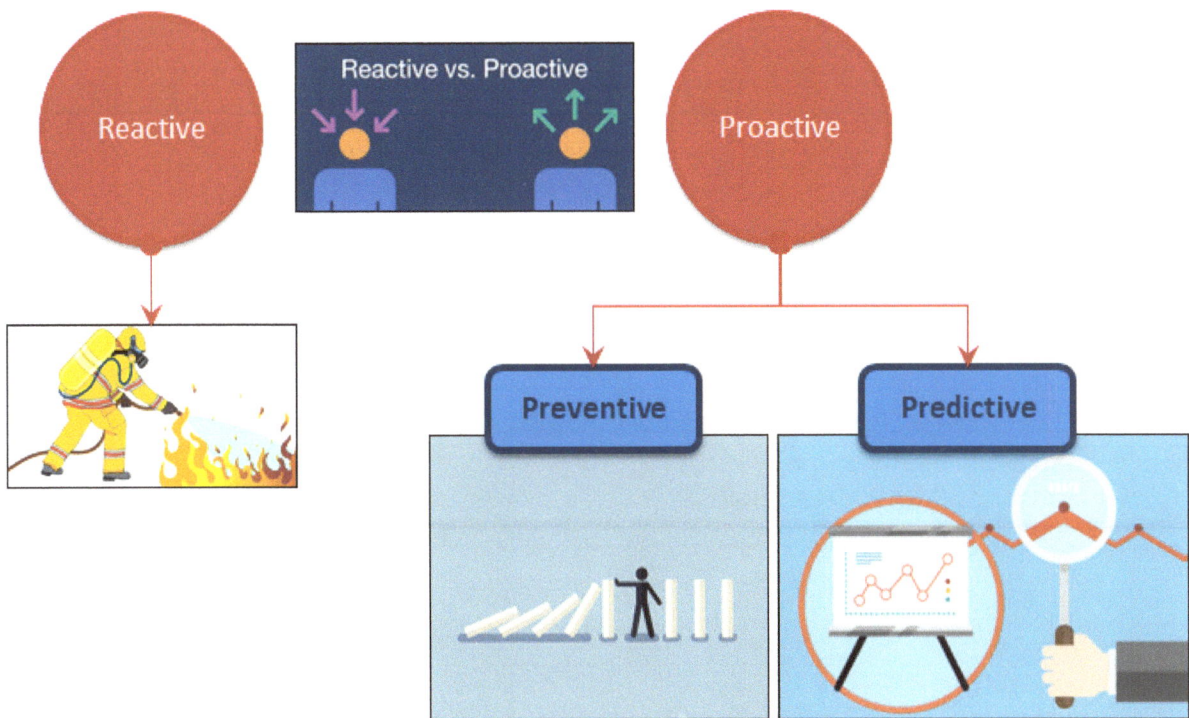

Fig. 2.1- Maintenance Methodologies

Reactive Maintenance:
Reactive Maintenance is also referred to as "Run-to-Failure", "Break Down", "On Demand", or "Fire Fighting". In this method, hydraulic components are repaired as they fail. In such case, a hydraulic system can possibly fail and shutdown at any time. Today, the cost of running a machine to failure is too high in terms of:
- Production downtime Cost.
- Parts and material inventory.
- Labor cost.
- Additional costs may be associated with hydraulic system failure and other liabilities.

Preventive Maintenance:

Preventive Maintenance contains selected maintenance tasks that are routinely scheduled to avoid sudden failures. The below list of preventive maintenance actions has the greatest effect on a hydraulic system lifetime and should be scheduled more frequently than others. If any improper symptoms are found, find the root cause of such symptom and take proper actions to resolve the problem.

- **Hydraulic Fluid:** Check oil level.
- **Filters Breathers, and Strainers:** Check clogging conditions and change accordingly.
- **Hydraulic Reservoir:** Check inside reservoir for sludge, bacteria, etc.
- **Operating Pressure:** Check and record operating pressure.
- **Operating Temperature:** Check and record operating temperature.
- **Plumbing:** Check tightness (not to the point of distortion), and leakage.
- **Outside Surfaces**: Check and keep it clean.
- **Pumps**: Check and record pump flow and suction pressure.
- **Actuators:** Check for external leakage, seals/wipers, and connections with load.
- **Valves:** Check spool shifting and electrical connections for EH valves.
- **Heat Exchangers:** Check for effectiveness.
- **Noise/Vibration/Odor:** observe if any of these conditions are found.

Predictive (Smart) Maintenance:

Predictive Maintenance program is based on routine inspections and tests to detect *root-causes* that may lead to future failures. So, problems are solved in advance to avoid unexpected failures. Examples of predictive maintenance actions are

- Hydraulic fluid contamination analysis (checking cleanliness level).
- Hydraulic fluid content analysis (contents of Copper, Iron, Silicon, H_2O).
- Hydraulic fluid conditions (viscosity, additives, and oxidation).
- Wear and failure analysis.
- Vibration measurement.

Proactive Maintenance:

Proactive Maintenance program is a combination of Preventive and Predictive maintenance. Proactive maintenance is most expensive and requires skilled and trained personnel. Considering savings from avoiding unexpected downtime, proactive maintenance is least expensive and found to reduce operational cost by 30% over a reactive maintenance. Because components last longer, and sudden failure and shutdown are minimized.

Unfortunately, with that all concepts known; Fig. 2.2 is developed to reflect a matter of fact. Many of the manufacturing facilities, while they follow conservative spending policies on reliability training and proactive maintenance, they spend majority of the maintenance budget just to respond to systems failures.

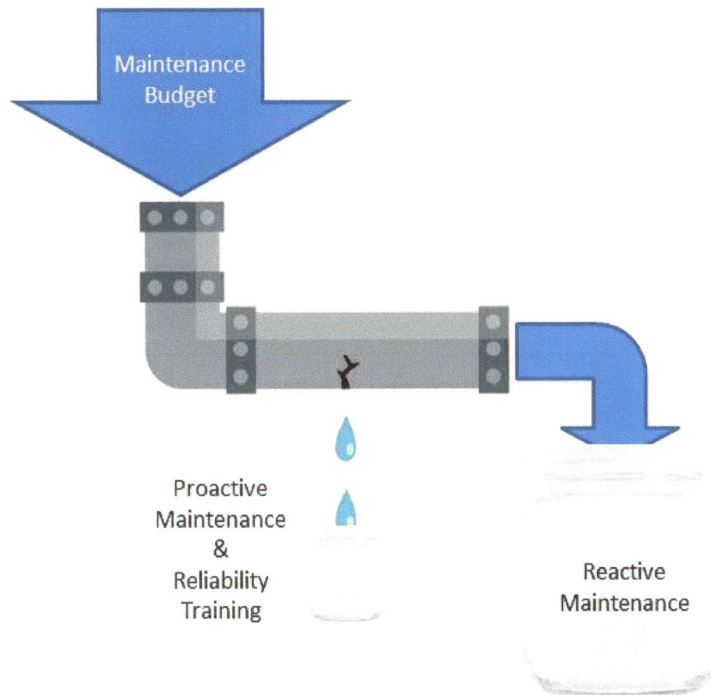

Fig. 2.2- Unwise Budget Spending

In this textbook: for each hydraulic component, some preventive maintenance actions are suggested and scheduled. The frequency of the suggested actions is based on the best understanding and the experience of the author under assuming that the system is working under normal operating conditions. This frequency may change based on several factors. Therefore, it is highly recommended to review the components and system manufacturer for further instructions.

2.3- Factors that affect the Frequency of Maintenance

As shown in Fig. 2.3, if maintenance tasks are scheduled based on number of hours, frequency of scheduling maintenance tasks for a hydraulic system is generally affected by the operating conditions under which the system works and how critical the application is. If maintenance tasks are scheduled based on calendar, then number of working hours per day will be considered and added to the factors shown in the figure.

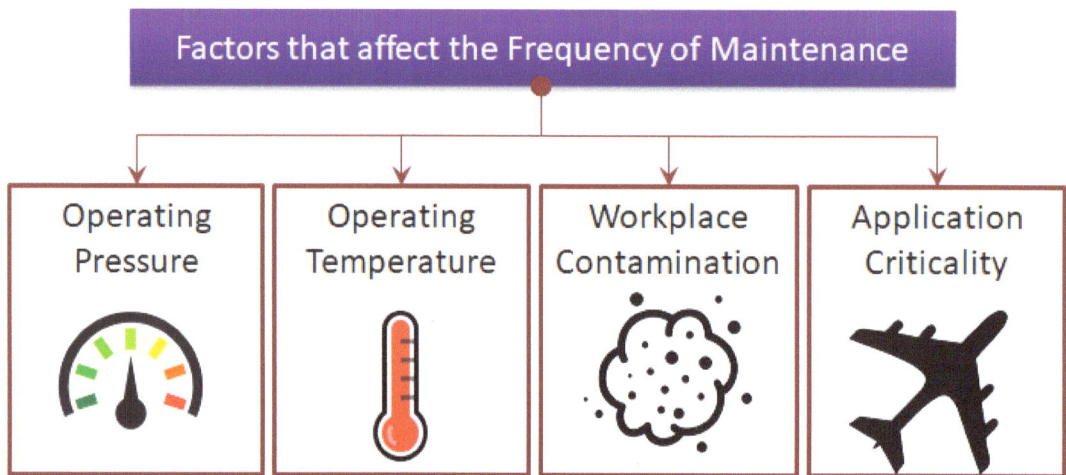

Fig. 2.3- Factors that affect the Frequency of Maintenance

❖ **Operating Pressure:**
Effect of abrasive contaminants under high pressure is much more severe than under low pressure. Therefore, systems that work under high pressure require more frequent checks.

❖ **Operating Temperature:**
High working temperature affects the properties of the life blood of the machine (hydraulic fluid) and accelerates failure of hydraulic systems. Therefore, such systems must be checked more frequently than others.

❖ **Workplace Contamination:**
Highly contaminated work environment increases the chances of oil contamination. Consequently, hydraulic systems are at high risk of failure.
- Example 1: cement factories have a lot of dusty air.
- Example 2: offshore workplaces have a lot of moisture.

❖ **Application Criticality:**
If a hydraulic system drives a very critical application, then maintenance program became very strict and more frequent.
- Example 1: hydraulic systems in aerospace industry for the sake of human safety.
- Example 2: hydraulic systems in expensive machines for the sake of reducing the downtime cost.

2.4- Hydraulic Systems Maintenance Record Keeping

Maintenance actions must be recorded for future investigations. Forms for collecting hydraulic system and component data should conform with ISO 4413. However, Table 2.1 shows a suggested record sheet for a hydraulic system maintenance.

Hydraulic Systems Maintenance Record		
Equipment #		
Location		
Service Person	Date	Action Performed

Table. 2.1- Suggested Record Sheet for Hydraulic System Maintenance

2.5- Skill Set for Maintenance and Troubleshooting Team

The person who is involved in hydraulic system maintenance and/or troubleshooting must maintain a good level of understanding of the following:
- **Safety:** Safety requirements for maintaining and troubleshooting of a hydraulic system.
- **Basic Principles:** force, work, pressure, flow rate, basic math, etc.
- **Hydraulic Components:** Construction and operating principle.
- **Hydraulic Systems:** Symbols and schematic.
- **Hydraulic Fluids:** Sampling, analysis, and interpretation of analysis report.
- **Filtration:** Filtration necessary to achieve the system's proper ISO contamination level.
- **Filters:** Types, application, clogging conditions, cleaning, and replacements.
- **Reservoir:** Cleaning and make up fluid.
- **Plumping:** Types, sizes, cleaning, fittings, replacement, tightening, etc.
- **EH Valves:** Construction, operation, null position, flow gain, and pressure gain.
- **Troubleshooting Principles:** Utilize "Root Cause Failure Analysis".
- **Commissioning:** Safety and startup to manufacturer's instructions.
- **Flushing:** Procedure and evaluation.

2.6- Setting Hydraulic Systems Stages of Services

As shown in Fig. 2.4, for extendable and effective servicing of a hydraulic systems, the following three lines of services are recommended.

First Stage of Service (On Spot):
- This means at the machine itself.
- At this level, proper tools should be available on spot to perform simple maintenance tasks such as filter change, make up oil, transmission line replacement, etc.

Second Stage of Service (Local Repair Shop):
- This means at a repair shop in a plant or at a dealer's service center.
- At this level, the repair shop should be equipped to perform intermediate repair tasks such as pump and cylinder rebuilds.

Third Stage of Service (Manufacturer):
- This means at the manufacture's site or certified service center.
- At this level, service centers are very well equipped to perform complex repair tasks that are restricted only for the manufacturer such as servo valve maintenance and reset, etc.

Fig. 2.4- Line of Services (Courtesy of Vickers)

2.7- Common Mistakes during Hydraulic System Maintenance

This section presents common mistakes that were found frequently repeated during hydraulic system maintenance.

- **Mistake 01-Tightening Torque:** *Tightening* hydraulic components, particularly valves, to unspecified torque leads to body distortion and spool seizure (Fig. 2.5) and added mechanical stress to the component.

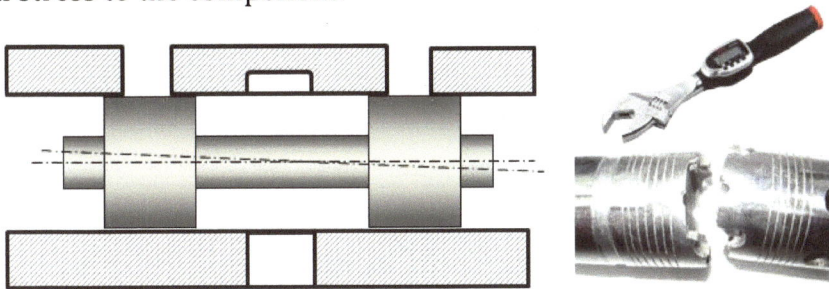

Fig. 2.5- Valve Body Distortion due to Over Tightening Torque

- **Mistake 02-Running a Component on Dry Conditions:** If a component needs to be tested, it shouldn't be tested on dry conditions. An example of that is a pump must never run without oil or an EH valve must not be signalized without oil (Fig. 2.6). Otherwise, if the component runs without oil, internal clearances are affected due to lack of lubrication.

- **Mistake 03-Use of Inadequate Washing Fluid:** Cleaning internal parts of a hydraulic components requires special cleaning fluids. Such cleaning fluids remove dirt and lubricate moving parts as well. As shown in Fig. 2.7, a common mistake is to use Benzene fuel to clean internal parts of hydraulic components. Such a fuel does not provide lubrication and will make it difficult to assemble these parts back. So, specified solvents must be used. Other related mistakes are using washing fluid with improper cleanliness level or chemically incompatible with the hydraulic components' material. Another common mistake is using such solvents in a way that risks the environmental safety.

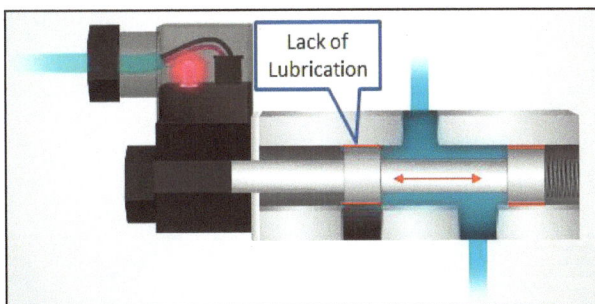

Fig. 2.6- Signalizing a Valve with no Oil affects the Internal Clearances

Fig. 2.7- Only Specified Solvents in Cleaning Internal Parts

- **Mistake 04-Use of Inadequate Cleaning Towels:** As shown in Fig. 2.8, only industrially specified and lint-free towels are used for cleaning. Using woolen or lint type tissues to wipe and dry hydraulic components after washing them is a common mistake. The hair from such towels will find its way inside the components.

- **Mistake 05-Search for Leakage with the Operator's Hand:** As shown in Fig. 2.9, a common mistake that machine servicers used to do is to use their hands to check *oil leakage*. It is very dangerous to do that because this method most likely will result in *oil injection* into the servicer's hand. Alternatively, special tools must be used to search for leakage.

Fig. 2.8- Cleaning Towel

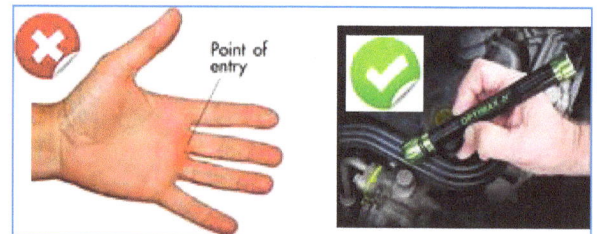

Fig. 2.9- Oil Injection

- **Mistake 06-Improper Line Routing:** As shown in Fig. 2.10, a common mistake a machine servicer used to do is to route the transmission lines improperly. *Routing* transmission lines improperly results in increasing energy losses and premature failure.

- **Mistake 07-Long hoses Unfastened:** Leaving hard tubing or long hoses *unfastened* causes noise, vibration, and rubbing of hoses against sharp edges. Figure 2.11 shows that long hoses must be fastened to a fixed frame.

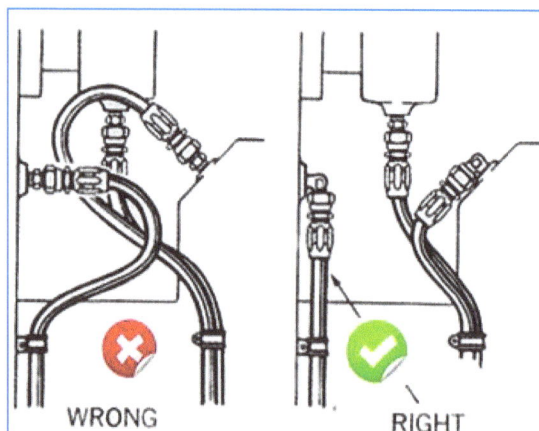

Fig. 2.10- Proper Routing of Hoses

Fig. 2.11- Fastening Long Hoses

- **Mistake 08-Wrong Identification of a Pump Suction Port:** Usually, as shown in Fig. 2.12, a unidirectional pump has a *suction port* that is larger than the discharge pump to avoid pump cavitation. A pump can drastically fail if it was improperly connected to the reservoir.

- **Mistake 09-Wrong Identification of the Direction of Rotation:** A pump *direction of rotation* is related to the position of the suction port of the pump. If a pump was driven in the wrong direction, even if it was connected properly to the reservoir, it will be cavitated immediately.

- **Mistake 10-Shutoff Valve on a Pump Suction Line is Left Unlocked:** In some systems, a *shutoff valve* may be placed on a pump suction line so that the pump can be disassembled without draining the reservoirs. As shown in Fig. 2.13, such a valve must be locked open during machine operation. Otherwise, if left unlocked and mistakenly closed while the pump is running, the pump will fail immediately.

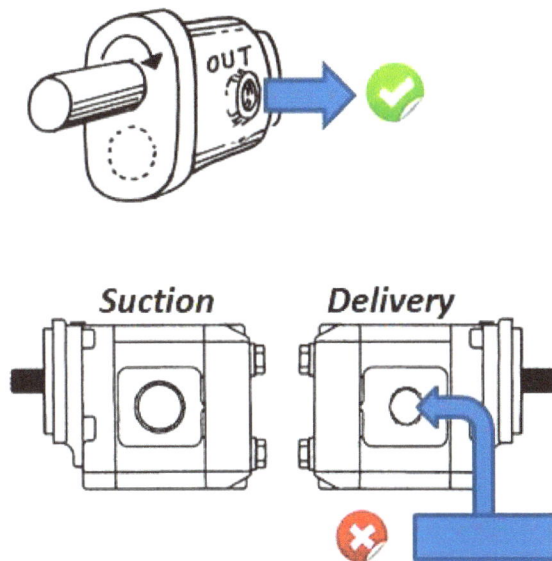

Fig. 2.12- Incorrect Pump Connection   **Fig. 2.13- Lockable Shut Off Valve**

- **Mistake 11-Mixing Different Types of Hydraulic Fluids:** A common mistake is to use a different hydraulic fluid to make up the fluid in the reservoir. Mixing different types of hydraulic fluid results in unfavorable products.

- **Mistake 12-Make Up Oil Without Filters:** A common mistake is making up hydraulic fluids directly to a reservoir without pre-filtration. Oil in the reservoir may be cleaner than the unused oil in the barrel. A portable filtration unit with appropriate filter rating must be used to transfer the oil from the barrel to the oil tank.

- **Mistake 13-Changing Component Location:** As shown in Fig. 2.14, in a *Meter-In* speed control system, the flow control valve was originally placed on the line that connects the directional valve to the piston side of the cylinder. Just moving the flow control to the line that connects the pump to the directional valve will make the machine perform totally different. The flow control valve in its original place controls only the extension speed of the cylinder. The flow control valve in its new location controls the speed of the cylinder in both directions, this may put the machine operation out of sequence. As a result, possible damage to the machine could occur. However, before changing the location of a hydraulic components, the circuit operation must be reanalyzed to make sure that it will work safely and reliably.

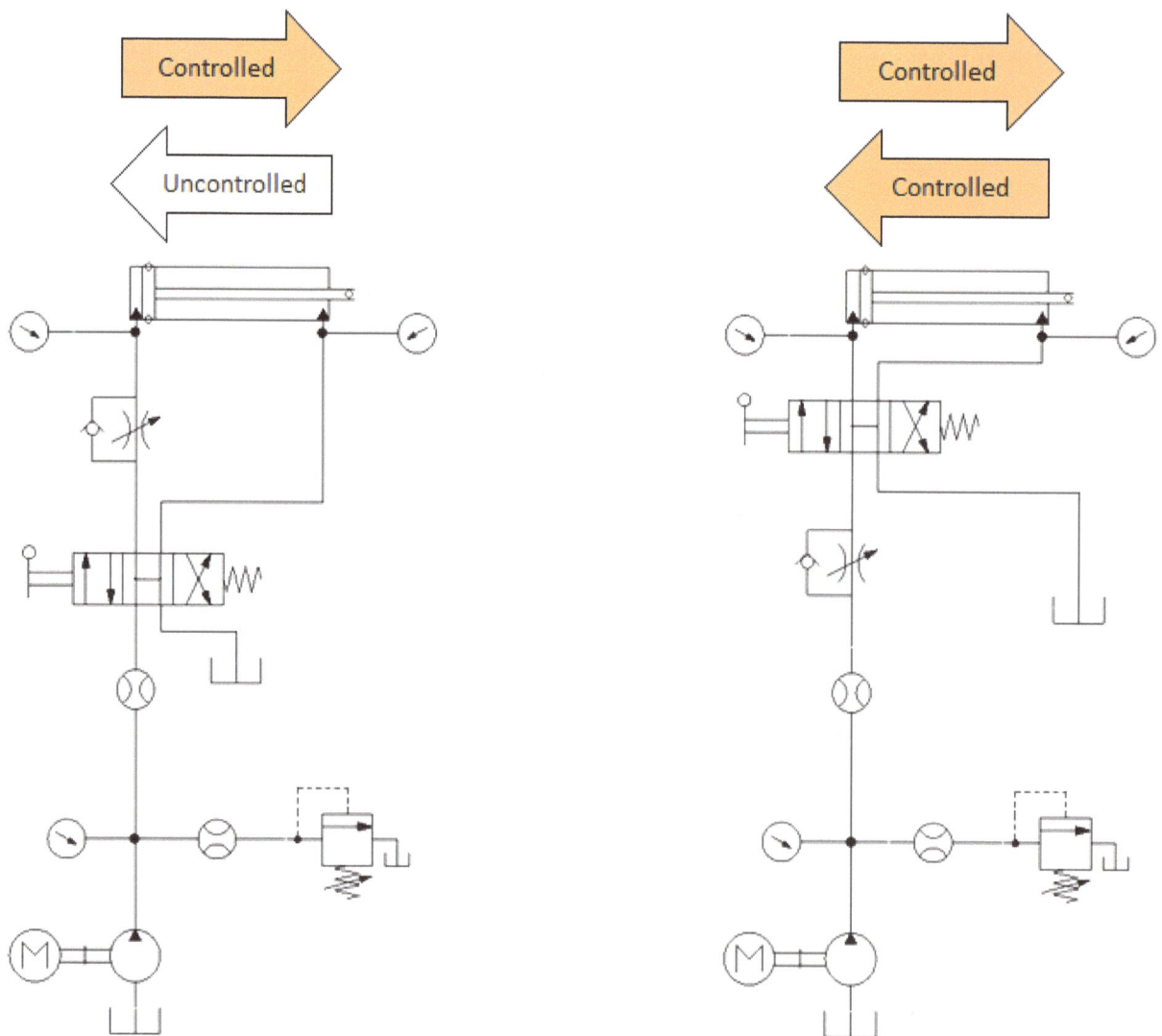

Fig. 2.14- Meter-In Speed Control

- **Mistake 14-Changing Component Specification:** When a hydraulic problem occurs, usually one component has failed. It is essential to match the part numbers between the new and old components. Hydraulic pumps and valves that look alike are not necessarily the same. An example of that is the hydraulic-driven cooling system shown in Fig. 2.15. The system contains a pilot-operated, float-center directional valve. Float-center valve is selected so that when the system is shut off, the fan inertia doesn't fracture the motor shaft. Replacing the original directional valve by one that has a closed center may result in a motor shaft or line failure. The probable cause is that the service person was confused between the symbol of the pilot stage and the main stage. Regardless the central position of the main stage, pilot stage should be of a float type in order to make sure that the main stage is perfectly centered when the pilot stage is de-energized.

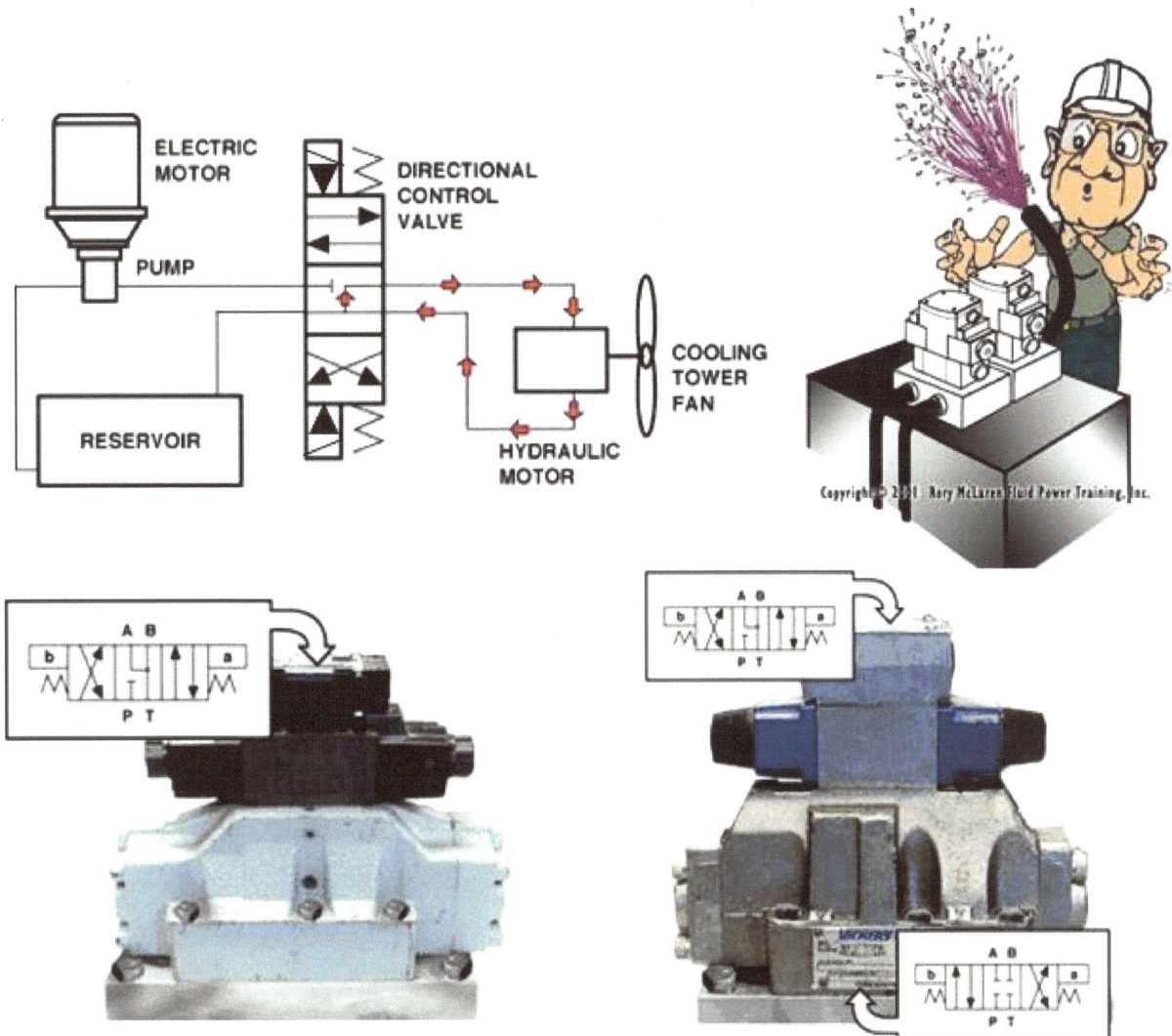

Fig. 2.15- Hydraulic Driven Cooling System

▪ **Mistake 15-Changing Component Settings:** As best practices for safe servicing a hydraulic-driven machine, after maintaining adjustable hydraulic components, they must be properly reset. Some of the settings are critical to prevent unsafe operation of the system. Because of that, only trained and authorized personnel are permitted to make components adjustments or settings. Valves with lockable knobs are recommended for critical adjustments. The following are examples of adjustable components:

o **Sequence Valve**: Improper setting of a sequence valve may result in machine or work piece destruction.

o **Pressure Relief Valve:** Setting a PRV low results in inoperative hydraulic actuators. Setting a PRV high may result in machine or work piece destruction.

o **Variable Displacement Pump:** Variable pumps have different controllers as follows:
   - Power-Controlled Pump set Low → loss of power + inoperative system.
   - Power-Controlled Pump set High → prime mover stalls.
   - Pressure-comp. pump set Low → reduced pressure + stopped actuators.
   - Pressure-comp. pump set High → Increased pressure + damage.
   - Displacement–Controlled Pump set Low → reduced flow + slow actuators.
   - Displacement–Controlled Pump set High → increased flow and possible:
     ✓ Turbulent flow in the pipelines.
     ✓ Higher flow forces in the valves.
     ✓ Increased speed of actuators.

▪ **Mistake 16-Reuse of Sealing Elements:** *Sealing Elements*, such as *O-Rings* and *Cutting Rings* shown in Fig. 2.16, perform their job when they are plastically deformed and squeezed into their installation space. A common mistake is to reuse the sealing elements. That results in system leakage.

Fig. 2.16- Sealing Elements

- **Mistake 17-Incompatible Sealing Material:** Using a sealing element that is *incompatible* leads to seal physical and chemical deterioration, internal/external leakage, blocking control orifices, and stick-slip actuator motion. The following factors must be considered when choosing a seal material:

  o TEMPERATURE COMPATIBILITY: *Temperature Compatibility* is one of the prime considerations involved in seal material selection. In selecting a seal material, the user must determine the upper and lower extreme temperatures at which the seal will be expected to function and also the duration to which these extremes of temperature will be experienced, bearing in mind that compression set (and therefore seal-ability) is a function of the time/temperature relationship.

  o PRESSURE COMPATIBILITY: *Pressure Compatibility* is another consideration when selecting a seal material. This characteristic is determined by the durometer (hardness) of the material and by its internal chemistry.

  o FLUID COMPATIBILITY: *Fluid Compatibility* is very important because all elastomers are affected to some degree when they get in contact with system fluid.

- **Mistake 18-Improper Pump Priming:** Pump *priming* is an essential step before starting up a new system or after pump major repair. Failing to prime a pump may lead to cavitation and premature failure.

- **Mistake 19-Improper Air Bleeding:** Cylinder *air bleeding* is an essential step before starting up a new system or after cylinder major repair. Failing to bleed air may lead to sudden or erratic movement.

- **Mistake 20-Potential System Contamination and Environmental Pollution:** Maintenance procedures must be structured to minimize potential contamination of the hydraulic system and environmental pollution. Common mistakes during hydraulic component disassembling are to allow oil spilling on the ground and to leave disassembled component ports uncovered.

2.8- Reasons to Void Warranty During Hydraulic System Maintenance

Usually manufacturers of hydraulic components provide warranty for components against manufacturing defects. The validation of this warranty is conditioned by the manufacturers' instructions. The following are examples of cases when warranty is voided.

- **Reason 1:** Oil contamination level is higher than what is specified by the manufacturer.

- **Reason 2:** Use of unspecified fluid. For example, using regular fluid instead of Extreme-Pressure (EP) fluid that is specified by the manufacturer.

- **Reason 3:** Use of non-original spare parts to recondition a hydraulic component.

- **Reason 4:** Abuse a component by using it out of the recommended maximum operating conditions; namely pressure, temperature, and speed.

- **Reason 5:** Improper installation of a component with disrespect to the manufacturer's instructions.

- **Reason 6:** Setting components that are specified to be set only by the manufacturer or certified agent. Figure 2.17 shows an example of a proportional valve in which the *Dither* amplitude is to be adjusted only by instructed specialists.

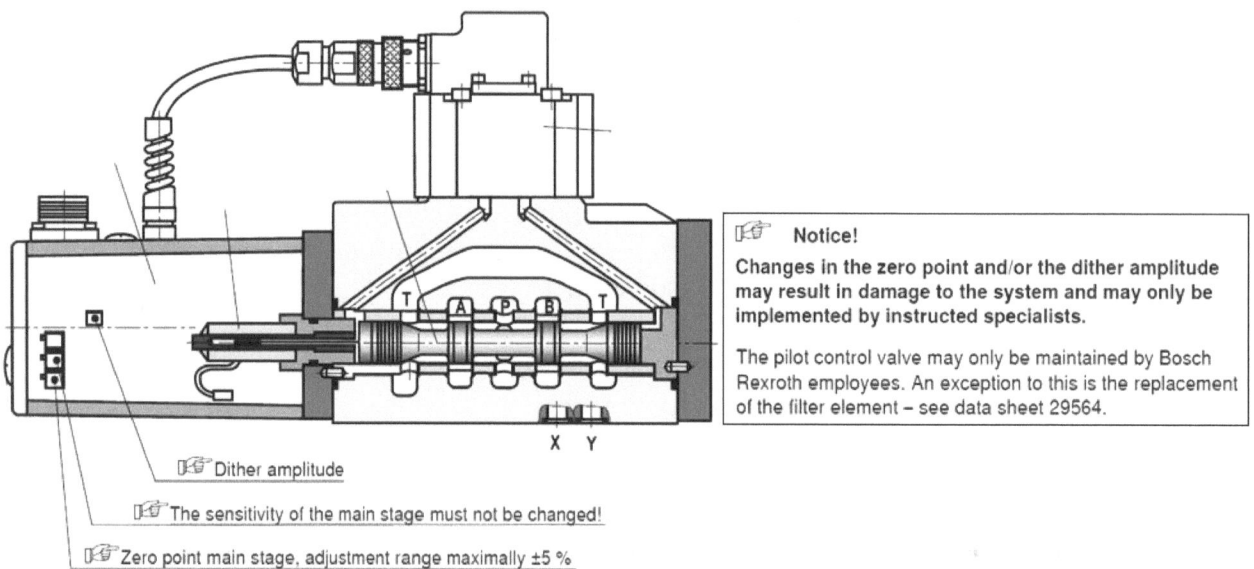

> **Notice!**
> Changes in the zero point and/or the dither amplitude may result in damage to the system and may only be implemented by instructed specialists.
>
> The pilot control valve may only be maintained by Bosch Rexroth employees. An exception to this is the replacement of the filter element – see data sheet 29564.

☞ Dither amplitude

☞ The sensitivity of the main stage must not be changed!

☞ Zero point main stage, adjustment range maximally ±5 %

Fig. 2.17- Instructions for Setting a Proportional Valve

2.9- Standard Tests for Hydraulic Components

2.9.1- Reasons for Tests

Hydraulic components are tested for one or more of the following reasons:
- Engineering Validation.
- Product Qualification.
- Quality Assurance at Production.
- Failure Mode Analysis.

2.9.2- Testing Standard Sources

Various sources contribute in developing standards, the following sources are the most common:
- American Society of Mechanical engineers (ASME).
- Society of Automotive engineers (SAE).
- National Fluid Power Associations (NFPA).
- International Organization for Standardization (ISO/TC 131).
- Application Specific.
  - MIL Standards.
  - Manufacturers.
  - Other Countries.

2.9.3- Type of Tests

The following are typical tests performed to evaluate the capabilities of hydraulic components:
- Pressure Rating (Burst, Fatigue, and Impulse).
- Temperature-Viscosity.
- Contamination (Wear & Sensitivity).
- Efficiencies (Volumetric, Mechanical, and Overall).

2.9.4- Testing Requirements

Test process must be thoroughly designed to obtain accurate results. Designing a test process requires paying attention to the following requirements:
- Maintaining high accuracy and repeatability.
- Define proper instrumentation, scale, and monitoring devices.
- Proper design of fixtures and loading mechanisms.
- Cost.

Chapter 3

Hydraulic Measuring Instruments

Objectives

This chapter provides an overview of the common measuring devices used in hydraulic systems including devices for measure pressure, flow, temperature, oil level, and load cells. The chapter introduces the difference between a meter, a switch, and a sensor. The chapter also discusses the best practices for measuring devices selection & replacement, maintenance scheduling, installation & maintenance, and standard tests & calibration.

Brief Contents

3.1-Classification of Measuring Instruments
3.2-Pressure Measuring Instruments
3.3-Flow Measuring Instruments
3.4-Oil Level Measuring Instruments
3.5-Oil Temperature Measuring Instruments
3.6-All-In-One Measurement Instruments
3.7- Load Cell
3.8-Other Measuring Devices
3.9-Hydraulic Data Logger
3.10-Wireless Sensors and IoT in Hydraulic System Maintenance
3.11-BP-Measuring Instruments-01-Selection and Replacement
3.12-BP-Measuring Instruments-02-Maintenance Scheduling
3.13-BP-Measuring Instruments-03-Installation and Maintenance
3.14-BP-Measuring Instruments-04-Standard Tests and Calibration

Chapter 3: Hydraulic Measuring Instruments

3.1- Classification of Measuring Instruments

As shown in Fig. 3.1, there are various types of measuring instruments for condition monitoring of hydraulic systems. Depending on the type of the measuring instrument, they contain mechanical or electro-mechanical mechanisms to provide the dial reading or digital reading on a small screen.

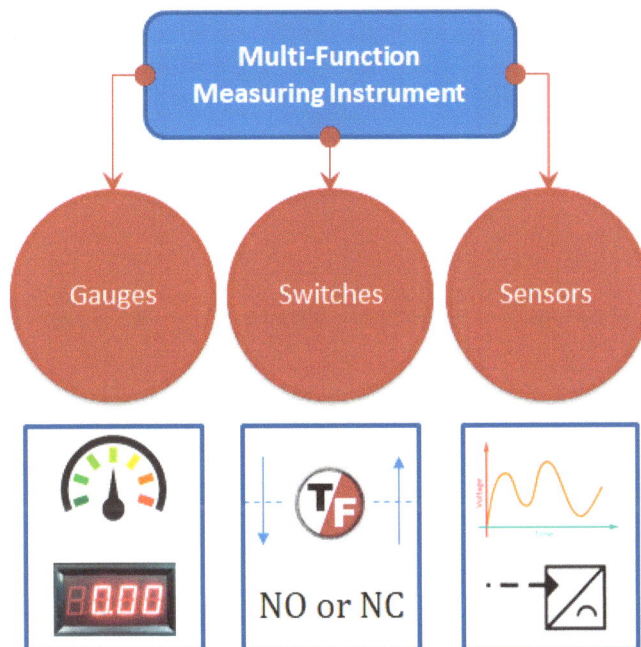

Fig. 3.1- Classification of Measuring Instruments

Gauges (Meters): *Gauges* are used for visual readings of the specific parameters that reflect working conditions of hydraulic systems. Gauges are also referred to as *Meters*.

Switches: *Switches* are devices used for *on-off* control mode. A switch provides a binary (TRUE or FALSE) electrical signal when the controlled parameter reaches a preset value while it is raising or falling. They can be selected as *normally-open* or *normally-closed*. They can be equipped with a light indicator or buzzing sound.

Sensors (Transducers): *Sensors* are devices used for *continuous* control mode. Sensors contains electro-mechanical part, referred to as *Transducers*, that converts the measured parameter into a proportional electrical signal in voltage or current format.

Multi-Function Measuring Instruments: The *multi-function* measuring instrument is a device that can be set to work as a gauge, switch, sensor, or any combination of them.

3.2- Pressure Measuring Instruments

3.2.1- Pressure Gauges

Pressure Gauges are used for visual readings of the working pressure. Figure 3.2 shows the classification of pressure gauges based on the number of measured points and based on the working mechanism.

Any of the shown below pressure gauges offers different scales and units of pressure values. Some are filled with glycerin to suppress pressure shocks and protect the reading mechanism such as the needle. If a small-scale pressure gauge is used to measure high pressure or subjected to high pressure shocks, the working mechanism may plastically deformed and become be no longer usable.

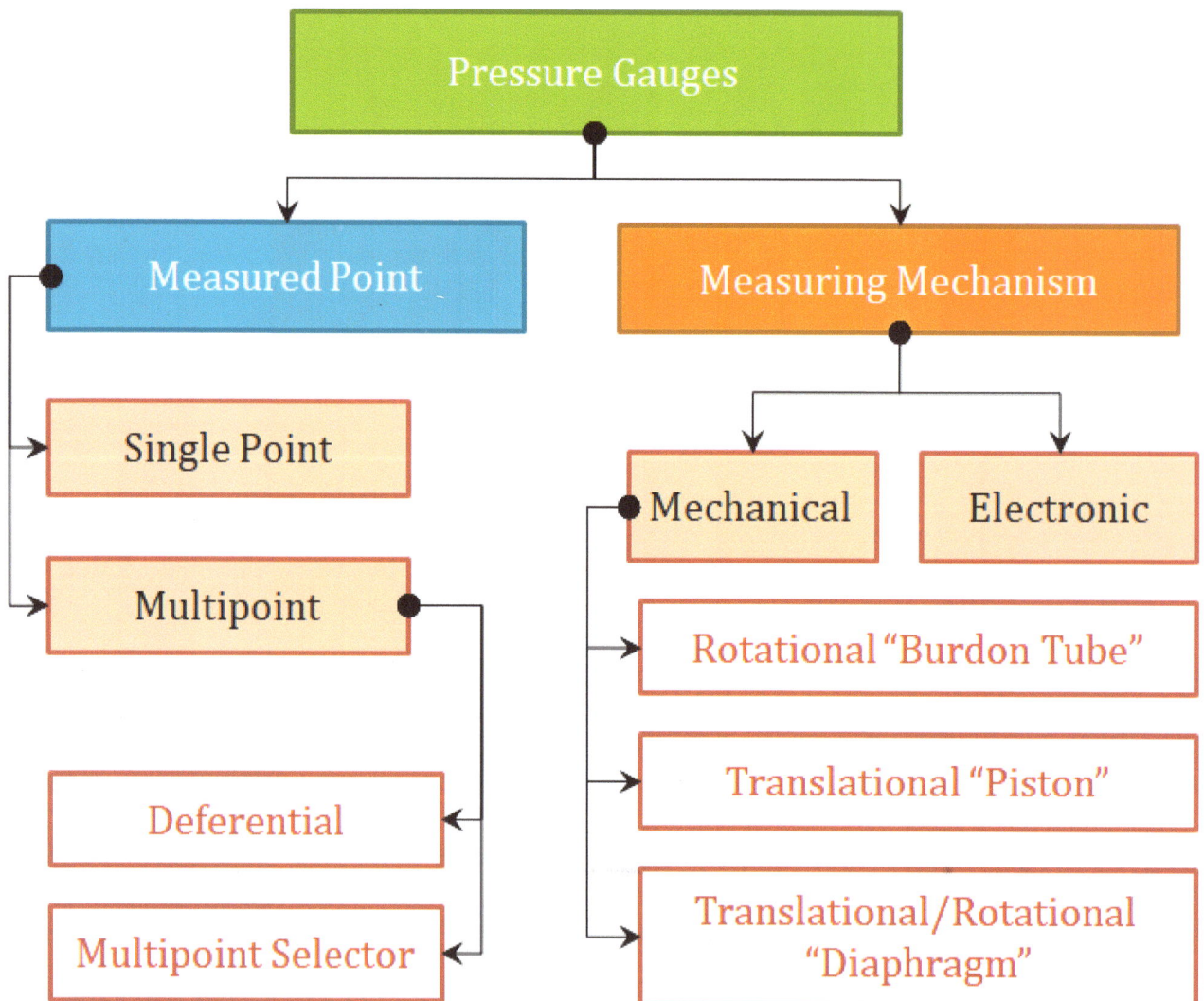

Fig. 3.2- Classification of Pressure Gauges

3.2.1.1- Burdon Tube Pressure Gauge

Burdon Tube pressure gauges are the most common for single point pressure measuring. As shown in Fig. 3.3, the working mechanism consists of a specially constructed and calibrated tube connected mechanically to a rotational mechanism. When the tube is expanded under the effect of pressure, it forces the connected mechanism to rotate a needle that points out to the corresponding pressure value. The figure shows a symbol for general pressure gauge.

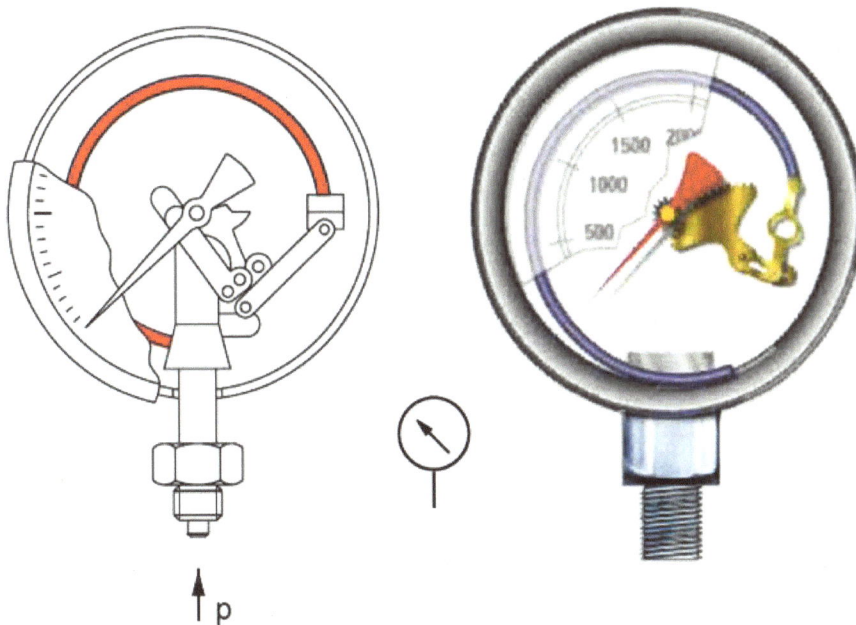

Fig. 3.3- Burdon Tube Pressure Gauge (Courtesy of Rexroth)

3.2.1.2- Piston-Type Pressure Gauge

Piston-Type pressure gauges are commonly used for a single point high pressure measuring. As shown in Fig. 3.4, the working mechanism is as simple as a piston subjected to pressure force from one side and supporting spring from the other side. When the pressure force exceeds the spring forces, the translational motion due spring compression is calibrated to provide the corresponding pressure value.

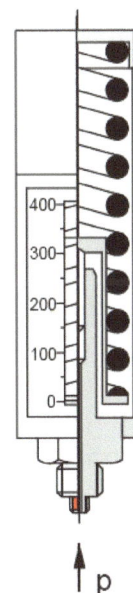

Fig.3.4- Piston-Type Pressure Gauge (Courtesy of Rexroth)

3.2.1.3- Diaphragm-Type Pressure Gauge

Diaphragm-Type pressure gauges are commonly used for a single point low pressure measuring. As shown in Fig. 3.5, the working mechanism consists of a translational part and a rotational part. A specially designed and calibrated diaphragm deflects translationally under the effect of pressure. As a result, the connected mechanism rotates a needle that points out to the corresponding pressure value.

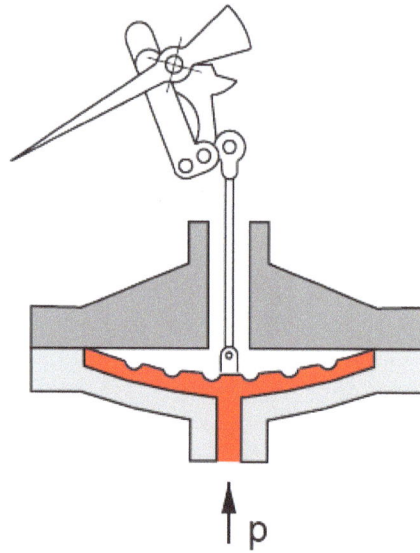

Fig. 3.5- Diaphragm Type Pressure Gauge (Courtesy of Rexroth)

3.2.1.4- Pressure Gauge with Digital Readings

As shown in Fig. 3.6, pressure gauges could be equipped to offer digital readings on a built-in small screen. In such a case, the motion of the mechanism is used to generate an electrical signal, which in turn is used to generate the digital reading on the screen.

Alternatively, instead of using any mechanical parts, a small piece of *strain gauge* is used. Strain gauge deflects under the effect of pressure. Hence, it produces an electrical signal proportional to the pressure. Strain gauge is much more reliable than mechanical mechanism.

Fig. 3.6- Pressure Gauge with Digital Reading (Courtesy of Parker)

3.2.1.5- Differential Pressure Gauge

As shown in Fig. 3.7, *differential* Pressure Gauges are made to measure the difference in pressure between two points. Such gauges are helpful in diagnosing proper functioning of hydraulic components and systems such as monitoring the clogging condition of a filter. As shown in the figure, pressure reading is offered either via a dial or digitally on a screen. The figure shows a symbol for general differential pressure gauge.

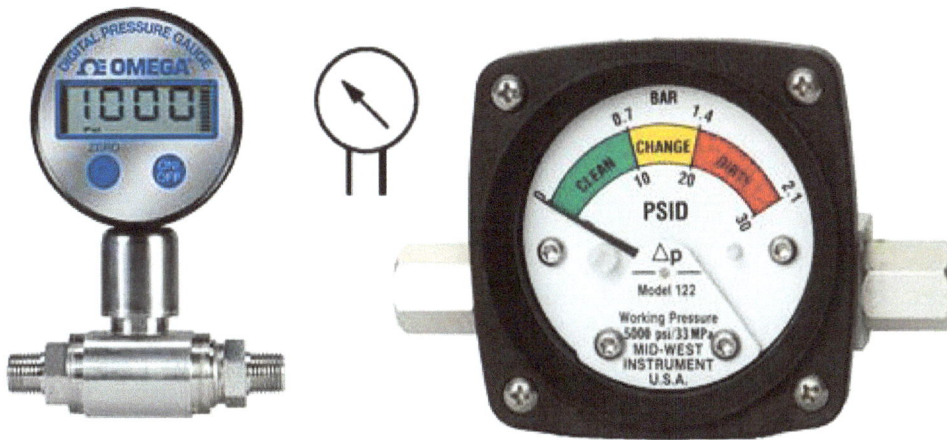

Fig. 3.7- Differential Pressure Gauge

3.2.1.6- Multipoint Pressure Gauges

Multipoint Point Pressure Gauges are used to measure pressure at more than two points. As shown in Fig. 3.8, a rotatable knob is used to select one reading at a time. Using such a gauge saves space for installing multiple gauges.

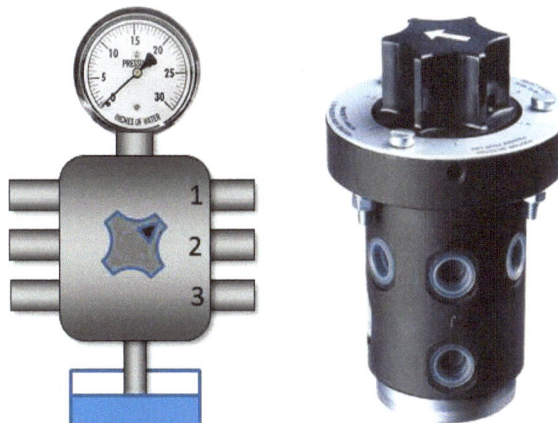

Fig. 3.8- Multipoint Pressure Gauge

3.2.2- Pressure Switches

Pressure *Switches* are devices used for *on-off* control mode. A switch provides a binary (TRUE or FALSE) electrical signal when the controlled parameter reaches a preset value while it is raising or falling. They can be selected as *normally-open* or *normally-closed*. They can be equipped with a light indicator or buzzing sound.

Figure 3.9 shows a Normally-Closed (NC) pressure switch. The switch has a socket with four pins, two for powering the switch and two for the signal. Since the micro switch (4) is in contact with the spring base (2), once the switch is powered, the signal is turned ON (TRUE). When the pressure force exceeds the force of the supporting spring (3), the push rod (1) displaces the spring base (2) against the supporting spring (3) separating it from the micro switch (4). Hence, the signal is turned Off (FALSE). Setting the pressure activation value is by changing the spring compression.

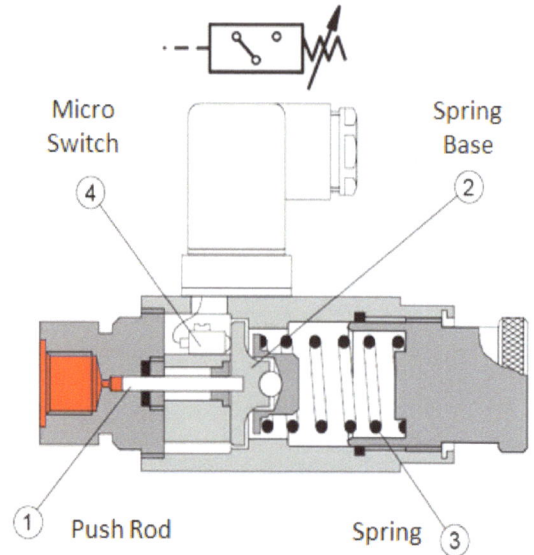

Fig. 3.9- Normally-Closes Pressure Switch (Courtesy of Rexroth)

Figure 3.10 shows a Normally-Open (NO) pressure switch. The construction and the function of this switch is the same as the normally-closed one except the signal is Off by default and switched On when the pressure activates the switch. For better troubleshooting and system condition monitoring, the switch shown in the figure is equipped with a light indicator.

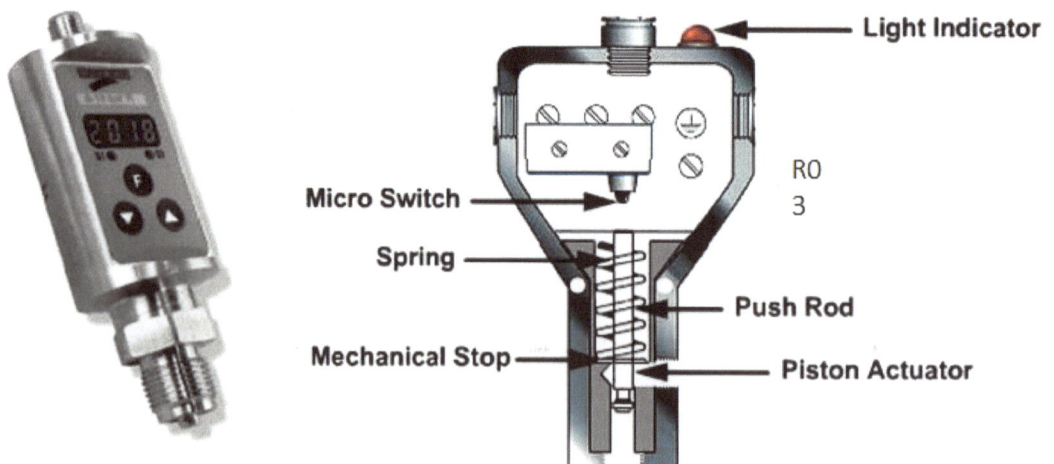

Fig. 3.10- Normally-Open Pressure Switch

3.2.3- Pressure Sensors

Pressure Sensors are devices used for *continuous* control mode. Pressure sensors contain an electro-mechanical part, referred to as *Transducer*, that converts the pressure into a proportional electrical signal in voltage or current format. Figure 3.11 shows an example of a pressure sensor and its relevant specifications.

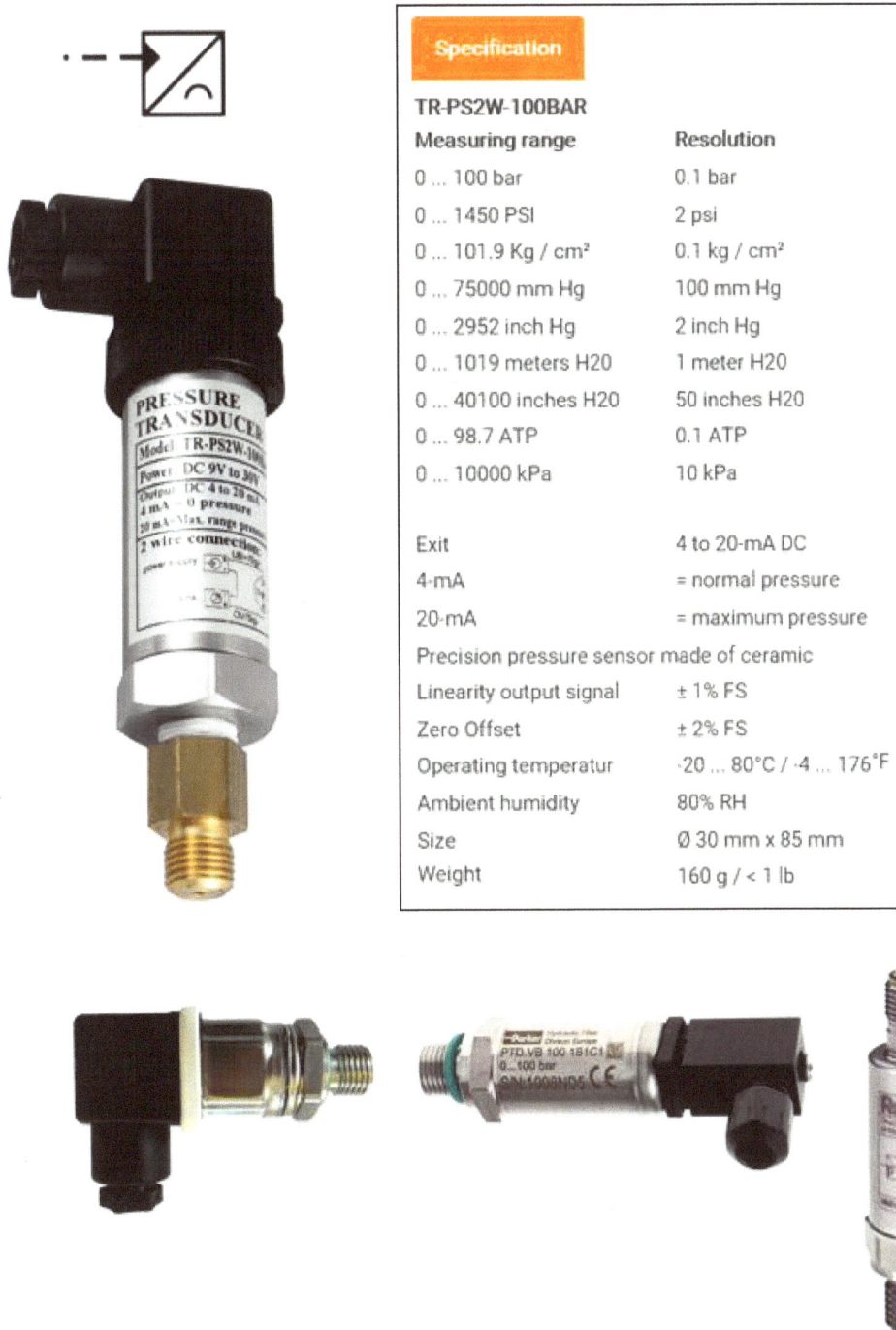

Specification	
TR-PS2W-100BAR	
Measuring range	**Resolution**
0 ... 100 bar	0.1 bar
0 ... 1450 PSI	2 psi
0 ... 101.9 Kg / cm²	0.1 kg / cm²
0 ... 75000 mm Hg	100 mm Hg
0 ... 2952 inch Hg	2 inch Hg
0 ... 1019 meters H20	1 meter H20
0 ... 40100 inches H20	50 inches H20
0 ... 98.7 ATP	0.1 ATP
0 ... 10000 kPa	10 kPa
Exit	4 to 20-mA DC
4-mA	= normal pressure
20-mA	= maximum pressure
Precision pressure sensor made of ceramic	
Linearity output signal	± 1% FS
Zero Offset	± 2% FS
Operating temperatur	-20 ... 80°C / -4 ... 176°F
Ambient humidity	80% RH
Size	Ø 30 mm x 85 mm
Weight	160 g / < 1 lb

Fig. 3.11- Examples of Pressure Sensors

3.2.4-Multifunction Pressure Measuring Instruments

As shown in Fig. 3.12, a *Multifunction* pressure measuring instrument functions as a pressure gauge, a pressure switch, and a pressure sensor as all-in-one device. It is a fully programmable device that offers a local display as well as electrical outputs to provide critical feedback to a control system.

PS+ Industrial Pressure Sensors with IO-Link

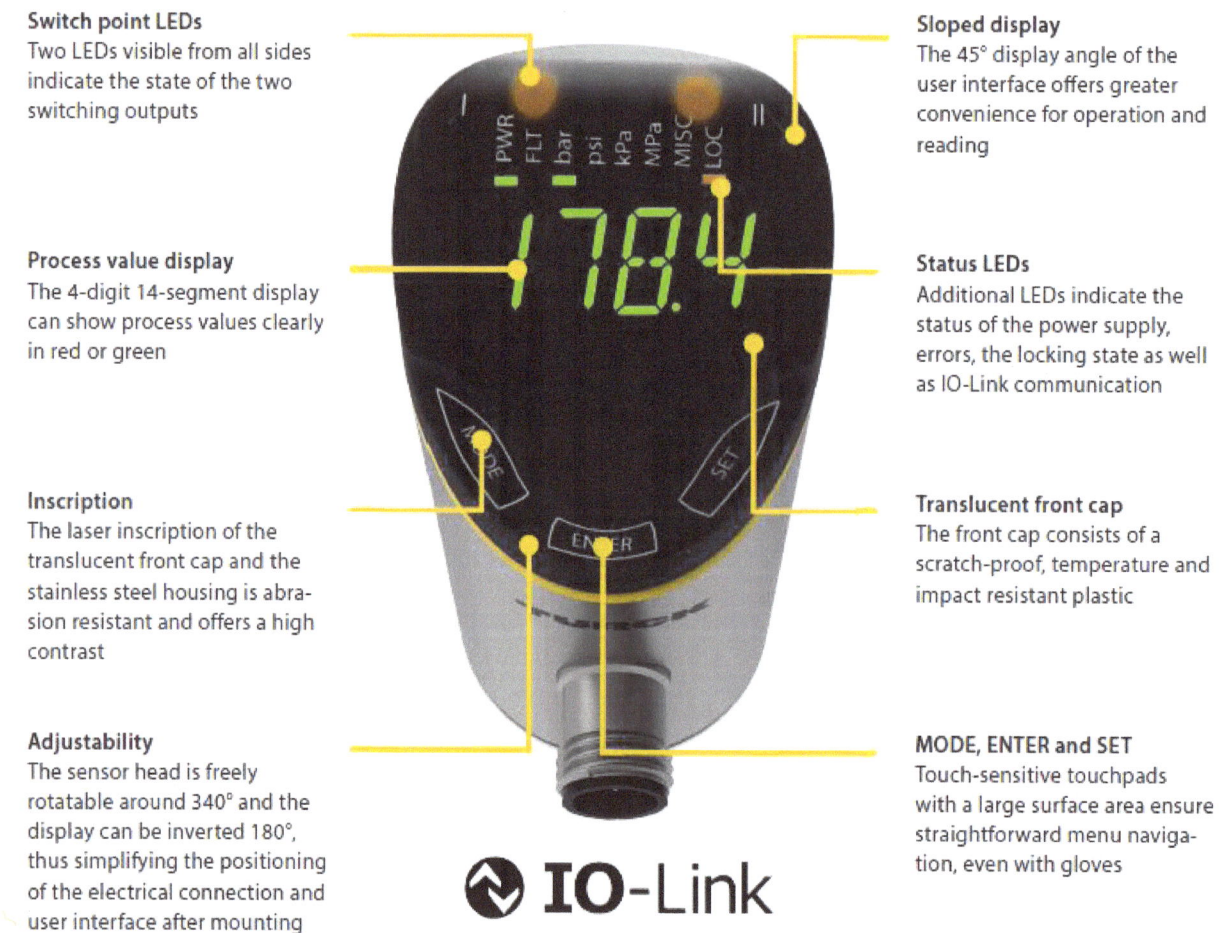

Switch point LEDs
Two LEDs visible from all sides indicate the state of the two switching outputs

Process value display
The 4-digit 14-segment display can show process values clearly in red or green

Inscription
The laser inscription of the translucent front cap and the stainless steel housing is abrasion resistant and offers a high contrast

Adjustability
The sensor head is freely rotatable around 340° and the display can be inverted 180°, thus simplifying the positioning of the electrical connection and user interface after mounting

Sloped display
The 45° display angle of the user interface offers greater convenience for operation and reading

Status LEDs
Additional LEDs indicate the status of the power supply, errors, the locking state as well as IO-Link communication

Translucent front cap
The front cap consists of a scratch-proof, temperature and impact resistant plastic

MODE, ENTER and SET
Touch-sensitive touchpads with a large surface area ensure straightforward menu navigation, even with gloves

Fig. 3.12- Multifunction Programmable Pressure Sensor (Courtesy of Turck)

3.3- Flow Measuring Instruments

3.3.1- Operating Principles for Flow Measurement

Flow rate can be measured by just a flowmeter. Flow *Switches* produce binary signal for remote condition monitoring and on-off control mode. Oil flow *Sensors* produce an analog signal for remote condition monitoring and continuous control mode. Figure 3.13 shows the general symbol for flowmeter and the classification of flow measuring devices:

❖ **Nonintersecting Flowmeters:** They are flowmeters that measure the flow without being placed into the flow stream. They use *Ultrasonic* technique to measure the flow and are commonly used for measuring large flows (1000's of gpm) such as water flow in pipelines.

❖ **Intersecting Flowmeters:** They measure the flow by being placed into the flow stream.
  ▪ **Without Moving Element:** Such flowmeters have accuracy sufficient (usually ±5%) for general reading purposes in.
  ▪ **With Moving Element:** Such flowmeters have better (usually ±1% or better) accuracy.

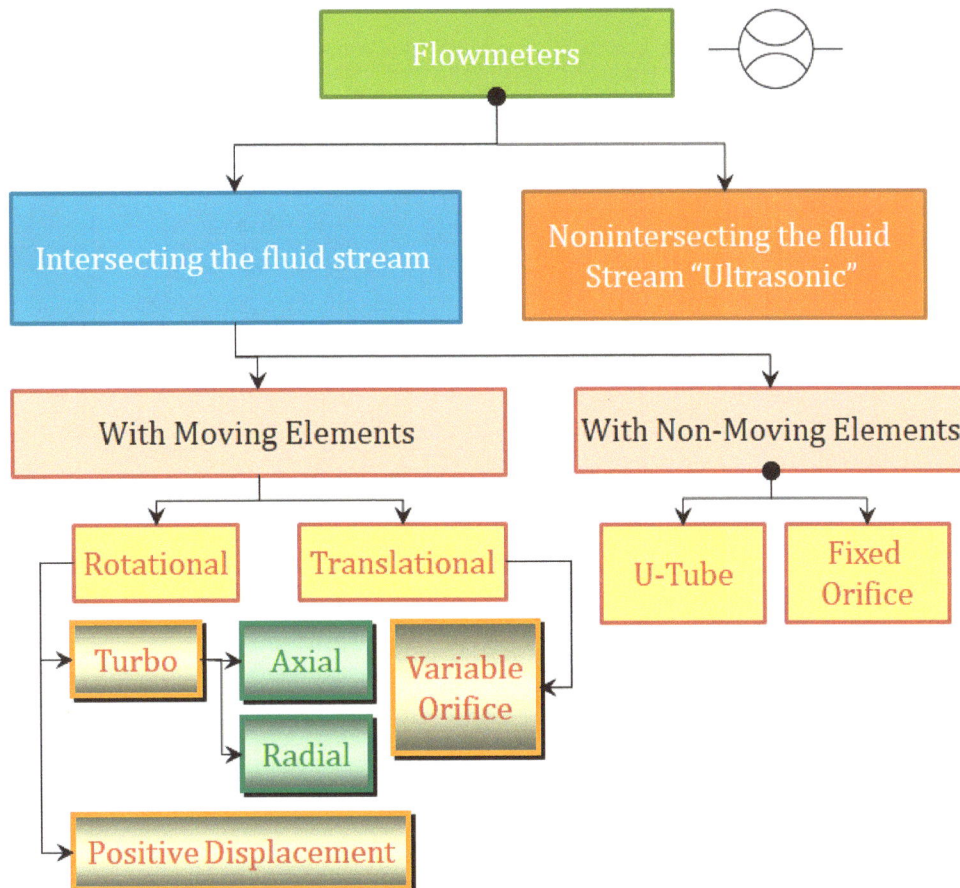

Fig. 3.13- Classification of Flowmeters

3.3.2- Fixed Orifice Flowmeters, Switches, and Sensors

As shown in Fig. 3.14.A, the operating principle of an *Orifice Flowmeter* is based on the fact that flow through a standard orifice is function of the differential pressure. So, differential pressure is calibrated to read flow rather than pressure. Obviously, there should be correction factors based on the density and viscosity of the fluid.

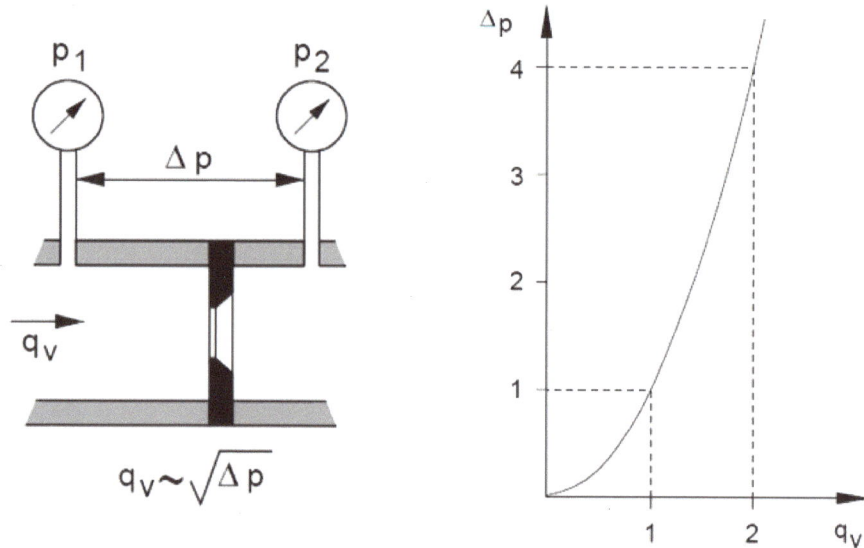

Fig. 3.14.A- Operating Principle of Orifice Flowmeters

Figure 3.14.B shows a typical example of orifice flowmeters. Such flow meters are commonly used for large flow rates of liquid, gas, and steam.

OPTIBAR DP 7060
with orifice meter run assembly

DP flowmeter for volume flow measurement of liquids, gases and steam

- For small line sizes

- Up to +400°C / +752°F; max. 160 bar / 2320.1 psi; (line pressure)

- Sizes: DN15...100 / ¾...2"

Fig. 3.14.B- Example of Fixed Orifice Flowmeters

3.3.3- U-Tube Flowmeters, Switches, and Sensors

As shown in Fig. 3.15, the operating principle of *U-Tube Flowmeters* is based on the fact that flow on top of a standard U-Tube is function of the differential pressure. So, differential pressure is calibrated to read flow rather than pressure. Obviously, there should be correction factors based on the density and viscosity of the fluid. The figure shows a typical example of U-Tube flowmeters. Such flow meters commonly used for large flows of liquid, gas, and steam.

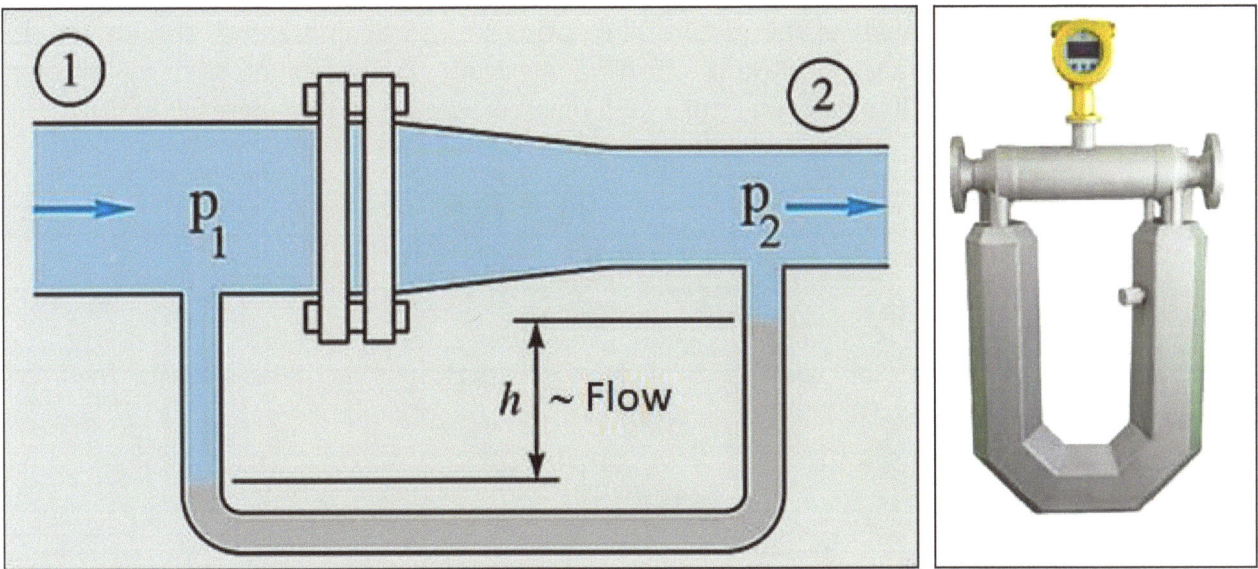

Coriolis Flow Meter 50mm, Stainless Steel Construction, LCD Display, Pulse, 4-20mA, RS485 Outputs
SKU: BF1001-50--L-DI-1-COM-NX-1-P-5

Coriolis Flow Meter 50mm
500 - 50,000 kg/hr Allowable range *
Mass Measurement Liquid & Gas
Stainless Steel Construction
LCD Display AC & DC options

Fig. 3.15- Operating Principle and Example of U-Tube Flowmeters

3.3.4- Variable Orifice Flowmeters, Switches, and Sensors

Variable Orifice Flowmeter is also known as *Float* Flowmeter. Figure 3.16 shows the operating principle and example from industry for variable area flowmeter. As the hydraulic oil or compressed air enters the meter, it is directed around a contoured metering cone located within the piston assembly. The pressure differential created by the flow around the cone causes the piston to move against the spring. The piston assembly carries a cylindrical magnet that is magnetically coupled to an easy-to-read linear flow scale that moves precisely and proportional to movement of the piston. The spring-loaded design allows the flow meter to operate in any orientation without affecting accuracy. Accuracy of such flowmeters is considered low to medium with error of $\pm$ (2-5) %. It is a one-way flowmeter.

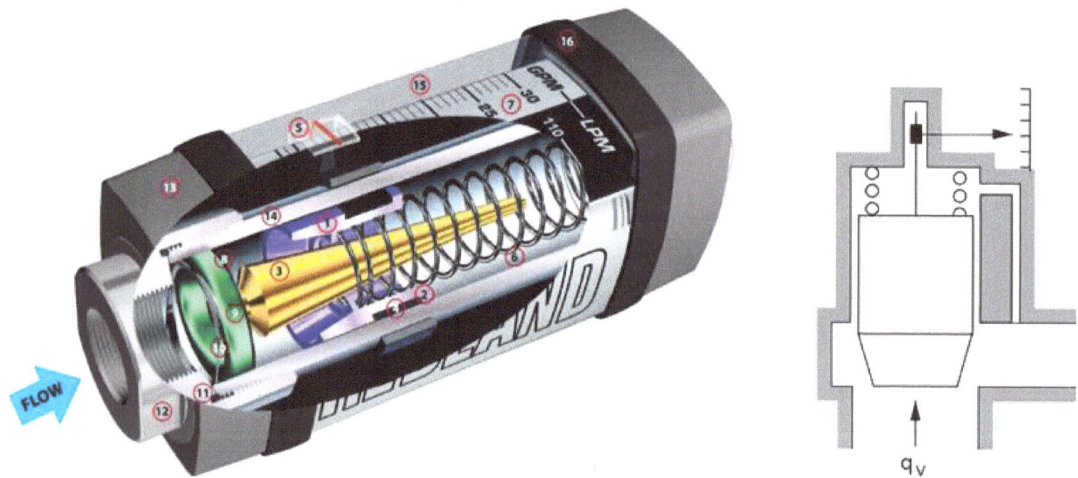

1	Orifice	9	Spider plate
2	Piston assembly	10	Retaining spring
3	Metering cone	11	Pressure seal
4	Internal magnet	12	End fitting
5	Flow indicator	13	End cap
6	Spring	14	Body
7	Flow scale	15	Guard
8	Retaining ring	16	Guard seal/bumper

Table 1: Meter components

HEDLAND

5 to 50 gpm Variable Area Mechanical Flowmeter

Aluminum Oil Flowmeters

• Max. pressure: 3500 psi Max. temp.: 240°F

  Accuracy: ±2%, except where noted Specific gravity:
  0.876 Viton® seals

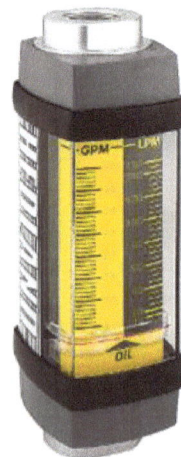

**Fig. 3.16- Operating Principle and Example of Variable Orifice Flowmeters
(Courtesy of Hedland)**

Another type of variable orifice flowmeter is shown in Fig. 3.17. The flow indicator consists of a sharp-edged orifice and tapered metering piston. The piston movement is directly proportional to the flow rate and the sharp edge orifice minimizes the effects of viscosity, so it has better accuracy than the float flowmeter. The piston is magnetically coupled to the rotary pointer assembly which registers on a clear 63 mm (2.5") scale displayed in lpm and USgpm. The FI750 flow indicators should not be installed in circuits where the flow is reversed. It has a built-in thermometer for temperature reading.

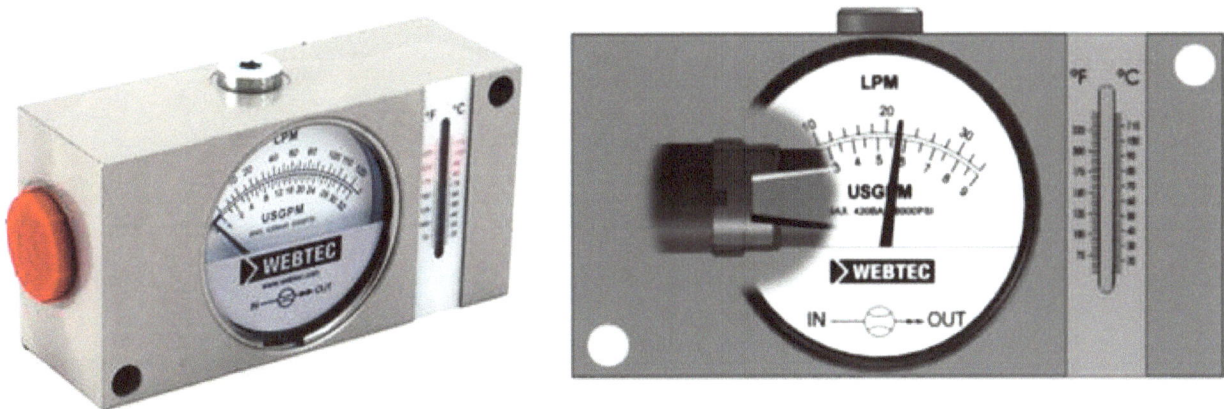

Specification

- **Ambient temperature**: -10 to 50°C (14 to 122°F)

- **Fluid temperature**:
  Continuous: 20 to 80°C (65 - 176°F)
  Intermittent use (less than 10 minutes) > 80 to 110°C (> 176 - 230°F)

- **Humidity**: 10 - 90 % RH

- **Fluid**: see model configuration

- **Seals**: FKM as standard (EPDM available on request)

- **Accuracy**: 4% of full scale (calibrated at 28 cSt)

- **Pressure**: 420 bar (6000 psi)

- **Flow range**: see model configuration

Fig. 3.17- Operating Principle and Example of Variable Orifice Flowmeters (Courtesy of Webtec)

3.3.5- Turbine Flowmeters, Switches, and Sensors

As shown in Fig. 3.18, the operating principle of *Axial* or *Radial Turbine* flowmeters is based on the fact that the flow is proportional to speed of the turbine. So, the measuring unit converts the rotational speed (RPM) of a standard turbine into a measurement of flow rate. Accuracy of such flowmeters is considered medium to high with error of <2%. Fig. 3.19 shows a typical example of turbine flow sensors.

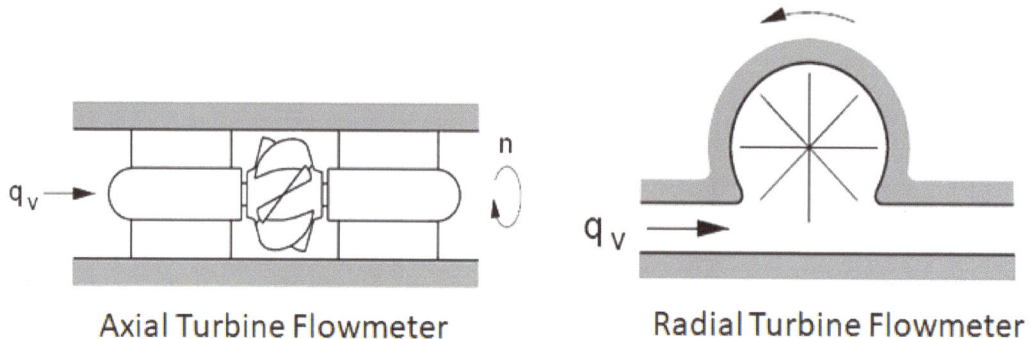

Axial Turbine Flowmeter

Radial Turbine Flowmeter

Fig. 3.18- Operating Principle of Turbine Flowmeters

INTERNATIONAL HYDAC
Flow Rate Transmitter
HFT 3100   Ex applications

| Turbine | High accuracy | Additional measuring connections |

Technical data:

Input data	
Measuring range and operating pressure	1.2 .. 20.0 l/min 420 bar
	6.0 .. 60.0 l/min 420 bar
	15.0 .. 300.0 l/min 420 bar
	40.0 .. 600.0 l/min 420 bar
Additional connection options [1]	2x G 1/4 female threads for pressure or temperature sensors with relevant approvals
Housing material	Stainless steel 1.4404
Parts in contact with fluid	Stainless steel: 1.4404, 1.4460, tungsten carbide
Output data	
Output signal, permitted load resistance	4 .. 20 mA, 2-conductor, with HART protocol $R_{Lmax} = (U_B - 12\,V) / 20\,mA\,[k\Omega]$ for HART communication min. 250 Ω
	HART communication acc. to HART 7 specifications
	HART Common Practice Commands, e.g. altering of measuring range limits (see table)
Accuracy	≤ 2 % of the actual value

Fig. 3.19- Example of Turbine Flow Sensors (Courtesy of Hydac)

The turbine flow meters, shown in Fig. 3.20, provides a precision solution to the measurement of flow in hydraulic systems on test stands, machine tools and other fixed or mobile applications. The flow meters can be installed anywhere in the hydraulic circuit for production testing, commissioning, development testing and analysis of control systems. The compact design of the flow meter allows them to be installed where space is limited. Various outputs are available such as 4 - 20 mA current loop, 0 - 5 V or 0 - 3 V. The figure shows a sectional view of the flowmeter in which an axial turbine is used. The turbine blade is designed to minimis the effects of variations in temperature and viscosity. The flow straightener before the turbine are used to increase the accuracy of the reading and to reduce flow turbulence and swirl after the turbine.

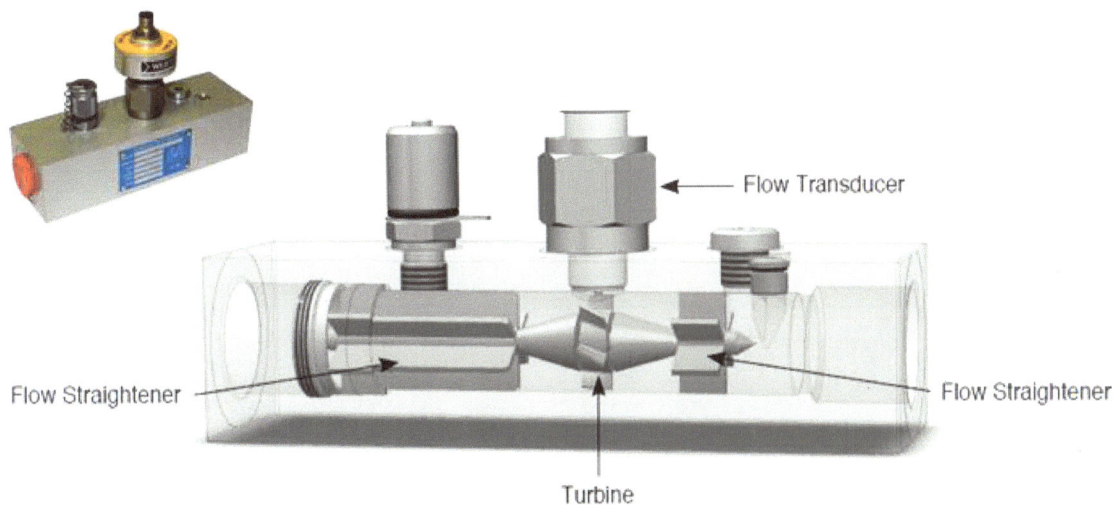

Fig. 3.20- Example of Axial Turbine Flow Sensors (Courtesy of Webtec)

Figure 3.21 shows the world's first turbine-type *Cartridge Flow Transmitter* that has the following features:
- Standard cartridge style design with turbine flow accuracy.
- Easy integration into hydraulic systems requiring no additional hardware or connections.
- Real-time data for predictive maintenance and remote troubleshooting.
- Continuous flow monitoring of all critical hydraulic functions.
- Configurable for standard 4-20mA output.

Fig. 3.21- Example of Axial Turbine Flow Sensors (www.dgdfluidpower.com)

3.3.6- Positive Displacement Flowmeters, Switches, and Sensors

Like turbine flow meters, the operating principle of a *Positive Displacement Turbine* flowmeter is based on the fact that the flow is proportional to the rotational speed.

Unlike turbine flowmeters, positive displacement flowmeters are the most accurate ones and are less affected by the working temperature or the fluid viscosity. They are used for measuring flow at a very high level of accuracy and for the purpose of automatic control of flow and hydraulic actuators speed.

As shown in Fig. 3.22, positive flowmeter comprises various mechanisms, the most common of them is the spur gear motor type.

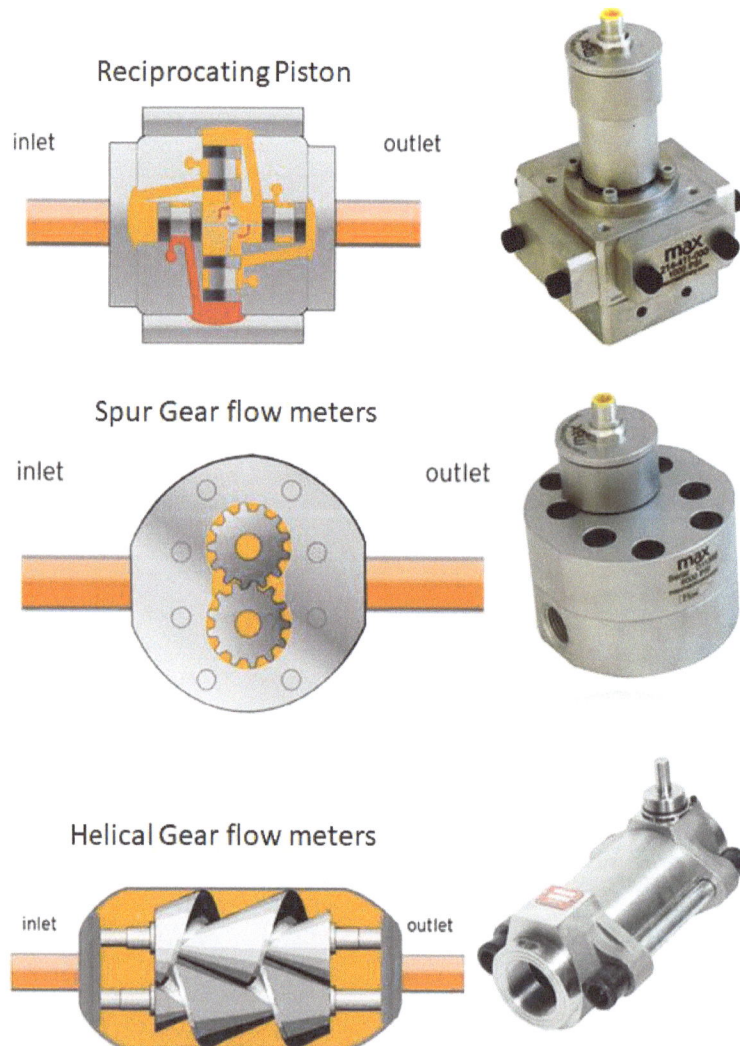

Fig. 3.22- Examples of Positive Displacement Flow Sensors (Courtesy of Max)

3.4- Oil Level Measuring Instruments

As shown in Fig. 3.23, Oil *Level* can be measured visually by just an indicator (1). Most oil level indicators are accompanied by a thermometer. Oil *Level Switches* (2) produce binary signal for remote condition monitoring and on-off control mode. Oil *Level Sensors* (3) produces analog signal for remote condition monitoring and continuous control mode.

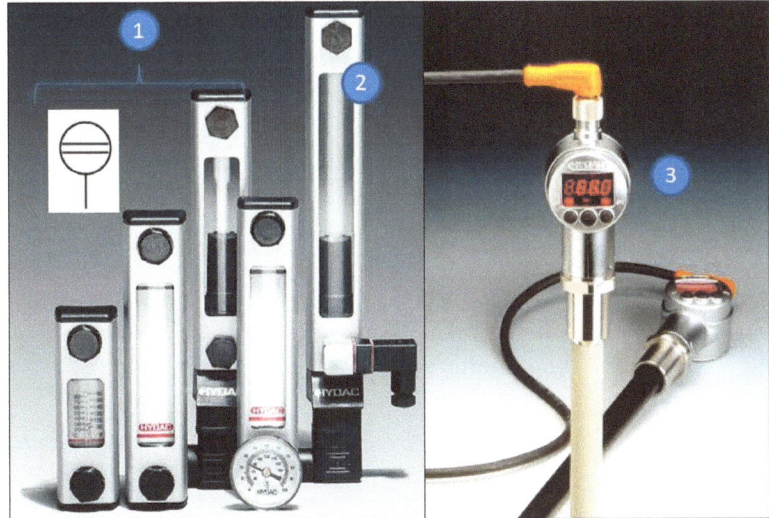

Fig. 3.23- Examples of Oil Level Indicators, Switches, and Sensors (Courtesy of Hydac)

Figure 3.24 shows other types of fluid level measuring devices as follows:
1. Visual and electrical level indicators.
2. Electrical float level indicator.

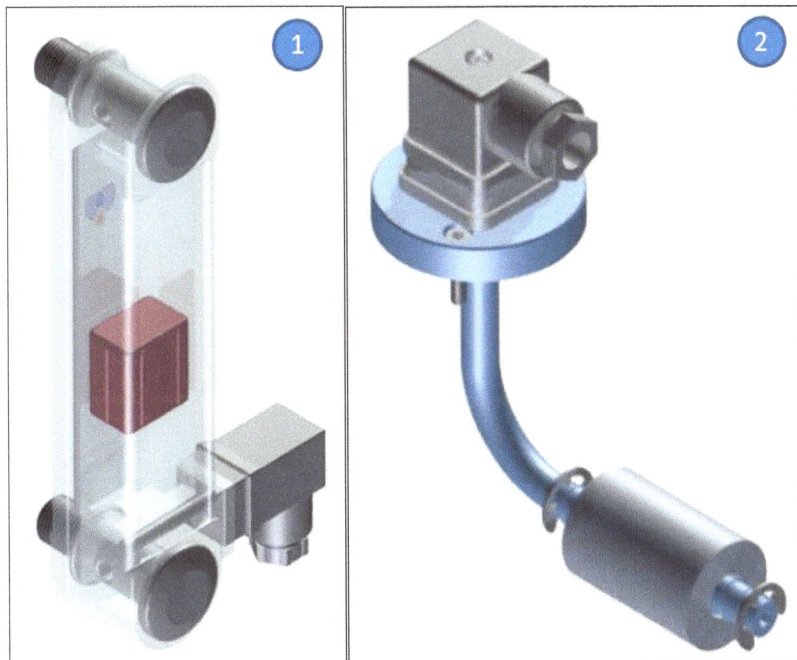

Fig. 3.24- Examples of Oil Level Indicators, Switches, and Sensors (Courtesy of MPFiltri)

3.5- Oil Temperature Measuring Instruments

As shown in Fig. 3.25, Oil *Temperature* can be measured visually by just an indicator (1). Oil *temperature Switches* (2) produce binary signal for remote condition monitoring and on-off control mode. Oil temperature *Sensors* (3) produces analog signal for remote condition monitoring and continuous control mode.

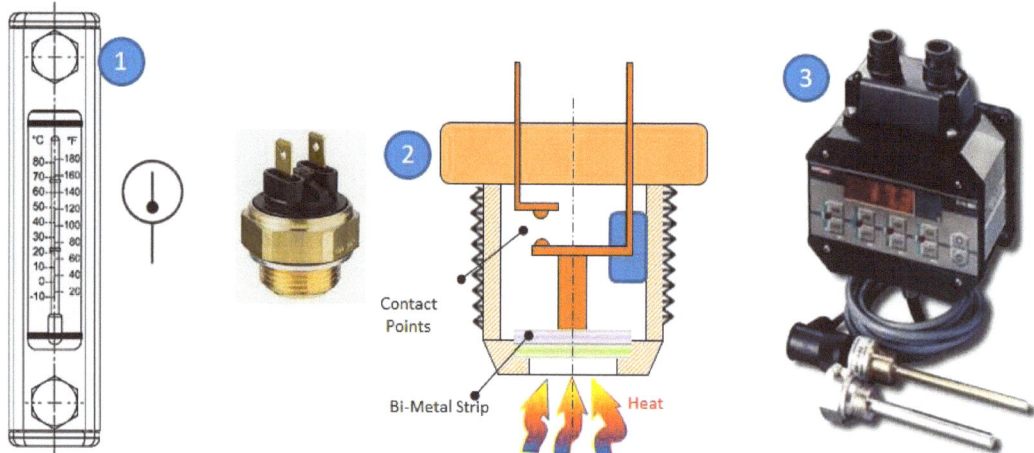

Fig. 3.25- Examples of Oil Temperature Indicators, Switches, and Sensors

3.6- All-In-One Measurement Instruments

For in field measurement and troubleshooting purposes, portable measuring instruments are available. Such units can measure multiple parameters simultaneously, provide visual readings, and produce electric signals for data acquisition and control purposes. Figure 3.26 shows a typical example from industry.

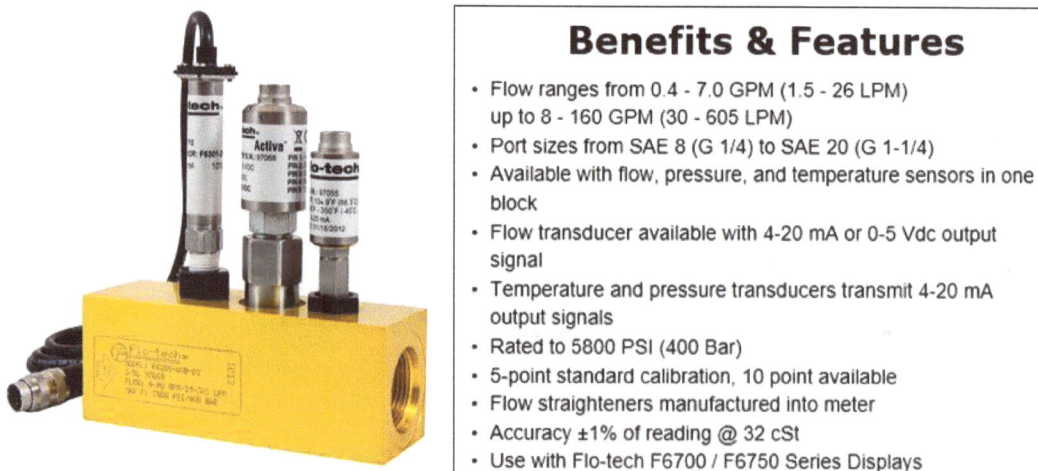

Benefits & Features

- Flow ranges from 0.4 - 7.0 GPM (1.5 - 26 LPM) up to 8 - 160 GPM (30 - 605 LPM)
- Port sizes from SAE 8 (G 1/4) to SAE 20 (G 1-1/4)
- Available with flow, pressure, and temperature sensors in one block
- Flow transducer available with 4-20 mA or 0-5 Vdc output signal
- Temperature and pressure transducers transmit 4-20 mA output signals
- Rated to 5800 PSI (400 Bar)
- 5-point standard calibration, 10 point available
- Flow straighteners manufactured into meter
- Accuracy ±1% of reading @ 32 cSt
- Use with Flo-tech F6700 / F6750 Series Displays

Fig. 3.26- All-In-One Flow, Pressure and Temperature Sensor (Courtesy of Flo-Tech)

3.7- Load and Torque Cells

As shown in Fig. 3.27, *Load and Torque Cells* are designed for measuring axial loads and bearing forces, and shaft torques on various hydraulic driven machine. The self-adapting piston and housing are constructed of high grade, corrosion resistant stainless steel and are available in standard or ring form. A high quality, highly accurate pressure gauge or transducer is attached for measurement indication.

SPECIFICATIONS

NOMINAL DIAMETER: 20 cm2
LOAD CELL HOUSING MATERIAL: Stainless steel
PISTON: Stainless steel
CONNECTING LINE: Direct connection – standard;
flexible tubing, capillary restrictor
RANGES: From 300 lb_f through 22,000 lb_f

MEASURING INSTRUMENT
PRESSURE GAUGE: 2-1/2" 300 Series,
one piece die cast brass case; dry or liquid filled;
4" 901 Series stainless steel case; dry or liquid filled
TRANSDUCER: 100, 200 or 615 Series transducer
OUTPUT SIGNALS: 4 mA to 20 mA, 2-wire:
0 Vdc to 5 Vdc, 0 Vdc to 10 Vdc, 1 Vdc to 5 Vdc,
1 Vdc to 6 Vdc & 1 Vdc to 11 Vdc, 3-wire
ACCURACY: ±0.5% full scale (BFSL) to ±0.125% full scale
OPERATING TEMP.: -4 °F to 140 °F (-20 °C to 60 °C)
AMBIENT TEMP.: -4 °F to 140 °F (-20 °C to 60 °C)

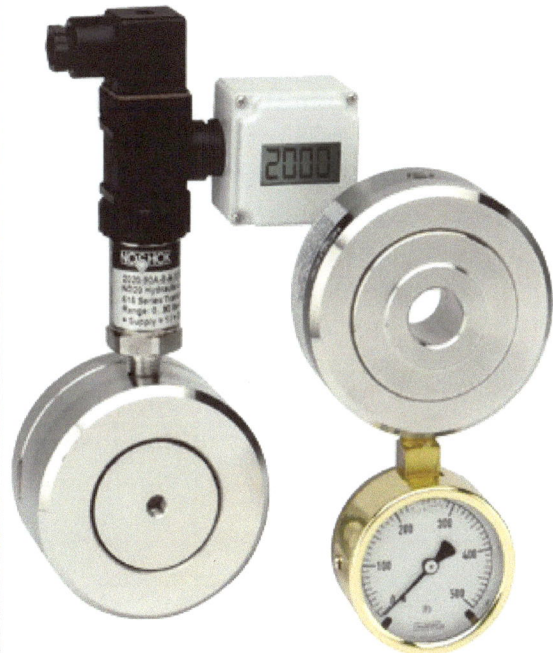

Fig. 3.27- Hydraulic Load Cells (https://www.noshok.com)

3.8- Other Measuring Devices

Operation of hydraulic systems still need additional measuring devices (such as to measure speed, vibration, contamination, etc.). these sensors are out of scope of this textbook.

3.9- Hydraulic Data Logger

As shown in Fig. 3.28, Hydraulic *Data Logger* is a portable multi-function measuring instrument used in hydraulic applications for testing and evaluation tasks. The device has the following features:

- **General Functions:** Measuring, monitoring, analyzing and saving data.
- **Measured Parameters:** Measuring pressure, temperature, and flow.
- **Number of Channels:** Measure up to 54 channels or up to 26 sensors.
- **Sensors Connections:**
  - Two electrically isolated CAN-bus networks (M12x1 connector).
  - Standard analogue inputs (Push-Pull plug).
- **Sensors Parameterization:** Automated parameterization of units and measuring ranges with automatic sensor-ID functionality (with up to 1 ms scanning rate).
- **Digital I/O:** One digital input and one digital output are also directly available.
- **Display:** The operator can select various display types, including numeric, bar graph, indicator gauge or curve chart.
- **Measurement Order**: Start/Stop measuring, point measuring and trigger measuring.
- **Measurement Memory:** Up to 4 million individual measured values. The entire measurement memory can contain more than one billion depending on the size of additional memory used such as microSD cards or USB drives.
- **Network:** A PC or Ethernet network can be connected using the USB and LAN ports.
- **Software:** PC-based analysis software for data analysis and visualization.
- **Safety:** All ports on the instrument are covered with rubber caps to protect them from being touched and from dust and moisture.

Fig. 3.28- Hydraulic Data Logger (Courtesy of Webtec)

3.10- Wireless Sensors and IoT in Hydraulic System Maintenance

3.10.1- Traditional versus Wireless Sensors

Traditional Sensors: Traditional condition monitoring means taking measurements on certain pieces of equipment or processes one at a time, either for diagnostics or performance analysis. Traditional condition monitoring has the following features:
- Use reliable wired sensors.
- Creates potential risk for workers in case of visual readings.
- High costs for setting up, wiring, and connectivity for remote condition monitoring.
- Labor-intensive process that consumes valuable man-hours.

Wireless Sensors and IoT-Enabled Solutions: Global competitiveness drives companies to find new ways to improve machine operation efficiency, product quality, and to reduce cost. As shown in Fig. 3.29, the Internet of Things (IoT) has changed the way manufacturing, condition monitoring, and machine servicing works. Nobody can afford to be left behind. Therefore, incorporating IoT-enabled solutions ensures your company is moving forward. *Wireless Sensors* and *IoT*-Enabled Solutions have the following features:
- Reduce potential risk for workers in case of visual readings.
- Reduce the cost of setting, wiring, and connectivity for remote condition monitoring.
- Advanced condition monitoring replaces the laborious, time-consuming process
- Improve access to condition monitoring from anywhere in the world.
- Help enforcing smart predictive maintenance and machine diagnosis.
- Reduce unplanned downtime.
- Battery powered wireless sensors require frequent battery charge.
- Cyber security is required for the safety of a machine operation.

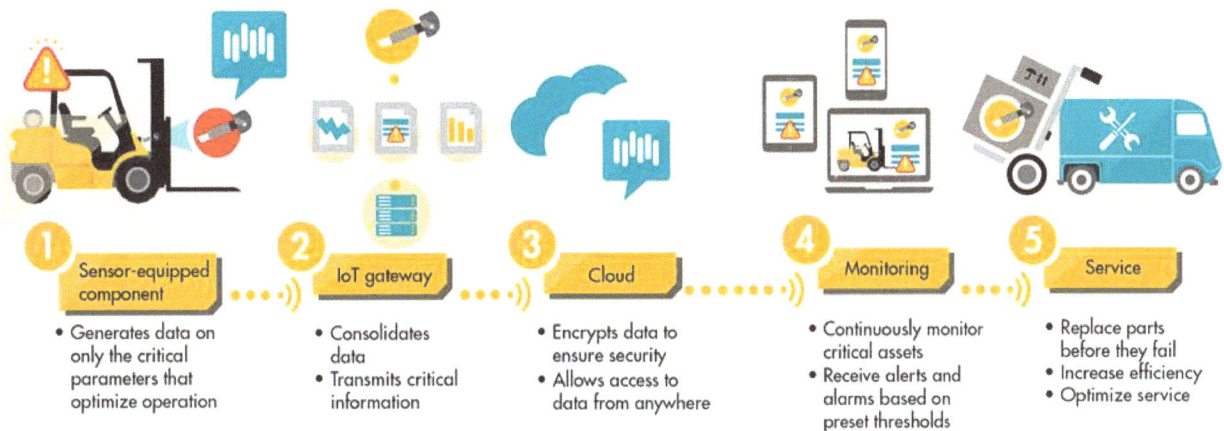

Fig. 3.29- IoT-Enabled Solutions in Industry (Courtesy of Parker)

3.10.2- Components of Wireless Sensors and IoT-Enabled Solutions

As shown in Fig. 3.30, *IoT*-Enabled solutions have the following components:

- Wireless Sensor: Transducer (1) + Wireless Transmitter (2) + Power Supply (3)
- Receiver (4): Nodes + Gateways.
- Network (5): Bluetooth or iCloud
- Software: PC Based (6) or Mobile Applications (7)

Fig. 3.30- Components of Wireless Sensors and IoT-Enabled Solutions (Courtesy of Parker)

3.10.3- Wireless Sensors

As shown in Fig. 3.31, wireless sensors are available to measure various variables such as pressure (1), temperature (2), humidity (3), flexible distance (4), vacuum (5), etc.

Wireless sensors are available with the following common features:
- User-definable measurement units (bar/psi), (F°/C°), etc.
- Port Options: Male NPTF, SAE, and others.
- Corrosion-resistant materials for challenging environments.
- User-selectable measurement and broadcast intervals.

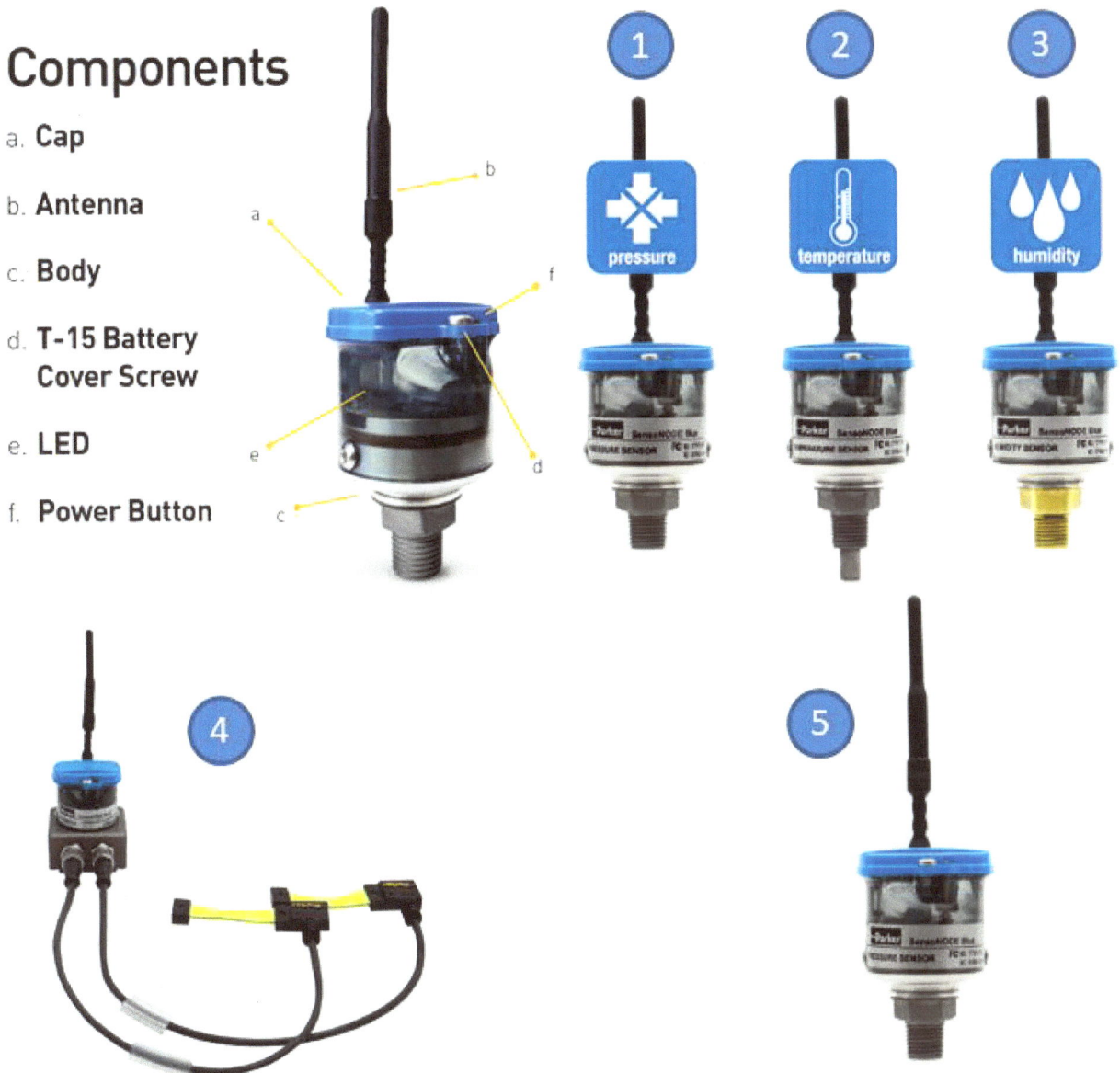

Fig. 3.31- Examples of Wireless Sensors (Courtesy of Parker)

3.10.4- Wireless Transmitters

Instead of replacing all sensors in a machine or a plant with fully assembled wireless sensors, a wired sensor can still be upgraded by connecting it to a *wireless transmitter*. With 4-20mA sensor measurements locked into many existing control systems, the 4-20mA Transmitter unlocks those measurements and allows customers to receive the information on their mobile device.

Figure 3.32 shows an example of a wireless transmitter. The figure shows that the wireless transmitter must be wired to a power supply and to the output of physical sensor.

Wireless Transmitters are available with the following features:
- Connects in line with any 4-20mA Sensor.
- Transmits wired sensor output including alarms and trend data.
- Threaded stud port or magnetic base for tool free mounting.

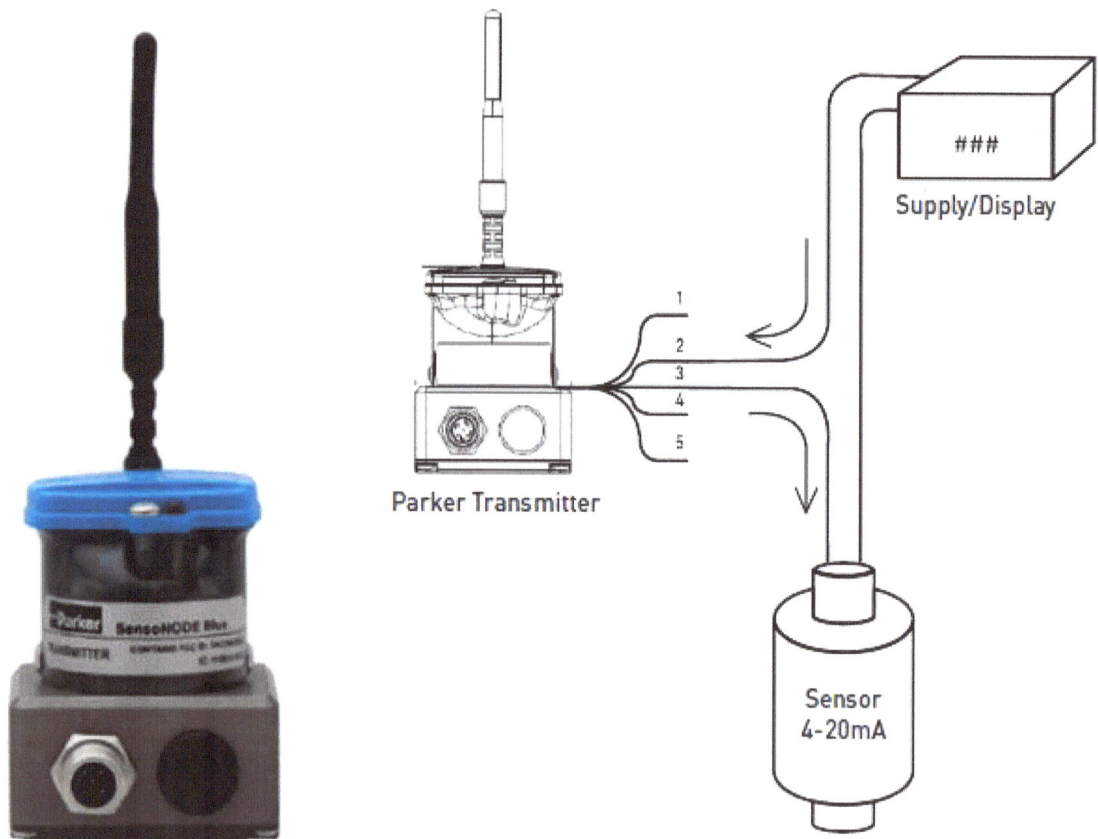

Fig. 3.32- Example of Wireless Transmitter (Courtesy of Parker)

3.10.5- Power Suppliers for Wireless Sensors

Instead of battery-powered operation of wireless sensors, as shown in Fig. 3.33, portable power suppliers are available with the following features:
- Supplies continuous power to sensors and eliminates the need for battery replacement.
- Used with IEC/UL 508 Class 2 power supply.
- Easy upgrade.
- Increases ambient working temperature range of sensors.

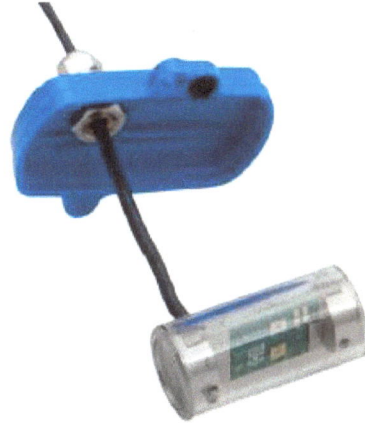

Fig. 3.33- Example of Power Supply for Wireless Sensors (Courtesy of Parker)

3.10.6- Mobile-Based Connectivity with Wireless Sensors

Transmitted data by the sensors must be received and analyzed by an associated software. As shown in Fig. 3.34, *Mobile Application* is used for diagnostics and condition monitoring for predictive maintenance. The app allows users to:
- Work under iOS and Android.
- Connect to multiple sensors concurrently to gather wide range of measurements.
- Offer immediate and historic trend information with analytical tools.
- Sensor setup and alarm setting.

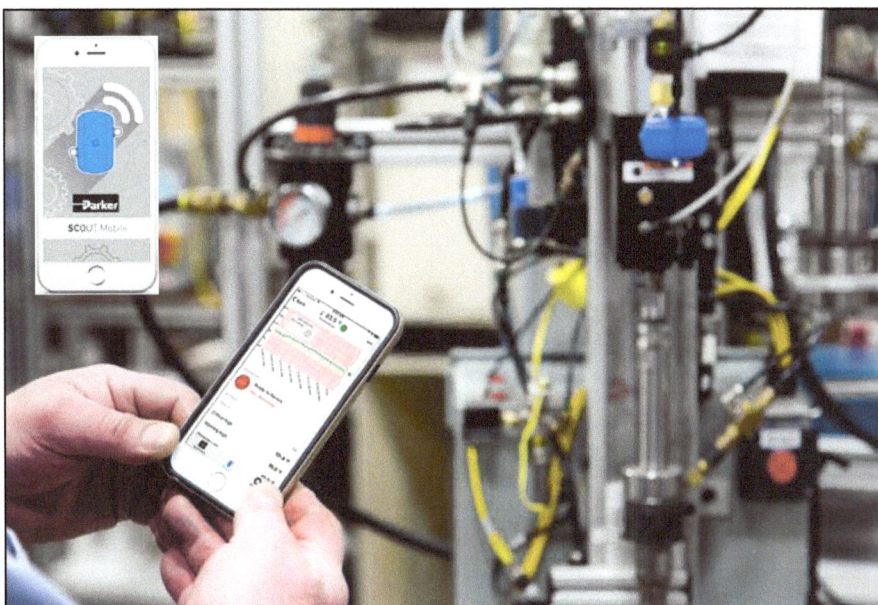

Fig. 3.34- Example of Mobile Connectivity for Wireless Sensors (Courtesy of Parker)

Figure 3.35 shows an example of applying Mobil-based connectivity for Pick & Place location.

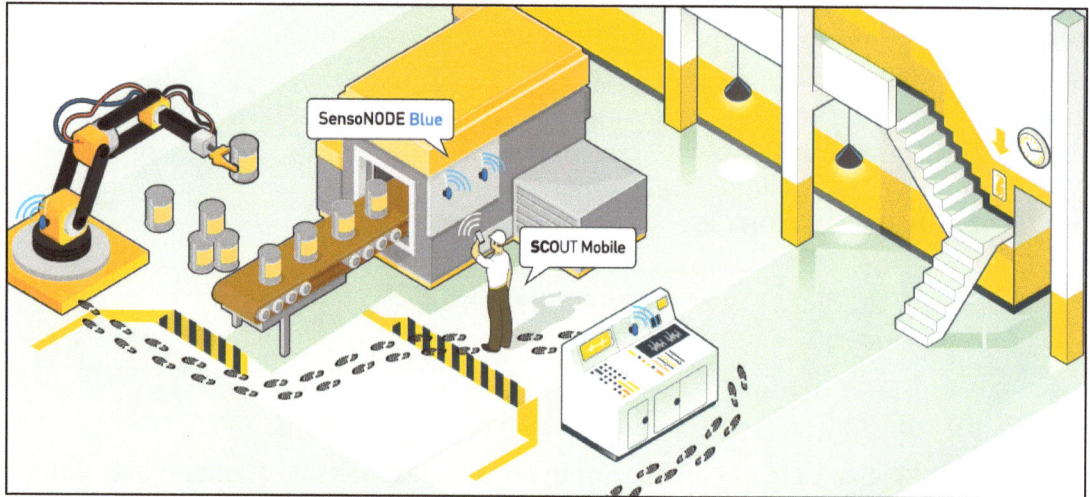

Fig. 3.35- Example of Mobile-Based Connectivity for Wireless Sensors (Courtesy of Parker)

3.10.7- iCloud-Based Connectivity with Wireless Sensors

For large scale operation, *iCloud-Based* connectivity is more adequate. Figure 3.36 shows the components required for such system that offers the following common features:
- Access to data anytime, anywhere.
- Multiple user access levels.
- Export and share data.
- Easy to use web-based interface, no software to download, and no updates.
- Better data visualization that makes most sense.

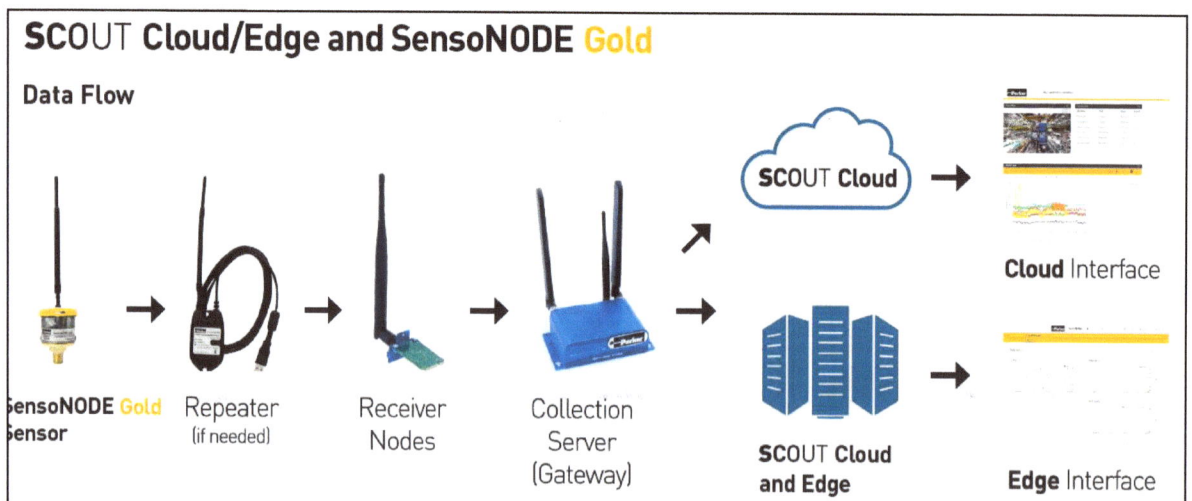

Fig. 3.36- Components for iCloud Connectivity for Wireless Sensors (Courtesy of Parker)

Figure 3.37 shows examples of applying iCloud-based connectivity with wireless sensors to manage the operation of material handling location, injection molding machine, and mining location simultaneously from one location.

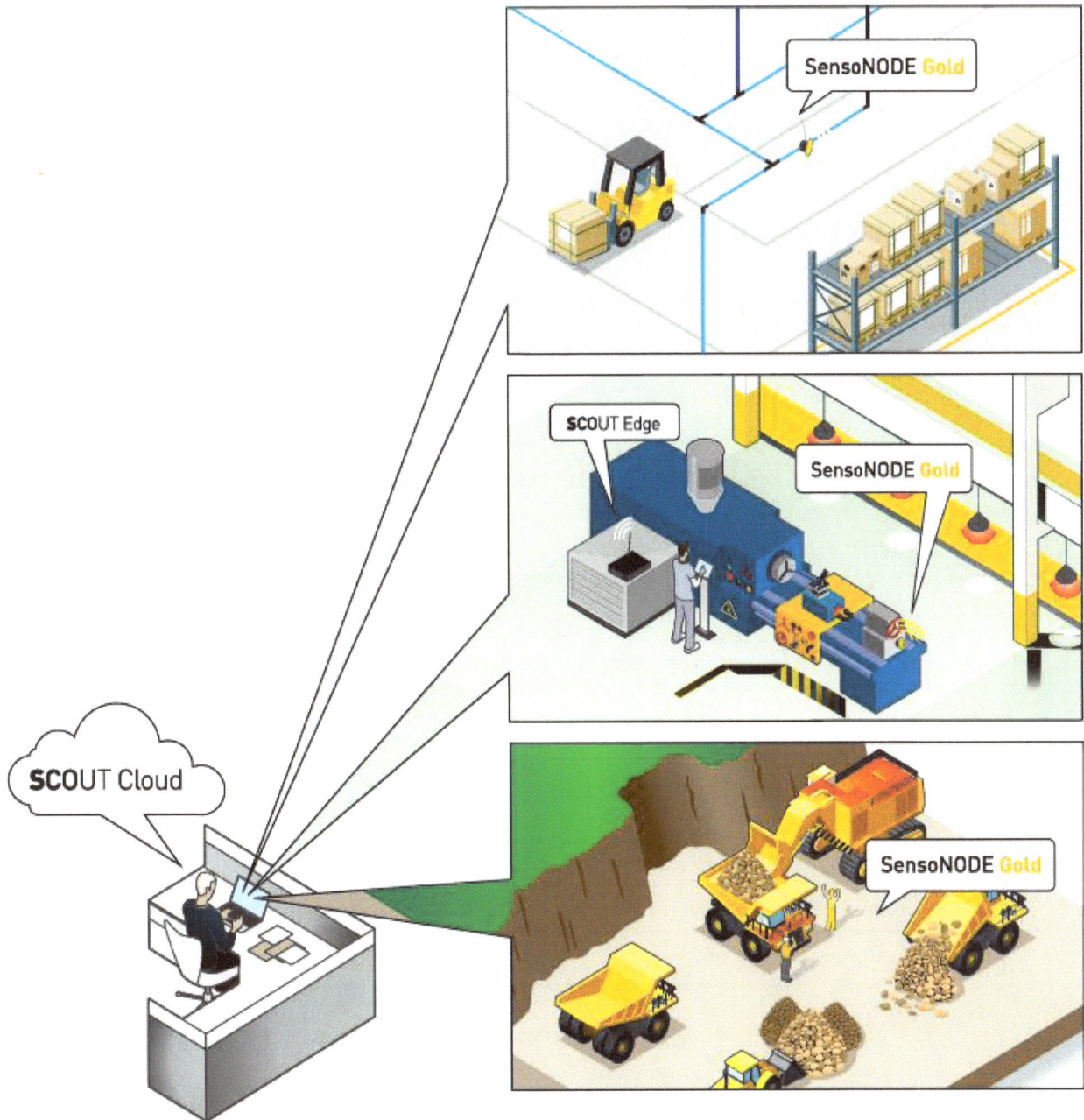

Fig. 3.37- Examples of iCloud-Based Connectivity for Wireless Sensors (Courtesy of Parker)

The following set of best practices provides general guidelines and may not be applicable for all cases. They are not intended to replace the instructions given by the component manufacturer manufacture. It is always strongly advisable to review and follow instructions provided by the component manufacturers in all cases.

3.11- BP-Measuring Instruments-01-Selection and Replacement

The following best practices list provides guidelines for selecting a new or replacing an existing measuring instrument.

BP-Measuring Instruments-01-Selection and Replacement:
- Review type/size of hydraulic connections to avoid using adaptors.
- Review the measuring scale and make sure the new one has the same scale.
- Review electrical connections and make sure to use standard connections.
- **Switches:** Review the status (NO or NC) and the adjustment range.
- **Sensors:** Review the output signal (Voltage and Current, 0-20 mA or 4-20 mA).

3.12- BP-Measuring Instruments-02-Maintenance Scheduling

Unless otherwise stated by the components and systems manufacturer, Table 3.1 provides guidelines for *scheduling* preventive maintenance actions for measuring instrument.

#	Preventive Maintenance Actions	Daily	Weekly	Monthly	Biannually	Annually
1	Clean around the outside surface		✔	✔	✔	✔
2	Check tightness and leakage around hydraulic connections			✔	✔	✔
3	Check electrical connections (if found)			✔	✔	✔
4	Check for offset from the zero point				✔	✔
5	Standard tests and calibration					✔

Table 3.1- BP-Measuring Instruments-02-Maintenance Scheduling

3.13- BP-Measuring Instruments-03-Installation and Maintenance

The following best practices provides guidelines for *installation and maintenance* of measuring instruments.

BP-Measuring Instruments-03-Installation and Maintenance:
- **General:** Avoid using adaptors.
- **Sensors:** Consider maximum distance between sensing element and receiver.
- **Sensors:** Consider sensor placement in most representative location for parameter being measured.
- **Pressure Gauges:** Protect pressure gauges from pressure surges **(See Note 1)**.
- **Pressure Gauges:** Select adequate installation method **(See Note 2)**.
- **Flowmeters:** Review if the flowmeter is unidirectional **(See Note 3)**.
- **Flowmeters:** Avoid placing flowmeters where the pressure is intensified **(See Note 4)**.
- **Flowmeters:** Inlet and outlet connections should have same size as the flowmeter ports **(See Note 5)**.

Note 1: As shown in Fig. 3.38, pressure gauges are installed on top of Snubbers to reduce pressure spikes and protect the needle. The traditional and cost-effective snubbing elements are conically shaped or disk-shaped elements made from porous material. Modern snubbing elements consist of a base with a built-in tunable restrictor.

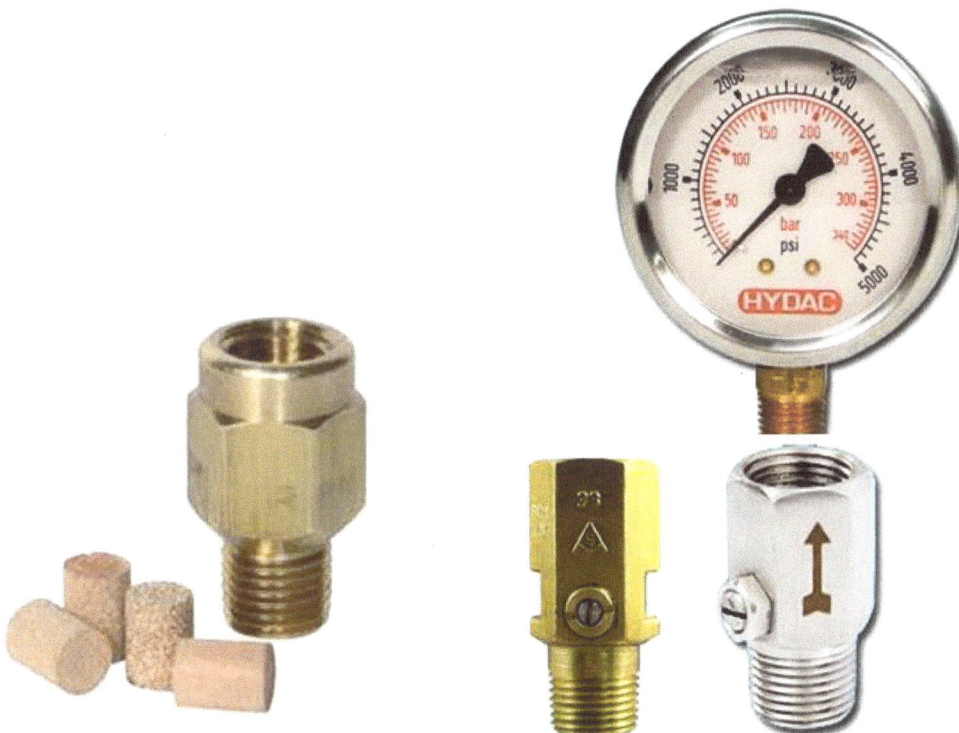

Fig. 3.38- Pressure Gauge Snubbers

Note 2: As shown in Fig. 3.39, pressure gauges are installed as follows:

1. A vented pressure gauge with a push button for pressure reading when needed. The button is pushed to read the pressure then released to vent and relax the gauge.
2. A pressure gauge with a knob to hold the last pressure reading. The knob is opened to read the pressure then closed to hold the last reading.
3. A pressure gauge is connected to multiple lines. A knob is used to select to read pressure of one line at a time.
4. A pressure gauge is mounted on a panel for permanent pressure reading.
5. A Panel-mounted pressure gauge is common for power units with front panels.
6. A test point equipped with a check valve so that a pressure gauge inserted for purposes of troubleshooting and diagnoses.

1-Continuous Measuring

2-Vented for measuring when needed

3-Keeps Last Measurement

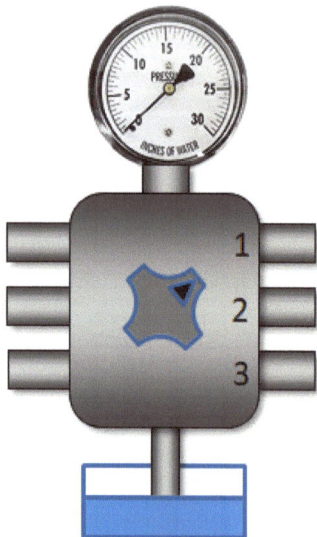

4-Multipoint Measurement

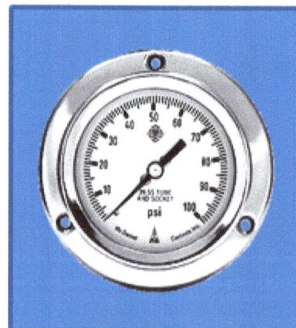

5-Panel-Mounted

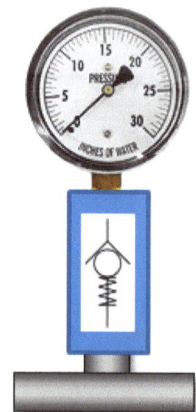

6- Test Point for Troubleshooting

Fig. 3.39- Pressure Gauge Installation

Note 3: As shown in Fig. 3.40, some flowmeters are unidirectional. So, make sure they are placed correctly in the direction of the flow. Otherwise, they read zero. As shown in the figure, some flowmeters have an integral check valve. So, if they are installed backwards, flow will be blocked and potentially resulting in unexpected machine movement, pressure intensification, catastrophic failure causing property damage, personal injury or death.

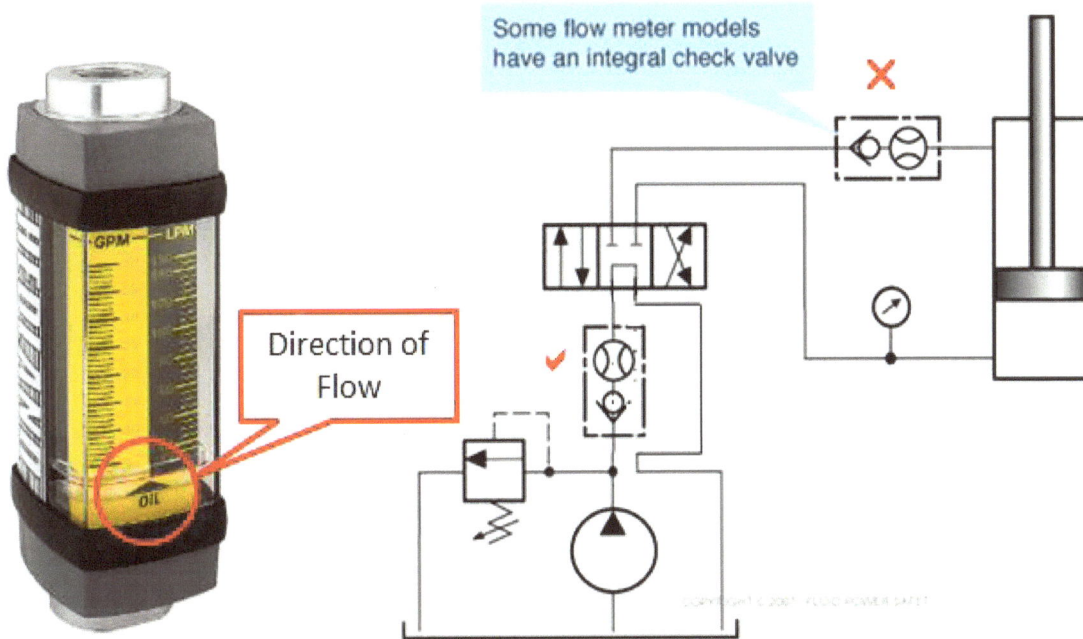

Fig. 3.40- Unidirectional Flowmeters

Note 4: As shown in Fig. 3.41, placing a flowmeter where the pressure is intensified may result in catastrophic failure of the flowmeter.

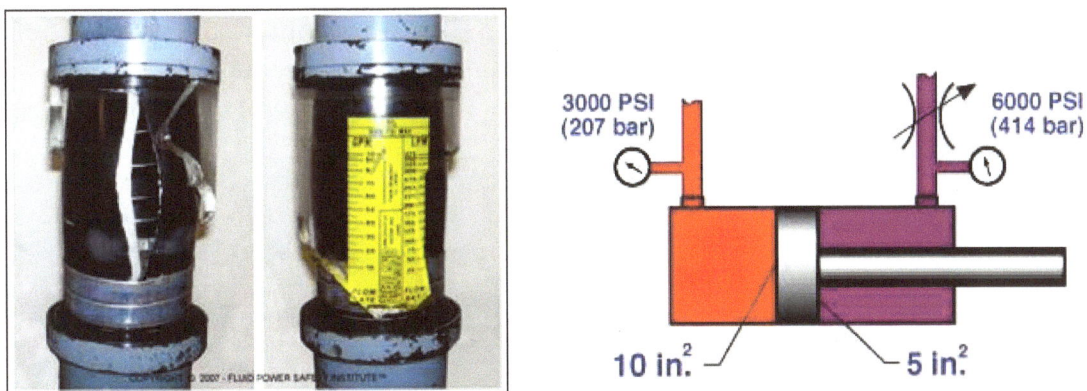

Fig. 3.41- Wrong Placement of Flowmeters (Courtesy of Fluid Power Safety Institute)

Note 5: As shown in Fig. 3.42, all hydraulic connections should be made by suitably qualified personnel. Inlet and outlet connections should always have a similar bore size to that of the flow meter to prevent venturi or contraction effects.

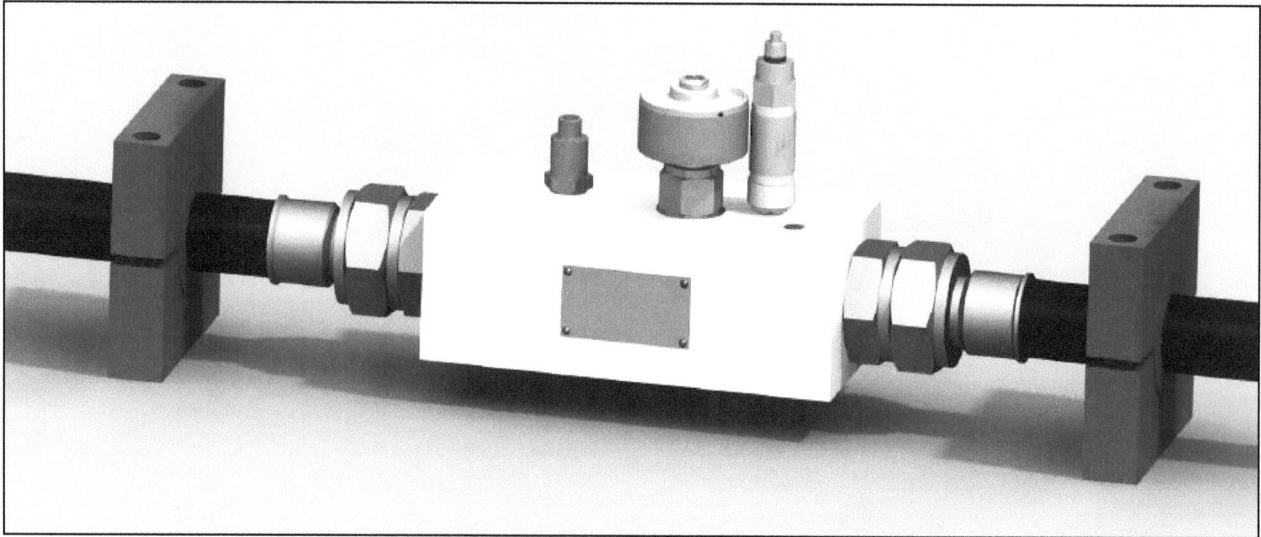

Fig. 3.42- Best Practices of Flow Meter Installation (Courtesy of Webtec)

3.14- BP-Measuring Instruments-04-Standard Tests and Calibration

Hydraulic components must be tested routinely within the scope of preventive maintenance in order to maintain reliable operation of the components/system.

Accuracy of measuring instruments are affected by the following factors:

- ❖ **Repeatability:** It is the variation that can occur over the time when measuring the same parameter repeatedly.
- ❖ **Response Time:** It is the transient time the instrument takes to provide steady measurement. Response time depends on the mechanism of the device.
- ❖ **Sensitivity:** It is the minimum value to trigger the measurement only at the beginning of the scale.
- ❖ **Resolution:** It is the minimum value to trigger the measurement within the range of measurement.
- ❖ **Full Scale (Range):** It is the maximum measurement value the device can detect.

What is meant by a sensor *calibration*?

When engineers design modern process plans, they specify sensors to measure important process variables such as pressure, flow, level, temperature, etc. Knowing these variables on a continuous basis are important to control the process and to ensure safe operation. The key element to maintain proper control is the sensor. To make sure that the sensor is working properly within an acceptable range of error, then the sensor must be tested and adjusted frequently. The process of testing and adjusting the sensor is referred to as "calibration".

What is the sensor *Error* and *Deviation*?
- Error = Measured Value – Ideal Value
- Deviation (%) = (Error/Ideal Value) x 100

What are the sources of errors in a sensor?

A sensor may give error in reading the measured variable because of one or combination of the following reasons:
- Offset from zero point.
- Mechanical wear or damage.
- Changing the scale of the operation.

The following list of best practices provides guidelines for *standard tests and calibration* of measuring instruments

BP-Measuring Instruments-04-Tests and Calibration:
- Review test and calibration instructions provided by the manufacturer.
- Test and calibration must be done by a trained person at a certified calibration center.
- Calibration and Standard Tests should be in compliance with **IEC/ISO 17025**.

What are the main steps of the calibration process?
- **Step 1:** Based on ideal readings from a reference device, develop the deviation curve. For example, for 5 points calibration at (0, 25, 50, 75, and 100) %.
- **Step 2:** Check if the calibration curve below or above the allowable deviation.
- **Step3:** if the calibration curve is above the allowable deviations, then perform the required adjustments to bring the device into compliance.
- **Step4:** Issue a certificate of calibration.

Chapter 4
Maintenance of Pumps

Objectives

This chapter provides guidelines for **pumps** selection, replacement, maintenance scheduling, installation, testing, storage and transportation. This chapter is supported by examples and figures provided by leading fluid power manufacturers.

Brief Contents

4.1-BP-Pumps-01-Selection and Replacement
4.2-BP-Pumps-02-Maintenance Scheduling
4.3-BP-Pumps-03-Installation and Maintenance
4.4-BP-Pumps-04-Standard Tests and Calibration
4.5-BP-Pumps-05-Transportation and Storage

Chapter 4: Maintenance of Pumps

The following set of best practices provides general guidelines and may not be applicable for all cases. They are not intended to replace the instructions given by the component manufacturer. It is strongly advisable to adhere to instructions provided by the manufacturer.

4.1- BP-Pumps-01-Selection and Replacement

4.1.1- Selecting or Replacing Pumps

The following best practices list provides guidelines for selecting a new or replacing an existing pump.

BP-Pumps-01-Selection and Replacement:
- Review maximum/optimum operating pressure.
- Review min/maximum/optimum operating speed.
- Review maximum overall efficiency at the optimum operating conditions.
- Review size (**See Note 1**) and displacement control requirements.
- Review type of fluid.
- Review contamination tolerance.
- Review noise level.
- Review initial cost.
- Review approximate service life.
- Review availability and interchangeability.
- Review maintenance and spare parts.
- Review physical size and weight.

Note 1:
If a pump is <u>oversized</u>, one or more of the following could occur:
- Actuators speed increases and may work out of sequence resulting in unsafe operation.
- Turbulent flow in the transmission lines resulting in all relevant consequences.
- Increased pressure drops across components, wasted energy and heat generation.
- Increased flow forces on spool valves resulting in improper valves operation.

If a pump is <u>undersized</u>, one or more of the following could occur:
- Actuators slowdown and may work out of sequence resulting in unsafe operation. Additionally, machines become less productive.
- Valves that are sized originally for large flow will lose controllability when receive low flow. For example, if the valve was originally a pilot operated valve, it will work improperly for low inlet flow.

4.1.2- Displacement Calculation of Legacy Pumps

When replacing a legacy pump, usually no information is available about the pump. The main challenge in replacing such legacy pumps is to find the pump size. The following sections show how to take some measurements so that pump size can be calculated.

External Gear Pumps:
The two gears of an *external gear pump* are identical. So, considering **N** = No. of teeth of one gear and the dimensions shown in Fig. 4.1, Eq. 4.1 shows how to calculate the pump displacement:

$$\textbf{Pump Displacement} \ = \ \textbf{L} \times \textbf{W} \times \textbf{H} \times \textbf{N} \times \textbf{2} \ = \textbf{L} \times \textbf{W} \times \frac{[\textbf{D}_1 - \textbf{D}_2]}{\textbf{2}} \times \textbf{N} \times \textbf{2} \qquad \textbf{4.1}$$

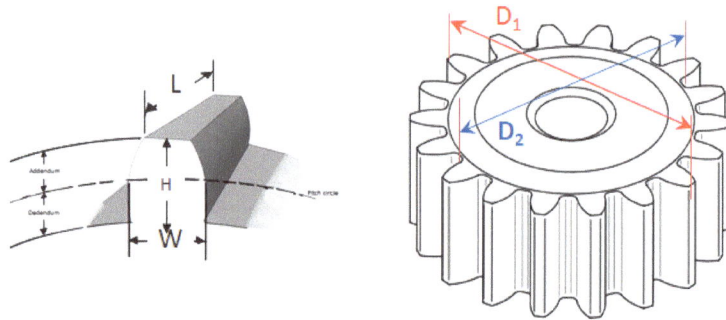

Fig. 4.1- Measurements for Calculating Size of External Gear Pump

Internal Gear Pumps:
For *internal gear pump*, Eq. 4.1 still applicable. But the drive and driven gear aren't identical. So, measurements are taken for the drive gear, which is the internal one (not the gear ring).

Gerotor:
As shown in Fig. 4.2, the rotor has one tooth less than on the inner stator. Equation 4.2 is used for *gerotor* where **N** = No. of inner gear teeth and **L** = width of the gear teeth.

$$\textbf{Pump Displacement} \ = \textbf{N} \times \textbf{L} \times [\textbf{A}_{\text{max}} - \textbf{A}_{\text{min}}] \qquad \textbf{4.2}$$

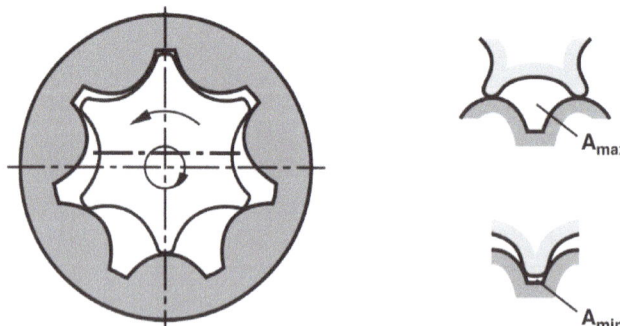

Fig. 4.2- Measurements for Calculating Size of Gerotor (Courtesy of Bosch Rexroth)

Screw Pump:

In a *screw pump*, the displacement chamber is formed between threads and housing. Considering the measurements shown in Fig. 4.3, Eq. 4.3 is used to calculate the pump displacement.

$$\textbf{Pump Displacement} = \frac{\pi}{4} \times [D^2 - d^2] \times s \times c \qquad \textbf{4.3}$$

Where **c** = correction factor < 1 that considers the interlocking of threads of both spindles.

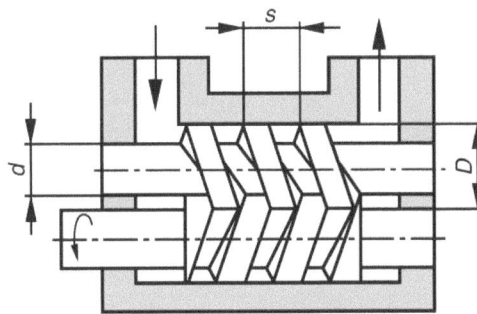

Fig. 4.3- Measurements for Calculating Size of Screw Pump (Courtesy of Bosch Rexroth)

Vane Pumps:

Figure 4.4 shows *unbalanced vane pump* (left) and *balanced vane pump* (right). Considering the dimensions shown in the figure, Eq. 4.4 and Eq. 4.5 show the calculation of unbalanced and balanced pump displacement; consequently.

$$\textbf{Pump Displacement} = 2 \times \pi \times b \times e \times D \qquad \textbf{4.4}$$

$$\textbf{Pump Displacement} = \frac{\pi}{4} \times [D^2 - d^2] \times b \times 2 \qquad \textbf{4.5}$$

Where **b** = Vane width and **e** = eccentricity.

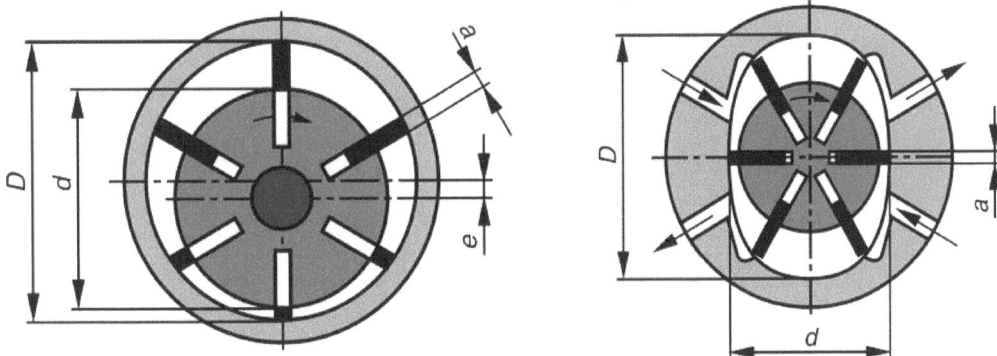

Fig. 4.4- Measurements for Calculating Size of Vane Pump (Courtesy of Bosch Rexroth)

Radial Piston Pumps:

Figure 4.5 shows *rotating cam piston pump* (left) and *rotating cylinder block pump* (right). Considering the dimensions shown in the figure, Eq. 4.6 shows the calculation of piston pump displacement.

$$\textbf{Pump Displacement} = 2 \times e \times z \times \left[\frac{\pi \times d_k^2}{4}\right] \qquad \textbf{4.6}$$

Where d_k = piston diameter, e = eccentricity, and z = number of pistons.

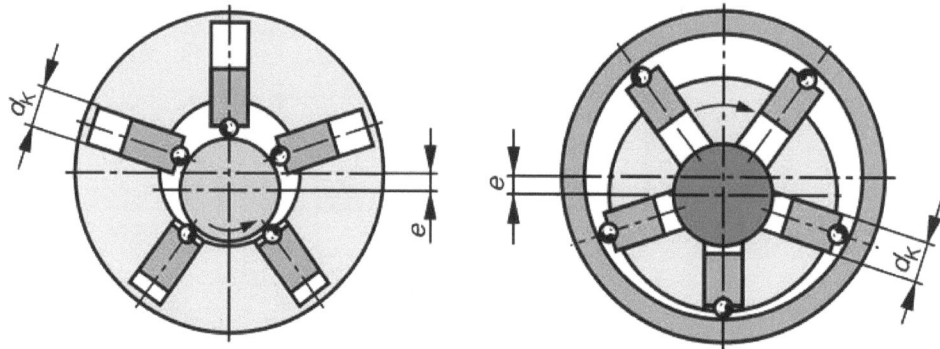

Fig. 4.5- Measurements for Calculating Size of Radial Piston Pump (Courtesy of Bosch Rexroth)

Swash Plate Pump:

Figure 4.6 shows *swash plate pump.* Considering the dimensions shown in the figure, Eq. 4.7 shows the calculation of swash plate pump displacement.

$$\textbf{Pump Displacement} = 2 \times r_h \times z \times \sin \alpha \left[\frac{\pi \times d_k^2}{4}\right] \qquad \textbf{4.7}$$

Where d_k = piston diameter, α = angle of inclination, and z = number of pistons.

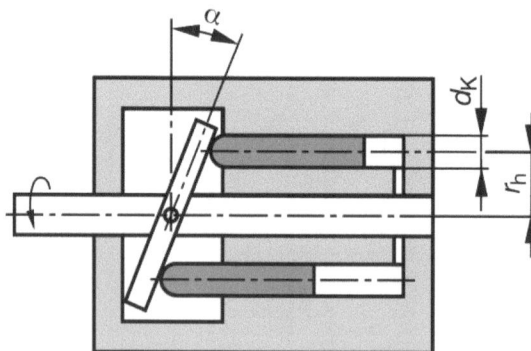

Fig. 4.6- Measurements for Calculating Size of Swash Plate Pump (Courtesy of Bosch Rexroth)

Bent Axis Pumps:

Figure 4.7 shows *bent axis pump.* Considering the dimensions shown in the figure, Eq. 4.8 shows the calculation of bent axis pump displacement.

$$\textbf{Pump Displacement} = 2 \times r_h \times z \times \tan \alpha \left[\frac{\pi \times d_k^2}{4} \right] \qquad \textbf{4.8}$$

Where d_k = piston diameter, α = angle of inclination, and z = number of pistons.

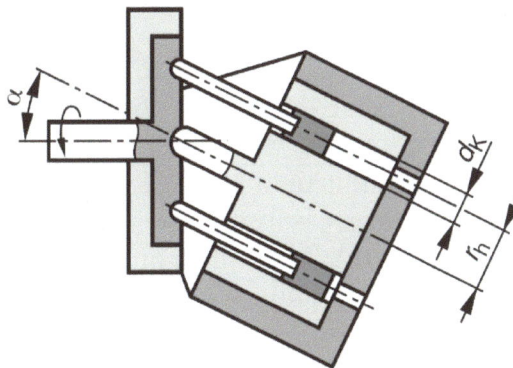

Fig. 4.7- Measurements for Calculating Size of Bent Axis Pump (Courtesy of Bosch Rexroth)

4.2- BP-Pumps-02-Maintenance Scheduling

Unless otherwise is stated by components and systems manufacturer, Table 4.1 provides guidelines for *scheduling* preventive maintenance actions for hydraulic pumps.

#	Preventive Maintenance Actions	Daily	Weekly	Monthly	Biannually	Annually
1	Clean around and outside surface	✔	✔	✔	✔	✔
2	Check for unusual sound	✔	✔	✔	✔	✔
3	Check temperature of the pump body	✔	✔	✔	✔	✔
4	Check tightness and leakage around hydraulic connections		✔	✔	✔	✔
5	Check electrical connections (if found)		✔	✔	✔	✔
6	Check for vacuum readings (if found)			✔	✔	✔
7	Check for vibration and condition of dampers			✔	✔	✔
8	Standard tests and calibration					✔

Table 4.1- BP-Pumps-02-Maintenance Scheduling

SUGGESTED MAINTENANCE ON HYDRAULIC POWER UNITS (Courtesy of Womack):

Daily Inspections:
- **Inspect for Oil leakage:** Visually inspect especially around shafts of pumps, cylinders, and hydraulic motors. Leakage often starts at these points just before seal breakdown. Tighten any fittings which leak.
- **Inspect for Oil Level:** Observe oil level in reservoir over a complete cycle of the machine. Make sure that oil level never falls below the LOW mark as cylinders extend. Add oil if necessary.

Weekly Inspections:
- **Housekeeping Inspection:** Should be made to observe general conditions around the machine. Clean up spilled oil. Clean dirt off top of reservoir.
- **Inspect Breather:** Inspect to see that filler and breather caps are in place. Check condition of breather; remove accumulations of lint and dirt. Breather must be cleaned, especially at higher altitudes where air pressure is lower.
- **Inspect Fluid and Water Content:** Tap a small quantity of fluid from one of the drain valves on the reservoir. Check color, odor, and water content. If water is found, suspect a water leak from a shell and tube heat exchanger. Water may be caused from condensation on inside walls of the reservoir due to atmospheric temperature changes around the reservoir. If this is the case, it will be necessary to tap off accumulated water every day.

500-Hour Intervals:
- **Filter Element:** Several sets of elements for paper-type micronic filters should be replaced. Wire mesh pump suction strainers seldom need replacing because they can be cleaned in most cases. Remove and clean pump suction strainers. Replace elements in micronic filters.
- **Hydraulic Oil:** Be sure to have a reserve supply of the same brand and type of oil used in the system. Usually a back-up supply of 10 to 25% of reservoir capacity should be kept as a minimum.
- **Hoses:** All hoses should be replaced after being in service approximately 5 years even though they appear to be in good condition.
- **Cylinders Overhaul:** Overhaul kits can be purchased from cylinder manufacturer which contains all replaceable gaskets, O-rings, and parts which normally wear out. These parts are usually packed in a sealed polyethylene bag. On receipt the bag should be dated. If not used within 2 years a new kit should be ordered to replace it. Rubber deteriorates by oxidation in a few years. If left sealed in an airtight bag it will deteriorate more slowly.
- **Pumps Overhaul:** Overhaul kits are sometimes available for pumps, and include all soft seals, gaskets, and sometimes bearings. Spare shaft seals should always be kept, and on important machines, entire pumps should be kept as spares.
- **Electrical Parts:** Keep replacement coils for solenoid valves, relays, and motor starters. Limit switches, fuses, lamp bulbs, and on important machines, spare relays should be kept.

4.3- BP-Pumps-03-Installation and Maintenance

Proper pump installation and maintenance are important processes for trouble-free operation of the system. DO NOT follow the crowed!!, mount the pump for life. The following best provides guidelines for *installation and maintenance* of pumps.

BP-Pumps-03-Installation and Maintenance:
1. Install the Pump to avoid Cavitation.
2. Identify Ports and Direction of Rotation.
3. Proper Placement with the Reservoir.
4. Proper Mounting for Direct Drive.
5. Proper Coupling for Direct Drive.
6. Proper Shaft Alignment for Direct Drive.
7. Proper Indirect (Side) Drive.
8. Adequately Damp Vibration.
9. Proper Oil Intake and Return.
10. Proper Priming.
11. Proper Case Drain.
12. Proper Oil Discharge.
13. Review Range of Driving Speed.
14. Review Range of Working Pressure.
15. Review Range of Working Temperature.
16. Review Compatibility with Working Hydraulic Fluid.
17. Review Prime Mover Overloading Conditions.
18. Proper Placement of Hydraulic Power Unit.
19. Proper Installation of Hydrostatic Transmission

The following sections provide detailed interpretations, with examples, for the actions listed in the previous best practices list.

4.3.1- Install the Pump to Avoid Cavitation

Every effort must be made to avoid developing cavitation. The following list shows actions that cause pump cavitation:
- Incorrect pump ports identification
- Incorrect pump direction of rotation.
- Improper pump placement with the reservoir.
- Improper pump oil Intake.
- Improper pump priming.
- Increased pump driving speed.
- Increased pump inlet temperature.
- Incorrect oil viscosity.

4.3.2- Identify Ports and Direction of Rotation

Suction and discharge ports of a hydraulic pump are relevant to its direction of *rotation*. If a pump is connected wrongly or rotates oppositely, the pump may fail in a very short time before somebody discover the mistake.

How to Identify Pump Ports: As shown in Fig. 4.8, Pump ports can be identified by one or more of the following signs:

1. knowing how a pump works and direction of rotation of the pump shaft.
2. Visual symbol on the name plate supposed to match the physical location of ports.
3. Suction port is larger than discharge port in unidirectional pumps.
4. Unbalanced pumps have bearing wear signs at the discharge side.

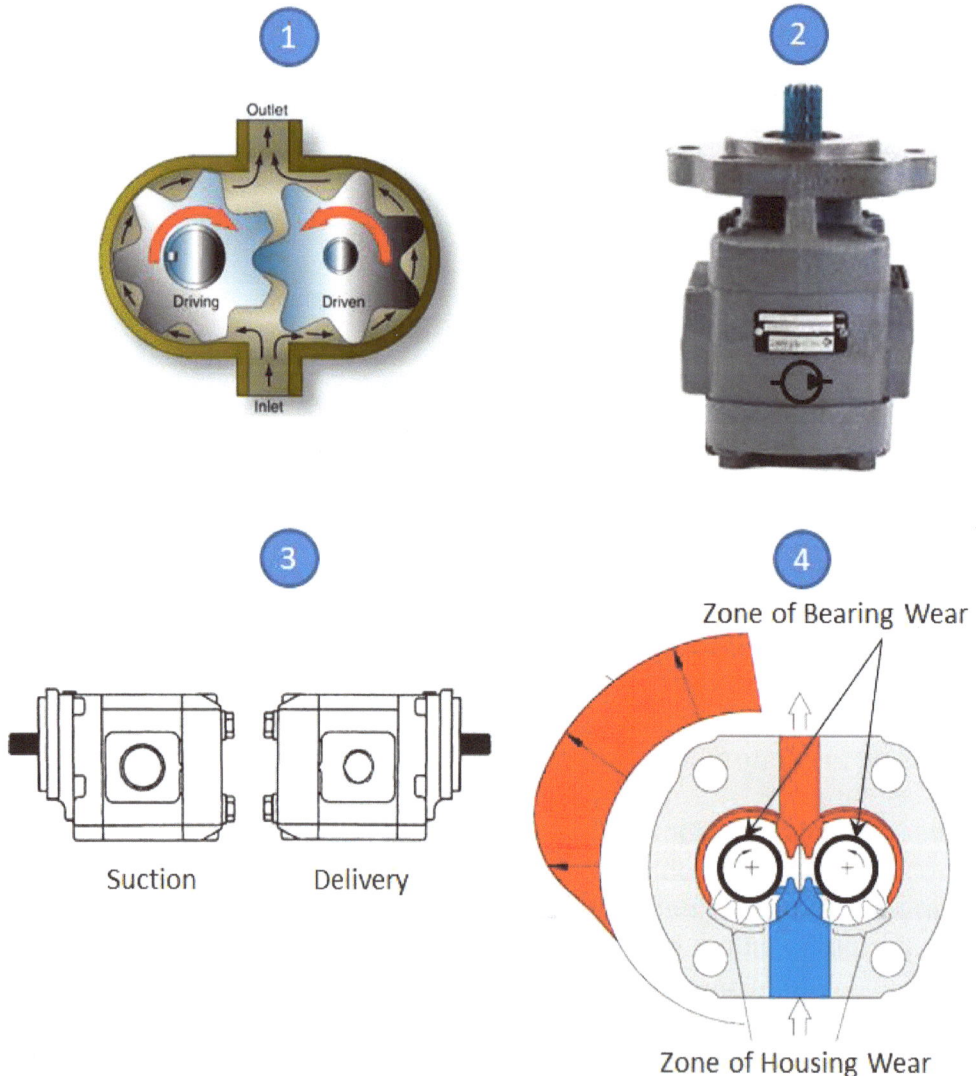

Fig. 4.8- Identification of Pump Ports

How to Identify Pump Direction of Rotation: A pump could be of *unidirectional*, *bidirectional* or *over-center* type. Over-center and unidirectional pumps have a predefined direction of rotation (CW or CCW). As shown in Fig. 4.9, Pump direction of rotation can be identified by one or more of the following signs:

1. knowing how a pump works and location of pump ports.
2. Visual symbol on the pump housing.
3. Review the ordering code or the name plate of the pump (**See Note 1**).

Note 1: A pump is called "Right or R" if it rotates CW and "Left or L" if it rotates CCW. These directions of rotation were assumed when facing the pump shaft or from the rear side of the prime mover. Pump users should specify the pump direction of rotation as part of the pump ordering code. In some cases, the letter **R** or **L** is written on the pump name plate.

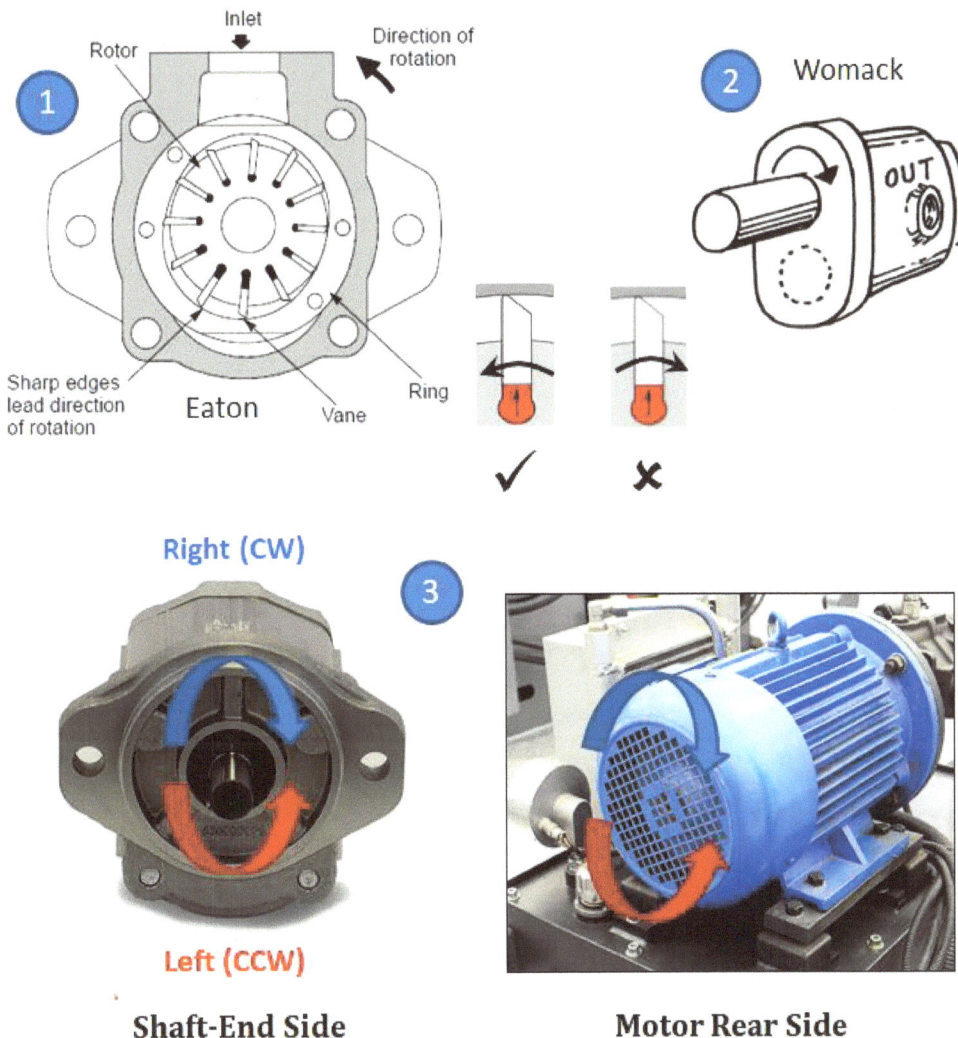

Fig. 4.9- Identification of Pump Ports and Direction of Rotation

Switching a Pump Direction of Rotation:

For a Bidirectional Pumps: As shown in Fig. 4.10, direction of rotation of such pumps can be switched. As shown in the figure, if the prime mover is an electric motor (1), direction of rotation can be switched by switching the motor rotation using appropriate motor controller. If a pump is driven by an engine (2) indirectly such as when a belt is used, the pump must be mounted on the other side to reverse its direction of rotation.

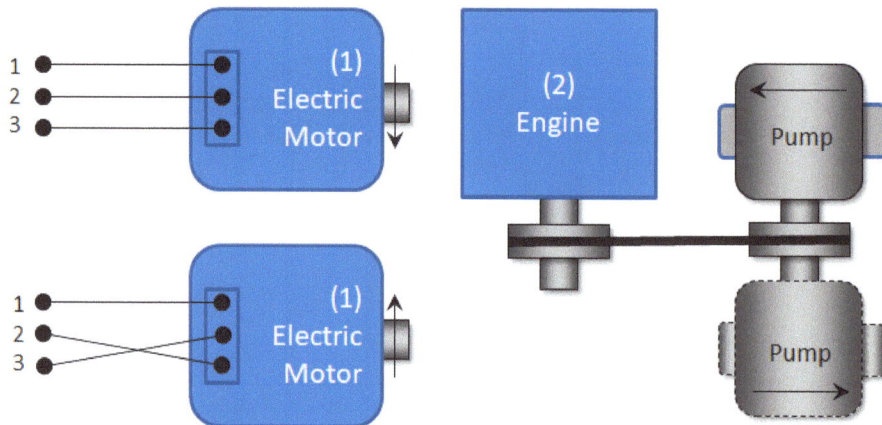

Fig. 4.10- Switching the Direction of Rotation

For a Unidirectional Pumps: Some of pump designs require disassembling and reconfiguring the pump as per the pump manufacturer before switching them to run oppositely. Figure 4.11 shows an example of instructions for switching the direction of rotation of a balanced vane pump.

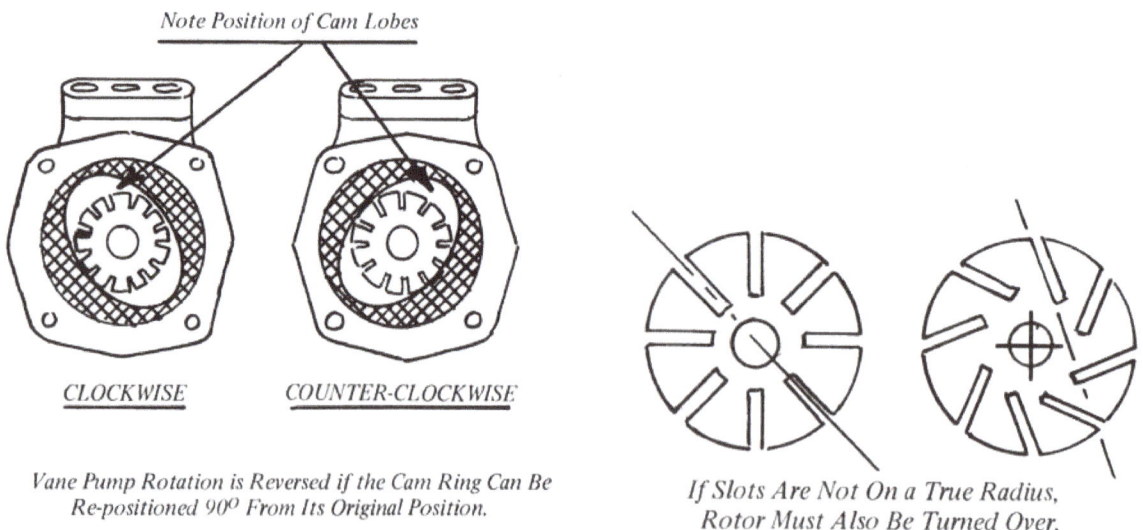

Note Position of Cam Lobes

CLOCKWISE   COUNTER-CLOCKWISE

Vane Pump Rotation is Reversed if the Cam Ring Can Be Re-positioned 90º From Its Original Position.

If Slots Are Not On a True Radius, Rotor Must Also Be Turned Over.

Fig. 4.11- Instructions for Switching A Vane Pump Direction of Rotation (Courtesy of Womack)

4.3.3- Proper Placement with the Reservoir

For better suction conditions and pump cavitation avoidance, pumps are recommended to be placed on a horizontal plane with the suction port is close as possible to the reservoir to minimize the number of bends on the suction line. As shown in Fig. 3.12, the following are the common scenarios of placing a pump with respect to the reservoir:

1- Small pumps usually are placed with the drive motors on top of the reservoir.
2- Medium size pumps are set aside of the reservoir on a L-shaped structure.
3- Larger size pumps are recommended to be placed underneath of overhead reservoirs to utilize the positive fluid pressure head to avoid cavitation.
4- Not as common, a pump is immersed in the oil inside the tank for better cooling conditions. However, this placement makes it difficult to monitor the pump condition.

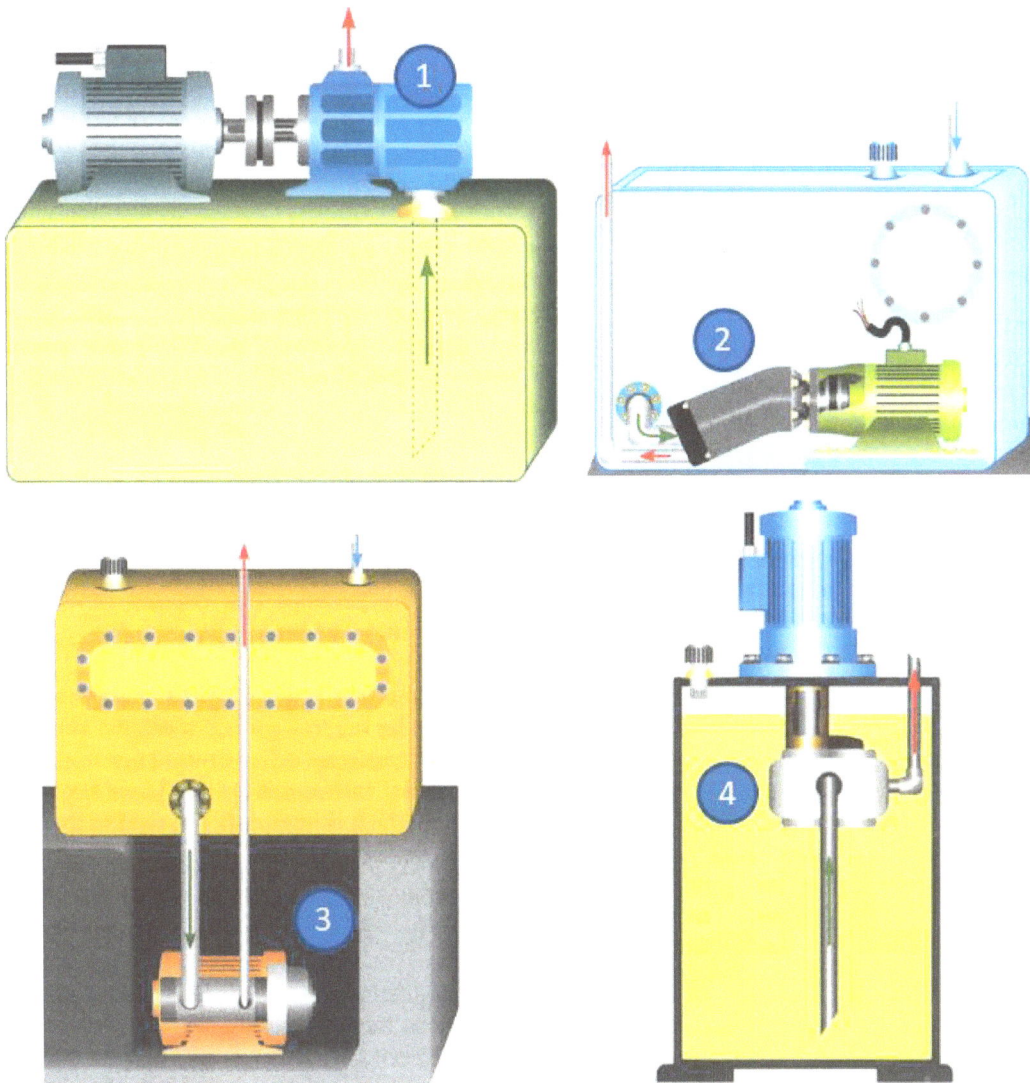

Fig. 4.12- Proper Placement with the Reservoir (Courtesy of Assofluid)

4.3.4- Proper Mounting for Direct Drive.

Generally, for safe operation, rotating parts should be covered and guarded to provide adequate protection against hazard and noise. *Direct Drive* is commonly applied in industrial applications where a pump is directly driven by an electric motor using a standard coupling. However, direct drive is also applied in mobile machines where a main pump is directly driven by an engine using a special mount and couplings. This section presents different configurations for mounting a pump for direct drive:

Foot-Mounting: As shown in Fig. 13, *Foot-Mounting* is a traditional method in which the pump and the motor could be mounted independently on two different frames of references. Such a method has the following features and components:

- **Supporting Base:** The pump and the drive motor are mounted on a rigid *Support Base*. The support base must be rigid enough to prevent any flexing. Otherwise, pump-motor misalignment occurs.
- **L-Bracket:** *L-bracket* is an *adaptor* used to align the pump-motor shafts.
- **Shaft Alignment:** Shaft *alignment* is challenging and must be redone every time the pump is removed from the bracket.
- **Coupling:** flexible couplings are recommended in such method of mounting to absorb vibration that could occur due to any misalignment when the pump operates.
- **Distance between Flanges:** It is flexible to adapt to the exact required distance.

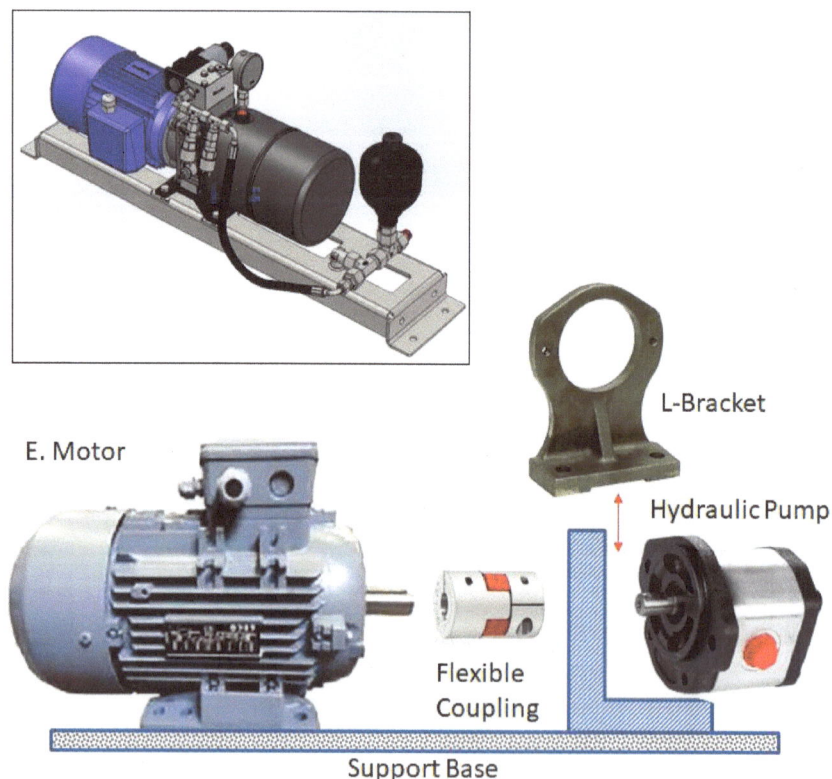

Fig. 4.13- Foot-Mounting for Direct Drive

Bellhousing Mounting: As shown in Fig. 14, *Bellhousing* is a common method in which both the pump and the motor are mounted on one frame of reference. Bell housing allows easy and a quick assembly of the prime mover to the pump. Such a method has the following features and components:

- **Supporting Base:** A large electric motor is foot-mounted to the supporting base and its flange supports the weight of the bell housing and the pump. For small electric motors, a bell housing is foot-mounted to the supporting base and it supports the weight of both the electric motor and the pump.
- **Bellhousing:** Because the electrical drive motor is approximately ten times the size of the pump, it's flange will be larger than the pump flange. Therefore, a bell-shaped housing is designed to integrate the pump with the drive motor. Bell housings are available in standard lengths to meet various pump-motor combinations.
- **Shaft Alignment:** Shaft alignment is considered in the design of the Bell Housing. Therefore, there is no need to redo the shaft alignment if the ump is removed.
- **Coupling:** flexible or rigid couplings are used. However, flexible couplings are still recommended.
- **Distance between Flanges:** It is fixed and must be considered in the Bellhousing design.

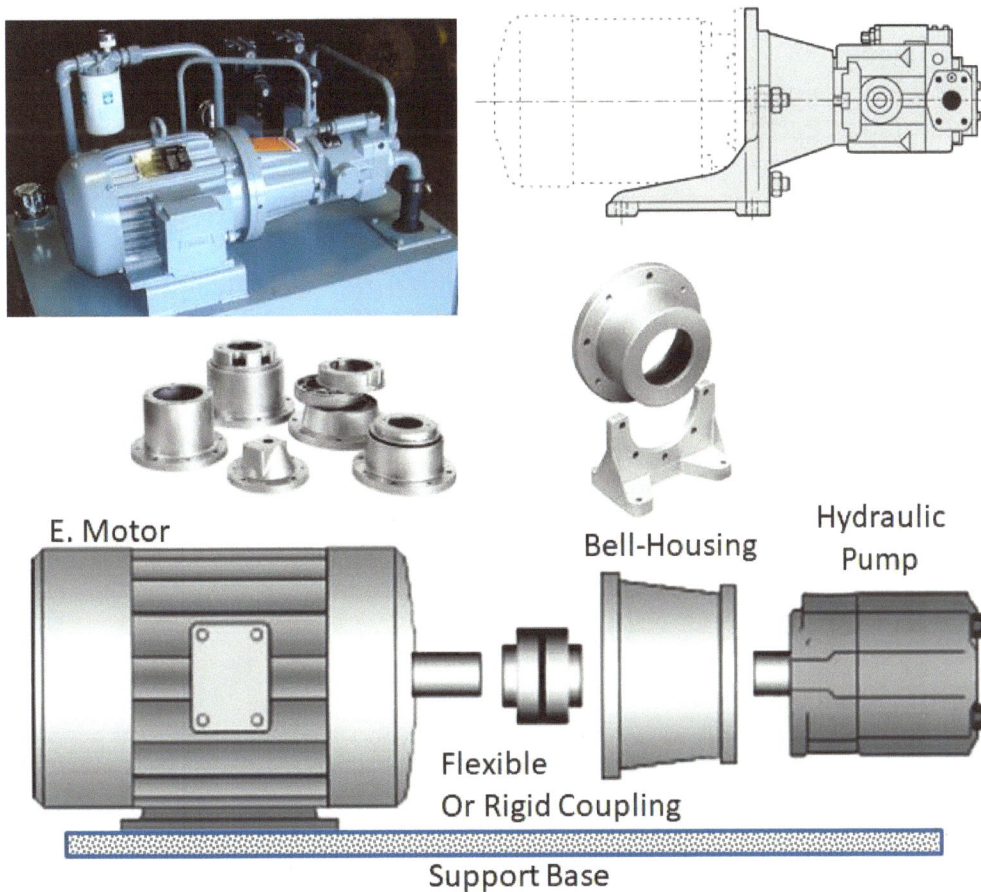

Fig. 4.14- Bellhousing Mounting for Direct Drive

4.3.5- Proper Coupling for Direct Drive

The following set of bullets explains the best practices for installing a coupling:
- Drive couplings and mountings shall be capable of continuously withstanding the maximum torque that can be generated under all conditions of intended use.
- Check the drive coupling for perfect fit on the pump and motor shafts. Loose fitting couplings cause accelerated wear of the drive shaft and should be replaced.
- DO NOT push hardly the coupling onto the pump drive shaft. If it is too tight, it may be necessary to heat coupling for installation.
- Drive couplings shall be provided with a suitable protective guard when the coupling area is accessible during operation of the pump.

Couplings for direct drives are available in different sizes and configurations:

Rigid Coupling: As shown in Fig. 15, a *Rigid Coupling* contains two metallic symmetric parts screwed together. Each part has a hole on the other side for a standard size shaft. Rigid couplings do not tolerate neither shaft misalignment nor shaft end backlash distance. Rigid couplings are commonly used with bellhousing method of pump-motor mounting. Rigid couplings have the following features:
- Compact design; low weight and mass moment of inertia.
- Independent of direction of rotation (suitable for reversing operation).
- Suitable for plug-in assembly in bellhousing installation.
- Suitable and certified for use in potentially explosive environments.

Flexible Coupling: The figure shows that, like a rigid coupling, a *Flexible Coupling* contains two metallic symmetric parts with one hole on the side of each part for a standard size shaft. Unlike a rigid coupling, a rubber-based flexible element is placed between the two halves of the coupling. Because of that, each size of flexible couplings allows a certain maximum permissible shafts misalignment and shaft end play distance. Flexible couplings are required for the foot-mounting method. Flexible couplings have the following features:
- The central part made of Nylon, Polyamide or Rubber.
- Rotational speed: max. 47,500 rpm.
- Compensation for shaft misalignments.
- Suitable for highly dynamic applications with proper central part material.
- Vibration-damping.

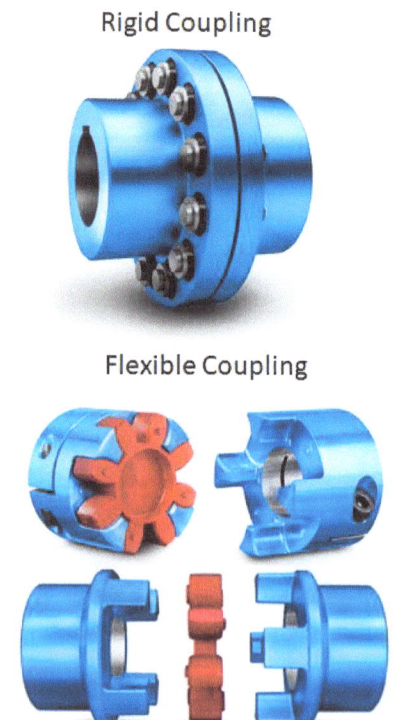

Rigid Coupling

Flexible Coupling

Fig. 4.15- Coupling Configurations for Direct Drive (www.flender.com)

4.3.6- Proper Shaft Alignment for Direct Drive

As shown in Fig. 16, ideally the pump-motor shafts should be coaxially connected and aligned to each other. Shaft misalignment causes:
- Increased vibration.
- Damage to bearings and couplings.
- Shaft seal failure.
- Wear of pump internal parts.
- Excessive heat and energy loss.

Therefore, pump-motor shaft alignment should be checked:
- During pump installation.
- Routinely during major maintenance.
- When there is any sign of increased noise and vibration.

As shown in the figure, misalignment could be *axial misalignment* (known also as *Offset*) or *angular misalignment* (known also as *Angularity*).

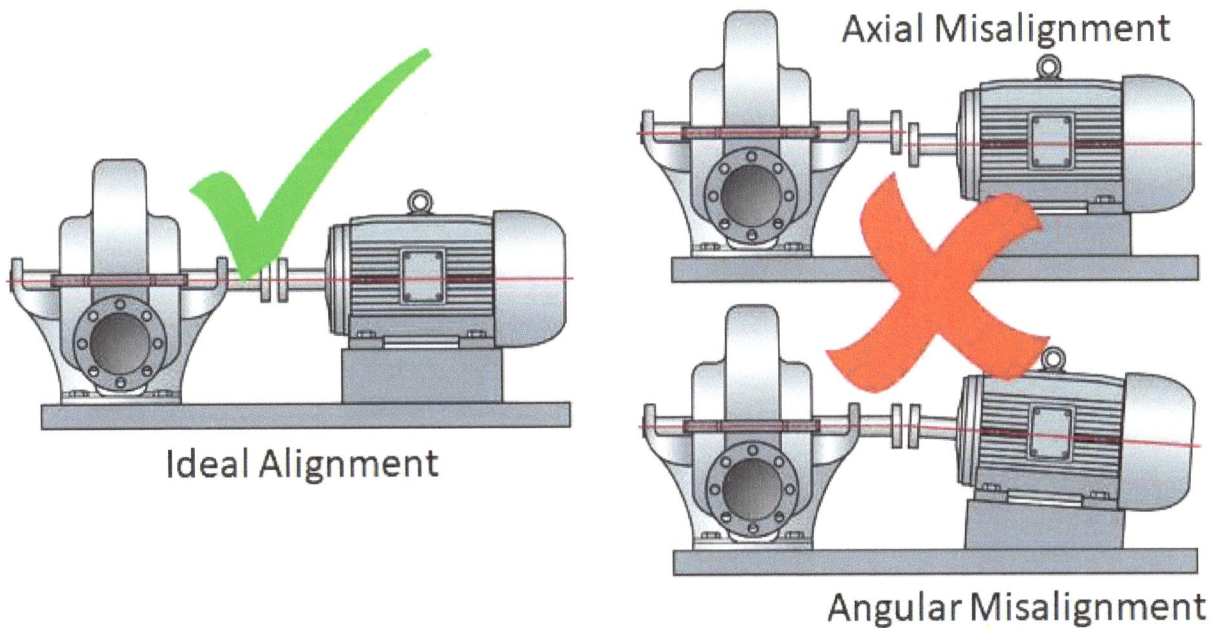

Fig. 4.16- Axial and Angular Shaft Misalignment (wastewater101.net)

Dial Indicator: As shown in Fig. 4.17, A dial indicator is a valuable tool that can be used for a variety of applications. One common application is checking shaft misalignment. A Dial indicator is using a precise gear mechanism that is driven by a plunger. That mechanism has a very fine measuring resolution up to a one thousand of an inch (0.001" = 1.0 mil or 1.0 thou). Dial indicators are also available with digital readings. Both should have means to adjust the zero point of the indicator.

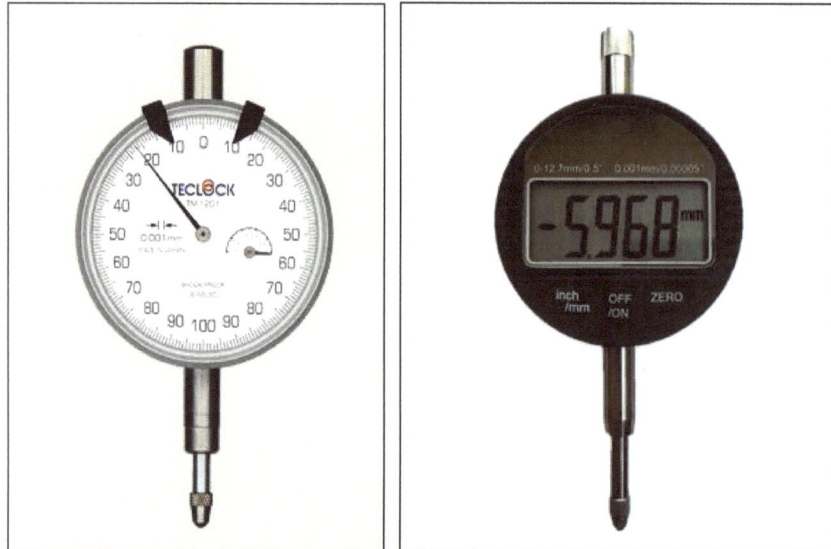

Fig. 4.17- Dial Indicators

Table 4.2 shows an example of how permissible misalignment is reported versus the rotational speed. Obviously, the permissible misalignment is inversely proportional to the rotational speed. While axial misalignment is measured in mils because it represents offset, angular misalignment is reported in mils/inch because it represents slope. Actual values must be reported by the component/system manufacturer or maintenance department.

Permissible Misalignment				
	Angular Misalignment (Mils)		**Axial Misalignment (Mils/Inch)**	
RPM	**Excellent**	**Acceptable**	**Excellent**	**Acceptable**
3600	0.3/1"	0.5/1"	1.0	2.0
1800	0.5/1"	0.7/1"	2.0	4.0
1200	0.7/1"	1.0/1"	3.0	6.0
900	1.0/1"	1.5/1"	4.0	8.0

Table 4.2- Example of Pump Shaft Permissible Misalignment

As best practices for measuring misalignment (Fig. 4.18):

- Rotate the pump shaft in the assigned direction of the pump rotation.
- Both the offset and angularity are measured on horizontal and vertical planes. However, they can be checked on any plane.

Checking Axial Misalignment (Offset): The standard reference point of measuring offset is at the center of the coupling. Therefore, axial misalignment is done by placing the dial indicator on the rim as shown in Fig. 18. That is why it is known as *Rim Check*. Make sure the plunger is in the middle of its stroke, then bring the zero pint on the scale at the place of the dial.

Vertical Offset:

- Start at 12 o'clock and turn the pump shaft 180 degrees till 6 o'clock.
- Vertical Total Indicator Reading (TIR) = indicator reading from top to bottom.
- Actual Vertical Misalignment Y = TIR/2 measured in mils.

Horizontal Offset:

- Start at 3 o'clock and turn the pump shaft 180 degrees till 9 o'clock.
- Horizontal Total Indicator Reading (TIR) = indicator reading from side to side.
- Actual Horizonal Misalignment X = TIR/2 measured in mils.

Checking Angular Misalignment (Angularity): As shown in the figure, angular misalignment can be checked by placing the dial indicator on the face of the coupling. That is why it is known as *Face Check*.

Vertical Angularity:

- Start at 12 o'clock and turn the pump shaft 180 degrees till 6 o'clock.
- Vertical Total Indicator Reading (TIR) = indicator reading from top to bottom.
- Actual Vertical Angularity = TIR/Coupling Diameter.

Horizontal Angularity:

- Start at 3 o'clock and turn the pump shaft 180 degrees till 9 o'clock.
- Vertical Total Indicator Reading (TIR) = indicator reading from side to side.
- Actual Vertical Angularity = TIR/Coupling Diameter (D).

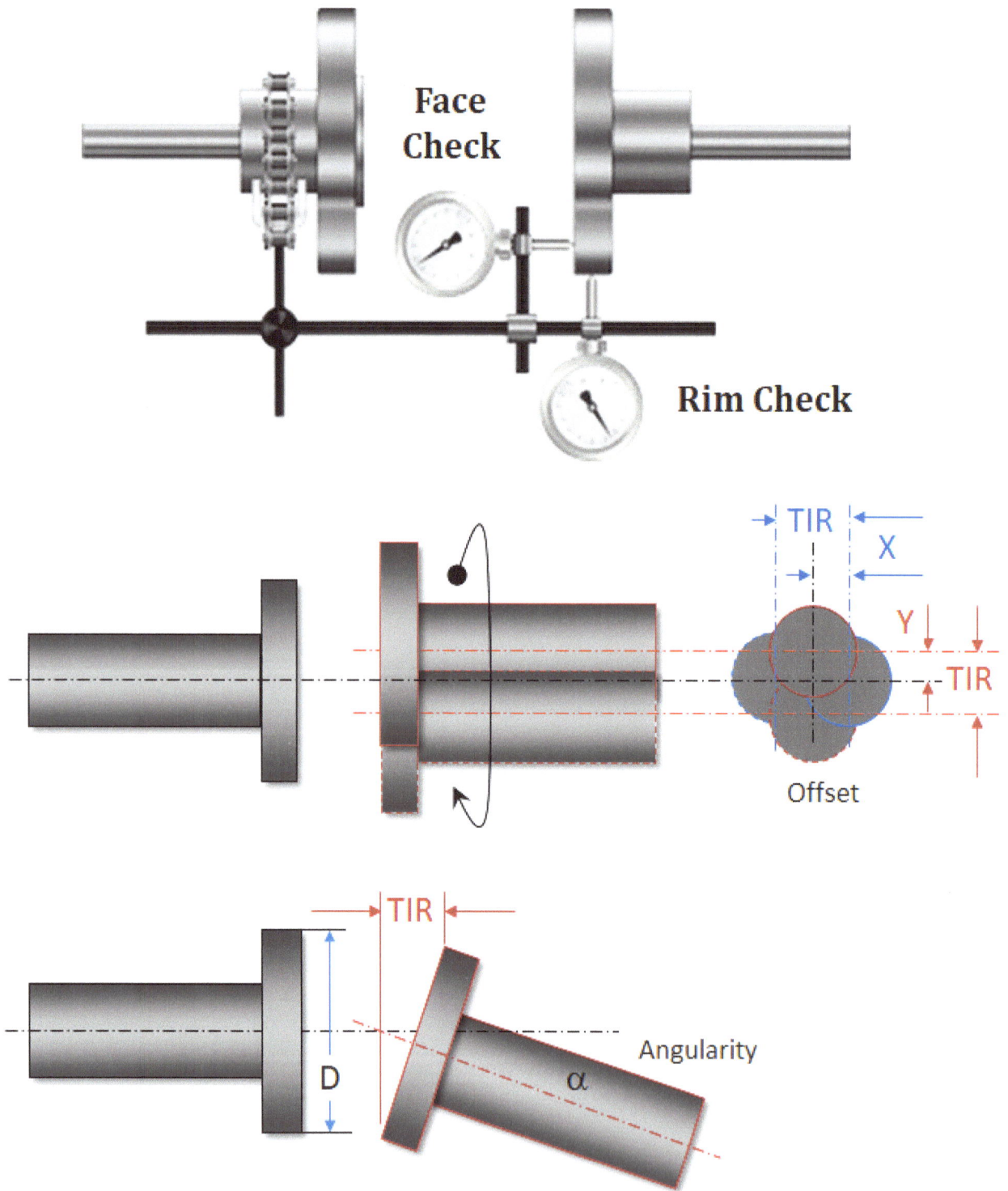

Fig. 4.18- Measuring the Offset and Angularity

4.3.7- Proper Indirect (Side) Drive

As shown in Fig. 4.19, *side drive* is commonly used in mobile applications when a pump is driven by an engine through an intermediate element such as belt, gear, chains, etc. However, it is still used occasionally in industrial applications when an electrical motor drives multiple pumps.

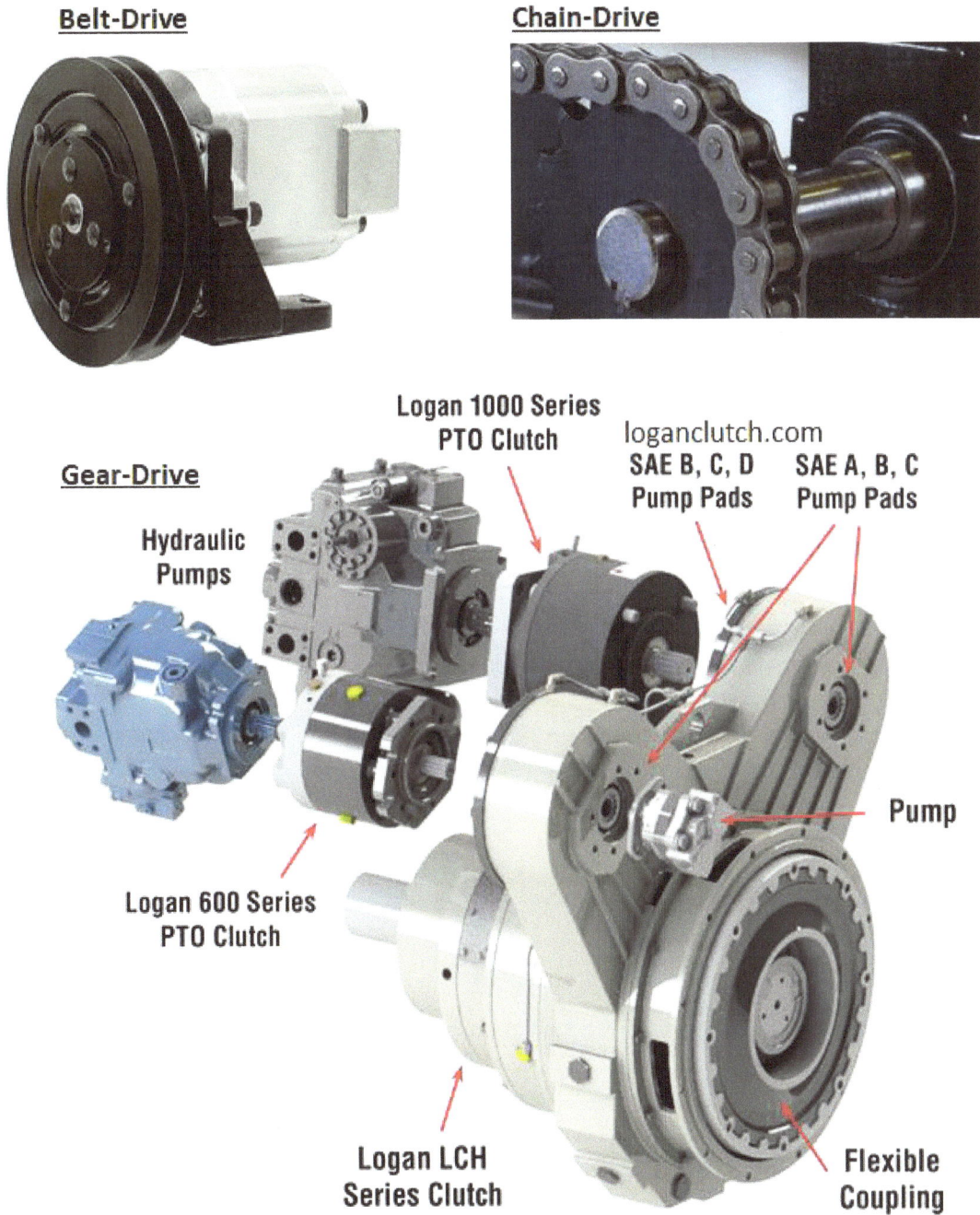

Fig. 4.19- Side Drive of a Hydraulic Pump

Side driving adds side load on shaft bearings. For each pump, there is permissible side load. Exceeding the permissible side load results in shaft misalignment and all the consequences mentioned previously.

Table 4.3 shows a typical example from industry for the permissible axial and radial loads for a pump shaft.

Axial Piston Fixed Pump A2FO

Technical data

Permissible radial and axial forces of the drive shafts

(splined shaft and parallel keyed shaft)

Size	NG		5	5[3]	10	10	12	12	16	23	23
Drive shaft	ø	mm	12	12	20	25	20	25	25	25	30
Maximum radial force[1] at distance a (from shaft collar)	$F_{q\ max}$	kN	1.6	1.6	3.0	3.2	3.0	3.2	3.2	5.7	5.4
	a	mm	12	12	16	16	16	16	16	16	16
with permissible torque	T_{max}	Nm	24.7	24.7	66	66	76	76	102	146	146
≙ permissible pressure Δp	Δp_{perm}	bar	315	315	400	400	400	400	400	400	400
Maximum axial force[2]	$+F_{ax\ max}$	N	180	180	320	320	320	320	320	500	500
	$-F_{ax\ max}$	N	0	0	0	0	0	0	0	0	0
Permissible axial force per bar operating pressure	$\pm F_{ax\ perm/bar}$	N/bar	1.5	1.5	3.0	3.0	3.0	3.0	3.0	5.2	5.2

Table 4.3- Example of Pump Shaft Permissible Side Loads (Courtesy of Bosch Rexroth)

Referring to Fig. 4.20, the following set of bullets explains the best practices for pump side drives:

- **Connecting Element (1):** Select proper connection (belts. gears or chains.
- **Shaft Bearings (2):** Pump shaft is supported by a pair of tapered bearing placed face-to-face to carry radial and axial Loads.
- **Neutralize the Side Load (3):** Operation of an unbalanced pump adds side load opposite to the flow. Therefore, if the drive motor is paced at the inlet side of the pump, the side loads will be concentrated on one side of the bearing. Then, it is recommended to locate the drive motor at the outlet side of the pump so that the side load from the driving element can partially balance the pressure side load.
- **Side Drive for Multiple Pumps (4):** If one prime mover is used to drive multiple pumps, they should be distributed to balance the side forces on the prime mover's shaft.

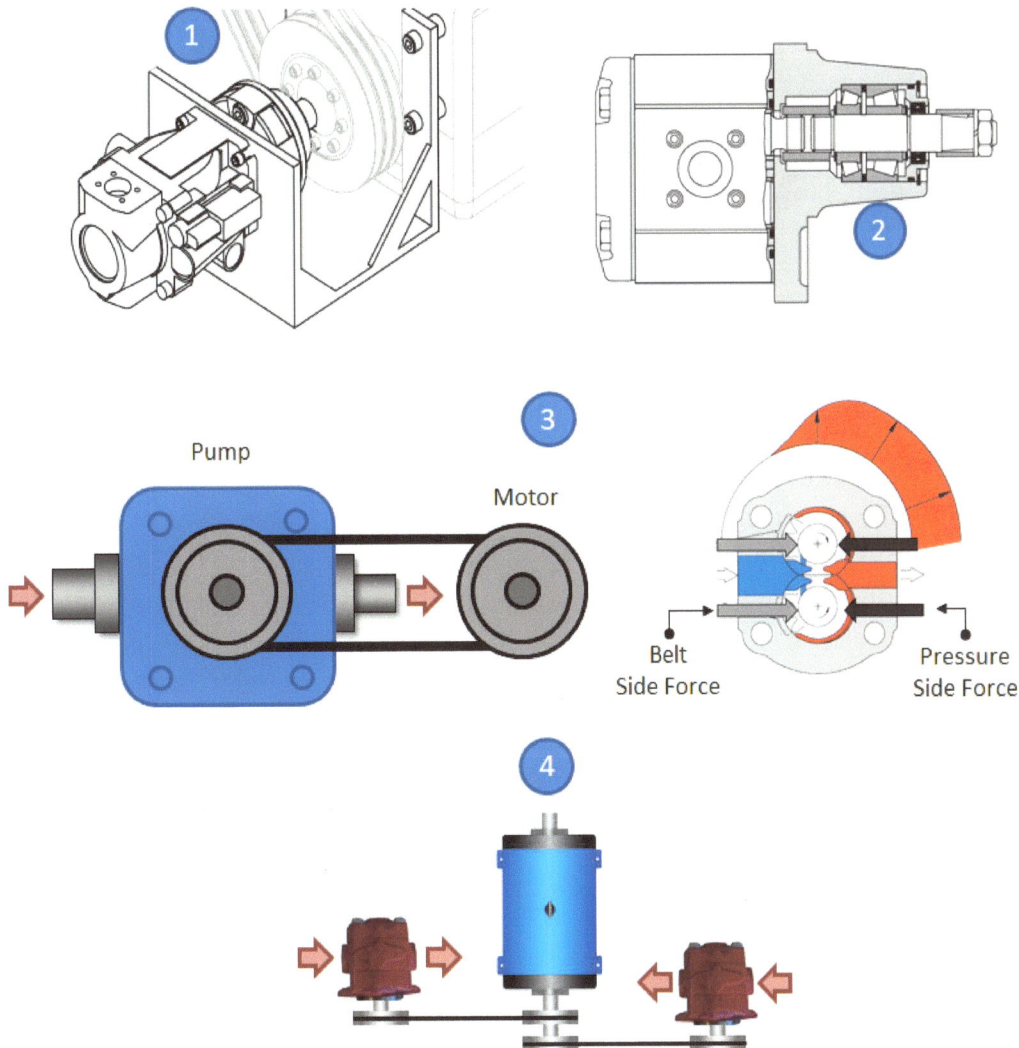

Fig. 4.20- Best Practices for Pump Side Drive

4.3.8- Adequately Dampen Vibration

As known for fact, operation of a positive displacement pumps is associated with pressure ripples and vibration due to flow pulsation. The vibration the pump and prime mover generate during operation causes noise that possibly exceeds recommended limits and possible resonance of the structure the pump is mounted on.

Vibration and noise control are topics that contain lots of details out of scope of this textbook. However, as shown in Fig. 4.21, during a pump installation and maintenance check conditions of vibration dampers (if found).

Fig. 4.21- Conditions of Vibration Dampers

To achieve favorable noise values, decouple all connecting lines from all vibration-capable components (e.g. tank) using elastic elements. As shown in Fig. 4.22, the flanges and the clamp bolts of the support surface must be covered with antivibration plugs, the hoses connected to the pump and the circuit must be secured by internally flexible collars and horizontal plates provided with rubber thru-bulkheads.

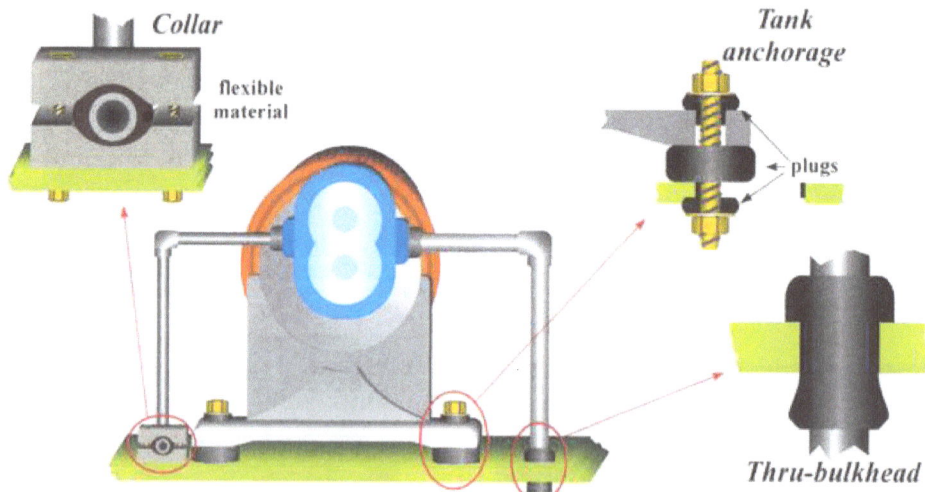

Fig. 4.22- Tools for Vibration Damping (Courtesy of Assofluid)

4.3.9- Proper Oil Intake and Return

Suction line of a hydraulic pump is the most critical one among the other transmission lines in a hydraulic system. Suction line design parameters supposed to be originally determined by the system designer. Details of the methods, rules of thumbs and calculations based on which these parameters are determined are presented in Volume 4 "Hydraulic Fluids Conditioning". Referring to Fig. 4.23, the following suction line design parameters SHOULD NOT be changed during system service, otherwise immediate pump cavitation and failure may occur:

Suction Line Diameter: It is predefined based on:
- Pump suction port size.
- Avoidance of turbulent flow in the line.
- Securing moderate flow speed in the line to minimize pressure losses and consequently avoid cavitation.

Suction Line Placement w.r.t. Baffle Plate: It is based on weather a baffle plate exists or not. if a baffle plate is used, suction and return lines are placed on the two sides of the baffle plat. If no baffle plate is used, they are placed on the diagonal of the reservoir. However, suction line should not in any case near the point of dirt collection.

Suction Line Ground Clearance: It is the distance defined based on the following rules of thumbs:
- Minimum volume of oil covers the bottom part of the intake line by at least 5 cm (2 in) to avoid air entraining to pump intake.
- Ground clearance = (2-3) x suction line diameter to minimize losses in suction line.

Suction Line Length: nether long nor short suction line are recommended. Suction line length It is defined based on:
- Proportional to the line diameter.
- How far the pump intake from the nearest turbulent point.
- Lowest permissible suction pressure at the pump suction port.

Suction Line Routing: That is also originally supposed to be decided by the system designer to minimize losses in the suction line.

Suction Line Type: When replacing a suction line, do not replace hard tubing suction lines by flexible hoses. However, if it was originally a flexible hose, make sure the new one is a suction hose (NOT a pressure hose). The reason is suction hoses are specially constructed with a spiral wire or supporting tube on the inner layers to prevent collapsing the inner layers due to negative pressure.

Suction Strainer: A suction strainer is used to be the first line of defense against the relatively large size contaminants. Suction strainers are then routinely must be cleaned. Figure 4.24shows a result of suction strainer replacement time overdue.

Return Line: Retrofitting the originally designed return line size, placement, and routing may result in developing back pressure and may bring turbulence to the reservoir.

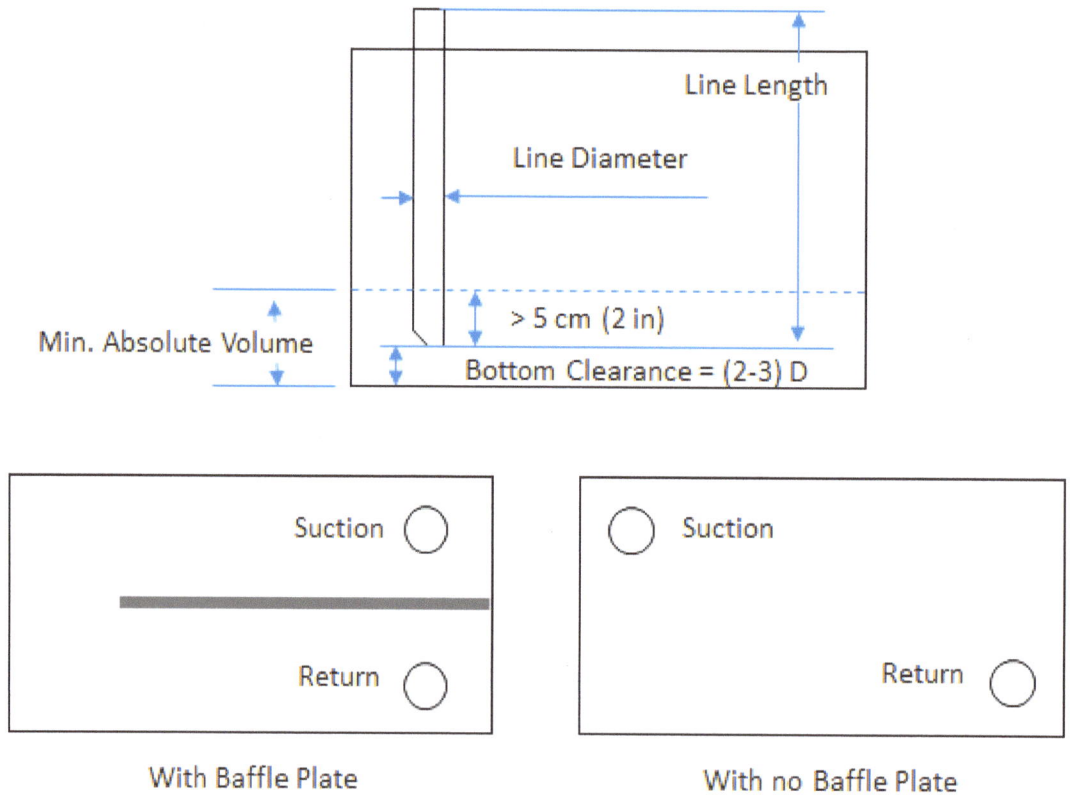

Line Length

Line Diameter

Min. Absolute Volume

> 5 cm (2 in)

Bottom Clearance = (2-3) D

Suction

Return

With Baffle Plate

Suction

Return

With no Baffle Plate

Fig. 4.23- Pump Suction Line Size and Placement

Fig. 4.24- Result of Suction Strainer Replacement Time Overdue

4.3.10- Proper Priming

The main purpose of *priming* hydraulic pumps is to bleed the air from inside the pump mechanism to avoid air lock and danger of dry run. Failure to do so will result in damage to the pump through inadequate lubrication on startup. Priming must be done during pump commissioning and following a relatively long standstill as the pump may drain back to the reservoir via the hydraulic lines. The process is simple. Fill the device with clean hydraulic fluid through the filling holes and simultaneously bleed the air from the *air bleed* port. The following set of bullets explains the best practices for priming a hydraulic pump:

- **Priming Fluid:** Priming hydraulic fluid must be same as used in the system and must be filtered before priming.
- **Identify Filling/Bleed Holes based on Pump Orientation:** Review manufacturer instructions to identify the filling port and the air bleed port versus pump orientation during priming. Figure 4.25 shows an example from a specific manufacturer. As shown in the figure, recommendation for installation position 8, a check valve in the drain line (cracking pressure 0.5 bar) can prevent draining of the pump housing. Units that are mounted vertically with the shaft up (4 and 8) require special attention to ensure that the fluid level in the case is high enough to lubricate the front shaft bearings.

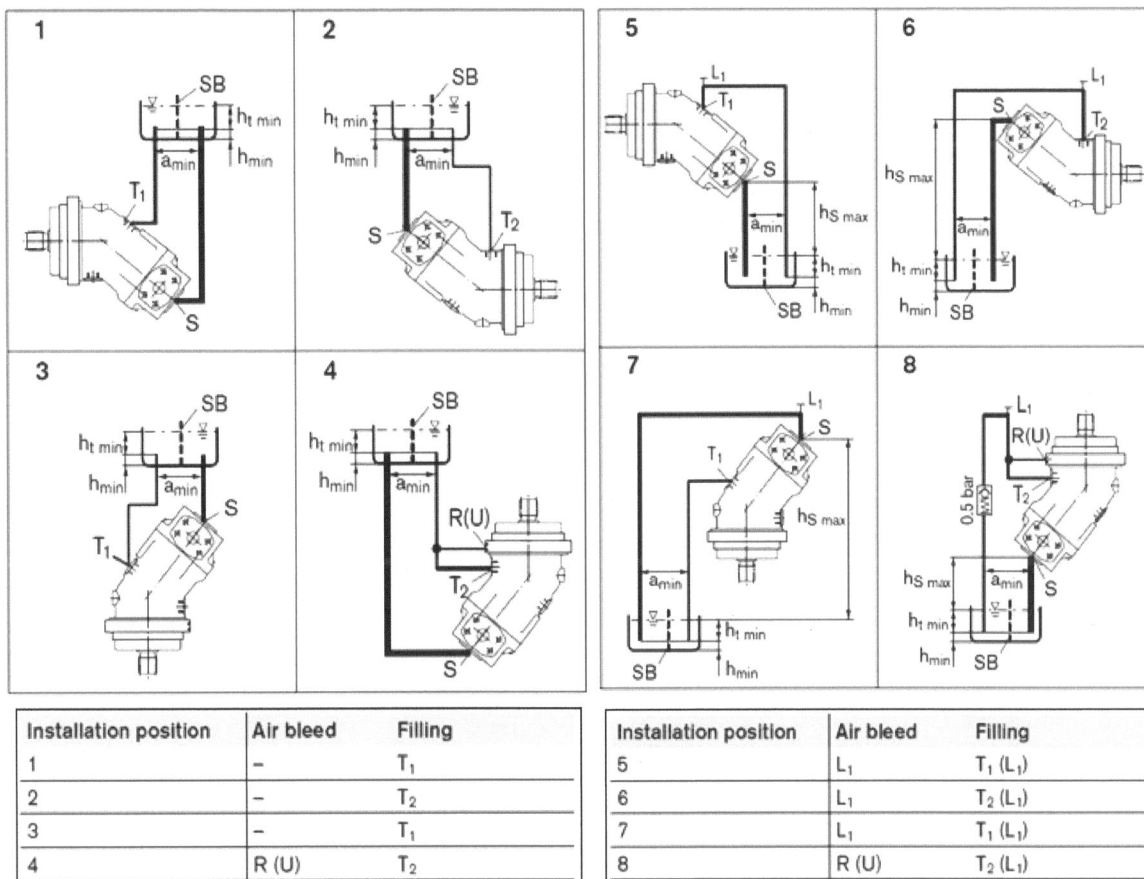

Installation position	Air bleed	Filling
1	–	T_1
2	–	T_2
3	–	T_1
4	R (U)	T_2

Installation position	Air bleed	Filling
5	L_1	T_1 (L_1)
6	L_1	T_2 (L_1)
7	L_1	T_1 (L_1)
8	R (U)	T_2 (L_1)

Fig. 4.25- Example for Identifying the filling/Bleed Ports (Courtesy of Bosch Rexroth)

- **Priming Procedure (Fig. 4.26):**
  - o Route the pump outlet to the tank.
  - o It is critical to ensure that the priming fluid and devices (i.e. funnel, container, etc.), used to transfer fluid into the pump are clean per pump manufacturer's recommendation.
  - o Fill the housing with clean priming fluid via the identified filling port.
  - o Rotate the pump shaft till it displaces bubble-free oil from the pump outlet.
  - o Close the pump priming hole and assemble the pump to the prime mover in the ready to work position.
  - o Run the pump and check if it displaces bubble-free oil in no less than 20 seconds. Otherwise, re-check the system.

Pour Oil Into Pump Case.

Rotate Pump by Hand With Fitting Cracked on Outlet Line.

Fig. 4.26- Hydraulic Pump Priming Procedure (Courtesy of Womack)

4.3.11- Proper Case Drain

The *case drain* allows drainage of leaking oil inside the pump housing back to the reservoir. Unidirectional Pumps/motors have one direction of leakage collected through internal cavities and directed to the pump inlet. In bidirectional and over-center pumps, since the leakage direction is switched, leakage is collected and routed separately to the reservoir. The following set of bullets explains the best practices for proper case drain:

- Route the case drain line so that it remains full of fluid (non-siphoning).
- Case drain pressure must be less than manufacturer's maximum value $\approx$ 1.7 bar (25 psi).
- For pumps that have more than one case drain port, case drain fluid must be directed to the reservoir via the highest available drain port (T1 or T2 shown on Fig. 4.25).
- Each drain line must be a separate line, unrestricted, full sized and connected directly to the reservoir below the lowest fluid level.
- Check if the drain line has a filter. Drain line filters can cause excessive case pressure, resulting in seal failure and mechanical damage.
- Avoid using a check valve in the case drain line.

4.3.12- Proper Oil Discharge

The following set of bullets explains the best practices for proper oil discharge:
- Avoid developing turbulent flow by sizing the discharge line to meet flow speed < 5 m/s. Recommended range is 2.1-4.6 m/s (7-15 foot/s).
- Select the discharge line to meet the maximum operating pressure of the system.

4.3.13- Review Range of Driving Speed

Efficiency and operation of a hydraulic pump is significantly affected by pump *driving speed*. Therefore, it is very important to make sure that the driving speed of a hydraulic pump is within the range recommended by the pump manufacturer. As Shown in Fig. 4.27, for proper selection of driving speed of a pump, the following should be considered:

Effects of operating a pump above the maximum speed:
- Reduced pump mechanical efficiency due to increased internal friction.
- Possible pump cavitation.
- Possible pump failure due to excessive friction.
- Reduced bearing lifetime. Life expectancy of a bearing running at 1800 rpm is double of the bearing running at 3600 rpm.

Effects of operating a pump below the minimum speed:
- Reduced pump volumetric efficiency.
- Increased leakage overheats the pump quickly.
- Some pumps, such as vane and radial piston, depend on centrifugal force for proper pumping action. Reduced speed for such pump reduces the pump ability to work under high pressure.

Optimum Speed: Optimum speed of a pump is at the maximum overall efficiency. If you select between two pumps, select the one that has higher overall efficiency at the current driving speed. If you were able to change the driving speed for an existing pump, change the driving speed to meet the maximum overall efficiency of the pump.

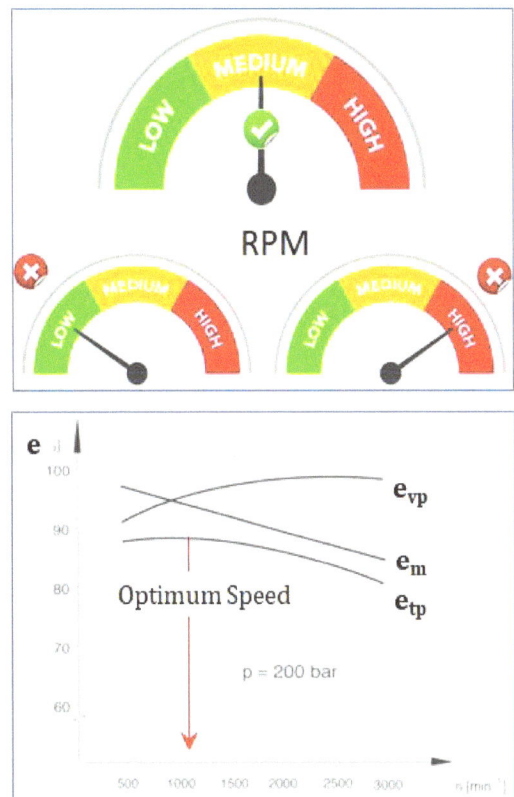

Fig. 4.27- Recommended and Optimum Driving Speeds of a Pump

4.3.14- Review Range of Working Pressure

For each pump mechanism and size, maximum *working pressures* should be defined by the manufacturer. As Shown in Fig. 4.28, for proper selection of working pressure of a pump, the following should be considered:

Effects of High Working Pressure:
- Maximum Pressure is the pressure under which the pump can still work but not as efficient as the optimum pressure.
- Reduced volumetric efficiency, increased leakage and heat generation.
- Mechanical failure.
- Reduced pump bearing life. As an example, life expectancy is increased by the cube of pressure decrease. Reducing pressure from 3000 to 1500 PSI increases bearing life expectancy by 8 times (the cube of 2).

Pressure Spikes:
- Pressure *Spikes* are intermittent pressure increase above the normal working pressure of the system. They occur when shifting valves, changing actuator direction or motion and starting and stopping actuators. Usually, they last for only few milliseconds. If this happens periodically, it causes fatigue failure of the pump and every other component of the system. Consult the pump manufacturer for the max intermittent and continuous operating pressure.

Effects of Low Working Pressure:
- Reduced mechanical efficiency due to lack of self-lubrication.

Optimum Pressure:
- Optimum pressure of a pump is at the maximum overall efficiency. If you select between two pumps, select the one that has higher overall efficiency at the current working pressure. If you were able to change the working pressure for an existing pump, change the working pressure to meet the maximum overall efficiency of the pump.

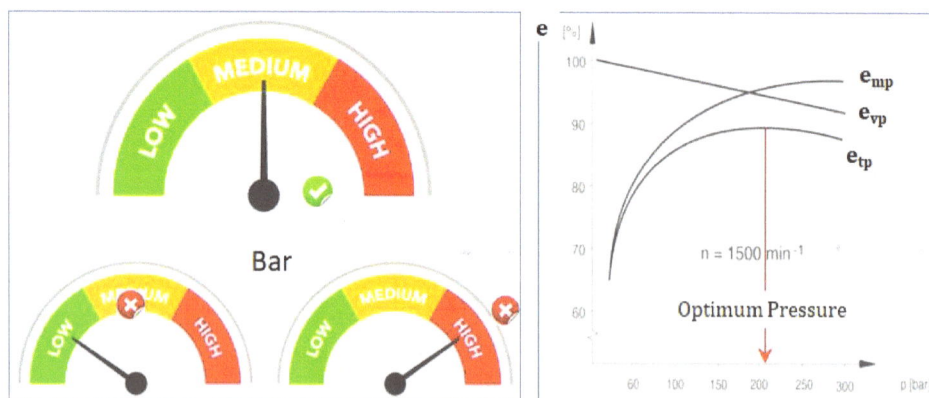

Fig. 4.28- Recommended and Optimum Working Pressure of a Pump

4.3.15- Review Range of Working Temperature

Working temperature mainly affects operation of hydraulic components including pumps as follows:

- Hydraulic fluid viscosity. Consequently, possibility of pump cavitation at cold start, increased leakage and lack of lubrication at high temperature.
- Hydraulic fluid additives and chemical properties. Consequently, the general performance of the pump.
- Internal clearances. Consequently, level of friction and internal leakage.
- Performance of the sealing elements.

Therefore, as shown in Fig. 4.29, pump manufacturers provide instructions about the recommended range of working temperature and/or oil viscosity that consider compromised effects on the pump operation.

If no information is found, should not exceed 60 $^{\circ}$C.

Viscosity and temperature of hydraulic fluid			
	Viscosity [mm²/s]	Temperature	Comment
Transport and storage at ambient temperature		$T_{min} \geq$ -50 °C T_{opt} = +5 °C to +20 °C	factory preservation: up to 12 months with standard, up to 24 months with long-term
(Cold) start-up[1]	v_{max} = 1600	$T_{St} \geq$ -40 °C	t ≤ 3 min, without load (p ≤ 50 bar), n ≤ 1000 rpm (for sizes 5 to 200), n ≤ 0.25 • n_{nom} (for sizes 250 to 1000)
Permissible temperature difference		$\Delta T \leq$ 25 K	between axial piston unit and hydraulic fluid
Warm-up phase	v < 1600 to 400	T = -40 °C to -25 °C	at p ≤ 0.7 • p_{nom}, n ≤ 0.5 • n_{nom} and t ≤ 15 min
Operating phase			
Temperature difference		ΔT = approx. 12 K	between hydraulic fluid in the bearing and at port T.
Maximum temperature		115 °C	in the bearing
		103 °C	measured at port T
Continuous operation	v = 400 to 10 v_{opt} = 36 to 16	T = -25 °C to +90 °C	measured at port T, no restriction within the permissible data
Short-term operation[2]	$v_{min} \geq$ 7	T_{max} = +103 °C	measured at port T, t < 3 min, p < 0.3 • p_{nom}
FKM shaft seal[1]		T ≤ +115 °C	see page 5
[1] At temperatures below -25 °C, an NBR shaft seal is required (permissible temperature range: -40 °C to +90 °C).			
[2] Sizes 250 to 1000, please contact us.			

Fig. 4.29- Example of Manufacturer's Instructions about Working Temperature for A2FO Pump (Courtesy of Bosch Rexroth)

4.3.16- Review Compatibility with working Hydraulic Fluid

Positive displacement pumps/motors (gear, vane, and piston types) are designed for use of *hydraulic fluids*, which lubricate the bearings and internal parts. They should never be used on fluids such as water, kerosene, fuel oil, jet fuel, gasoline, etc. because these fluids do not provide the necessary lubrication.

It is highly advisable to review the manufacturer's instructions about what type of fluid to be recommended. The oil recommendation will usually appear on the pump nameplate or in the instruction manual which accompanies the pump. Figure 4.30 shows an example of instructions that are provided by a pump manufacture about hydraulic fluids preferences. Additionally, if a special hydraulic fluid such as *synthetic* or fire-resistant are intended to be used, consult the manufacturer to make sure the internal seals and the material are compatible with such fluids.

If there is no oil recommendation, these general rules may be followed:

- **Petroleum-Based (Mineral) Fluids for Industrial Hydraulic Systems:** Most power units, except those built for *fire-resistant* fluids, will operate on petroleum-based oil (*mineral oil*) which has been compounded with the proper additives for use in hydraulic pumps. Recommended range of viscosity is from (22 – 100) cSt, i.e. (100 – 500) SSU. These viscosity ratings are at a 37°C (100° F).

- **Multi-Grade Fluids for Mobile Hydraulic Systems:** Such systems quite often operate on multi-grade automatic transmission fluids which, because of their high *Viscosity Index* requirements. will operate satisfactorily over a wide range of ambient temperatures.

- **Fire-Resistant Fluids for High Working Temperature:** Some applications in which the working temperature is higher than normal, and or the risk of fire is high, *fire-resistant* hydraulic fluids are specified. Example of applications are die casting and steel mills.

- **Synthetic Fluids for Special Applications:** *Synthetic fluids* are specified by system's manufacturers where some special fluid requirements are required. Aerospace is an obvious example of these systems where the safety of operation overrides all other requirements.

Technical data

Hydraulic fluid

Before starting project planning, please refer to our data sheets RE 90220 (mineral oil), RE 90221 (environmentally acceptable hydraulic fluids), RE 90222 (HFD hydraulic fluids) and RE 90223 (HFA, HFB, HFC hydraulic fluids) for detailed information regarding the choice of hydraulic fluid and application conditions.

The fixed pump A2FO is not suitable for operation with HFA hydraulic fluid. If HFB, HFC or HFD or environmentally acceptable hydraulic fluids are used, the limitations regarding technical data or other seals must be observed.

Details regarding the choice of hydraulic fluid

The correct choice of hydraulic fluid requires knowledge of the operating temperature in relation to the ambient temperature: in an open circuit, the reservoir temperature.

The hydraulic fluid should be chosen so that the operating viscosity in the operating temperature range is within the optimum range (v_{opt} see shaded area of the selection diagram). We recommended that the higher viscosity class be selected in each case.

Example: At an ambient temperature of X °C, an operating temperature of 60 °C is set in the circuit. In the optimum operating viscosity range ($v_{opt,}$ shaded area), this corresponds to the viscosity classes VG 46 or VG 68; to be selected: VG 68.

Note
The case drain temperature, which is affected by pressure and speed, can be higher than the reservoir temperature. At no point of the component may the temperature be higher than 115 °C. The temperature difference specified below is to be taken into account when determining the viscosity in the bearing.

If the above conditions cannot be maintained due to extreme operating parameters, we recommend flushing the case at port U (sizes 250 to 1000).

Selection diagram

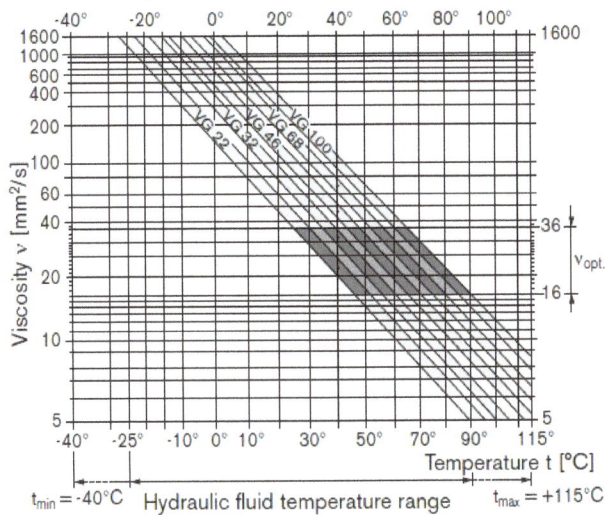

Fig. 4.30- Example of Manufacturer's Instructions about Hydraulic Fluid for A2FO Pump (Courtesy of Bosch Rexroth)

4.3.17- Review Prime Mover Overloading Conditions

Hydraulic pumps are driven by electric motors for industrial applications, and by engines for mobile applications. The following set of bullets provide some guidelines.

- **General Prime Mover:** Ideally, a prime mover should drive a pump at the maximum efficiency of the prime mover. However, practically, a prime mover should be able to comfortably (with an acceptable efficiency) satisfy the power needed by the pump in all phases of operation. Therefore, it is highly recommended to consult the manufacturer's datasheet. Review the power-efficiency curve to find the optimum loading condition of the prime mover.

- **Single-Phase Electric Motors:** They have less starting torque than 3-phase motors. It can be overloaded for short periods up 10% above its nameplate rated power before the circuit breaker trips the motor. Therefore, design of the hydraulic circuit should consider unloading the pump at start up.

- **Three-Phase Electric Motors:** They can be overloaded for short periods up 25% above its nameplate rated power before it stalls.

- **Engines:** Engine power is reduced over time due to engine aging. They can't work at a power above the rated power even momentarily. The Hydraulic circuit should be designed to unload the pump at start up to avoid stalling the engine.

4.3.18- Proper Placement of Hydraulic Power Unit

On most hydraulically powered machines the fluid power is generated by a unit assembly called a "*hydraulic power unit*". Usually the oil reservoir (tank) is used as a mounting base for electric motor, pump, and sometimes for other components. On machines operating in an industrial plant the hydraulic power unit is designed to set alongside or overhead near the machine, being connected to a cylinder or hydraulic motor on the machine through hose or steel tubing. On mobile equipment it is usually more convenient to separate the pump from the rest of the power unit and to mount it in the engine compartment to be driven from the engine power take-off shaft. However, the following are best practices for proper placement of power units:

- **Visual Inspection:** Before installation, check the hydraulic power unit for visible transport damage e.g. cracks, leaking seals, screws, protective covers.

- **Air Circulation:** A preferred location in the plant for the installation of a hydraulic power unit is where free air can circulate around all sides, including top and bottom. One of the important functions of the oil reservoir is to radiate heat which may accumulate in the oil. If the reservoir is installed too close to a wall, its heat radiating ability will be impaired and the entire hydraulic system may overheat.

- **Use of Forced Air Ventilation if Needed:** Do not install a power unit inside a cabinet, console, under a table or bench, or other closed space where air cannot freely circulate unless forced air ventilation is provided.

- **Shield from Heat Sources:** Power units should not be exposed to extreme temperatures, either hot or cold. Do not install in direct sunlight without a sunshade, and one which will not restrict vertical air circulation. Do not install near a furnace without interposing a heat shield unless a provision has been made to keep the oil cool with a heat exchanger.

4.3.19- Proper Installation of Hydrostatic Transmission (**Courtesy of Womack**)

Hydrostatic Transmission uses a hydraulic pump driven by an engine (in mobile applications) or electric motor (in industrial applications) and connected by means of a closed circuit to a hydraulic motor. A closed circuit is a circuit in which neither port of pump nor motor is vented directly to reservoir. Discharge oil from the motor re-circulates back to the inlet of the pump. A small volume, low-pressure, charge pump continually replaces about 10 to 15% of the circulating oil with fresh oil from the reservoir. The replaced oil is returned to the reservoir for filtering and cooling before being placed back in the circuit. Pressure is always maintained in both sides of the loop, usually about 4000 to 5000 PSI in the working line (high-pressure side) and from 150 to 500 PSI in the return line (low-pressure side) depending on the size of the transmission. Figure 4.31 shows the basic components found in the usual hydrostatic transmission. The following section provide guidelines for installation of hydrostatic transmission.

Charge Pump (Item 2): The charge pump is usually a gear or gerotor pump housed in the main pump rear cover (occasionally in the front cover). On some installations, if the internal charge pump does not have sufficient volume, the system may have been designed to use a separate external pump. The charge pump provides a small volume of fresh oil into the closed circuit,

Hydraulic Motor (Item 3): On most installations this will be a fixed displacement piston-type motor of approximately the same displacement as the pump, producing the same maximum speed as the pump driving the motor. On some installations the motor may be a slow speed radial piston motor of greater displacement than the pump, giving high torque output at slow speed. Case drain lines must be full size to avoid blowing out the motor shaft seal.

Charge Pump Filter (Item 4): Oil fed directly into the charge pump inlet port must be filtered. This will usually require a paper element filter. The filter should have sufficient filtering area to keep pressure drop very low - to 1 to 2 PSI - when the element is clean. If the filter becomes restricted the charge pump will cavitate. This may cavitate the main loop and may prevent the system from reaching maximum speed. A filter with an electrical or visual indicator is recommended.

Shut-Off Valve (Item 5): When the reservoir is at a higher elevation than the pump, this shut-off valve is necessary so the oil to the filter can be shut off while the element is being changed. It is recommend that a quarter turn plug valve which can be padlocked in the open position be used. Closure of this valve while the system is running will destroy the charge pump and might damage the main pump.

Heat Exchanger (Item 6): Many transmissions will require a heat exchanger. It must be placed in a low-pressure location. The only practical location on most transmissions is in the case drain lines from pump and motor. Joint case drain flow from pump and motor run to heat exchanger (after the filter if found). The discharge side of the heat exchanger should be piped to reservoir.

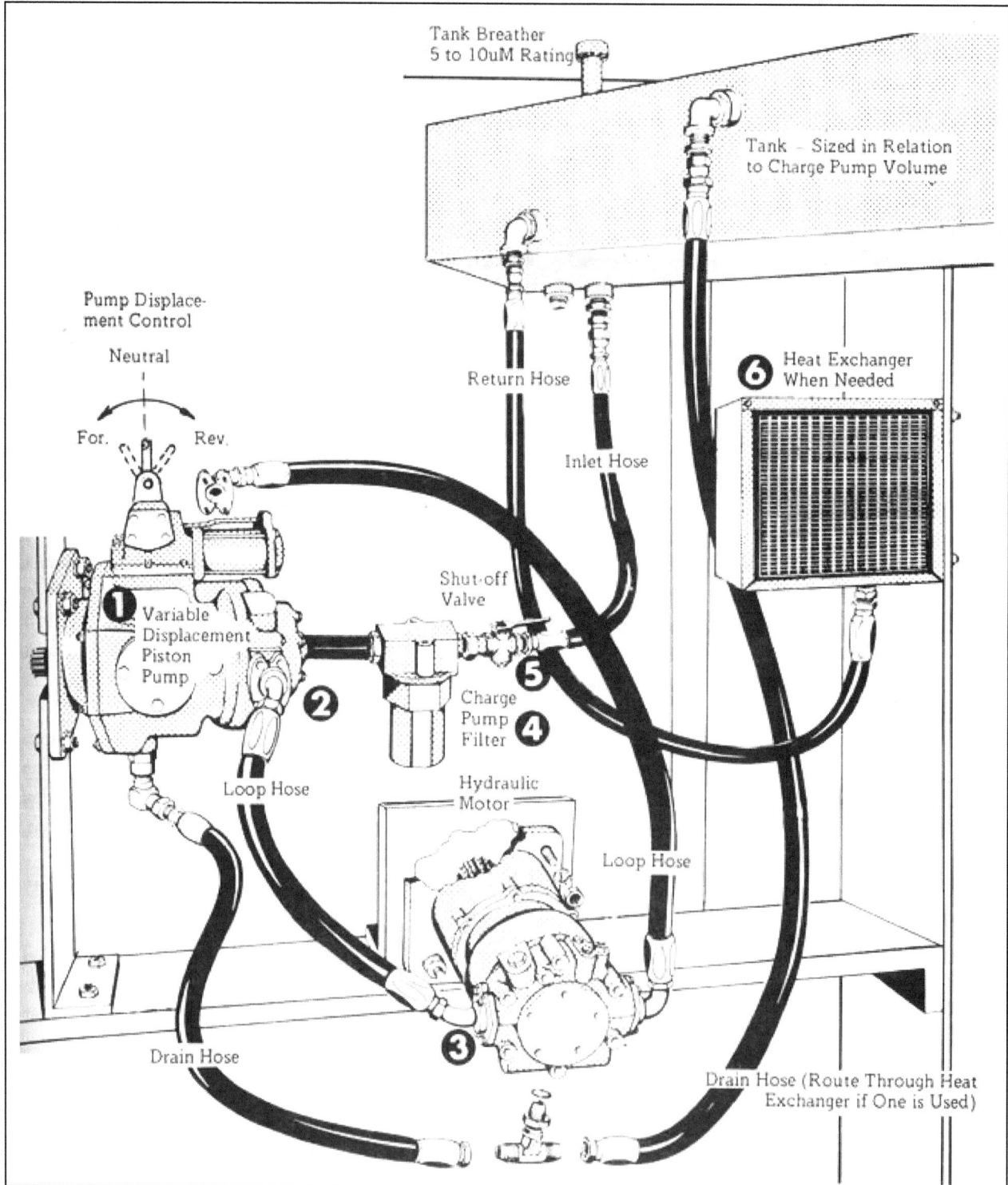

**Fig. 4.31- Proper Installation of Hydrostatic Transmission
(Courtesy of Womack)**

4.4- BP-Pumps-04-Standard Tests and Calibration

4.4.1- Testing of Fixed Displacement Pumps

Every new pump is factory tested to determine whether it meets acceptable performance standards. Pumps are also tested after overhauling or major maintenance. **ISO 4409:2007** specifies the requirements for test installations, test procedures under steady-state conditions and the presentation of test results. A positive displacement pump should discharge a theoretical flow **(Q_{th})** proportional to the pump size and driving speed, no matter what the working pressure is. As known, such a pump experiences internal leakage for the sake of self-lubrication. Internal leakage is affected by working pressure, driving speed, working temperature, and fluid viscosity. Figure 4.32 shows a pump characteristic curve developed by the pump manufacturer that plots actual flow of a brand-new pump **(Q_{new})** versus working pressure **(p_p)** at constant driving speed, working temperature, and fluid viscosity. Periodically a pump must be tested to check how healthy it is. So, testing a pump based on achieving maximum pressure only doesn't mean anything. A test characteristic curve **(Q_{test})** must be developed. For internally drained pump, we have no choice but measuring the pump outlet flow. For externally drained pump, we have the choice of measuring either by outlet flow of the pump or the case drain **(q_L)**. Plotting the test flow gives an idea about the volumetric efficiency. Similarly, plotting the input power curve **(P_{in})** gives an idea about the overall efficiency and the mechanical efficiency where:

$$\text{Volumetric Efficiency} = Q_{test} / Q_{th} = 1 - (q_L / Q_{th})$$
$$\text{Overall Efficiency} = P_{out} / P_{in} = (p_p \times Q_{test}) / P_{in}$$
$$\text{Mechanical Efficiency} = \text{Overall Efficiency} / \text{Volumetric Efficiency}$$

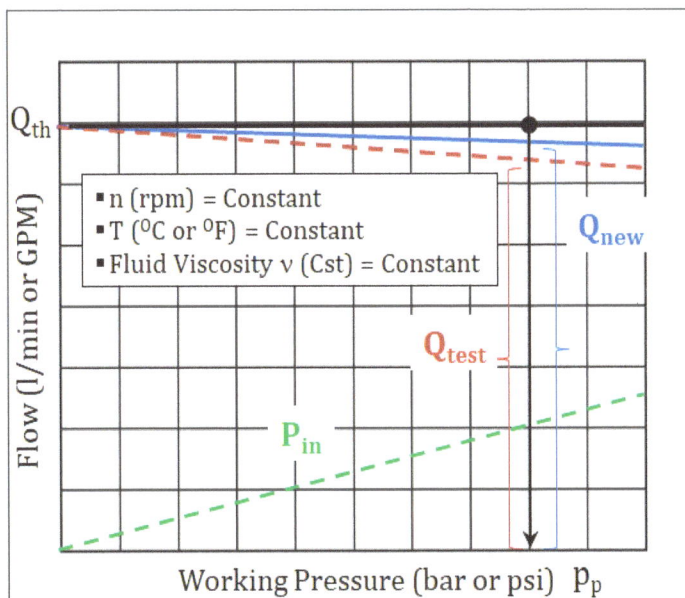

Fig. 4.32- Characteristic Curve for a Positive Displacement Pump

Figure 4.33 shows a circuit diagram for testing a pump by measuring the outlet flow (left) or by measuring case drain (right).

Test fixed-displacement pumps by checking the amount of flow through the flowmeter (Q_{test}). Turn the pump on and record the flow (Q_{test}) versus the pressure (p_p) that should be increased incrementally by closing the throttle valve.

Alternatively, if the pump is externally drained, an excellent method of monitoring the case-drain flow while operating is to permanently install a flow meter in the case-drain line.

If no flow flowmeter were found, check the flow out of the case drain line by porting the line into a container and timing it. Secure the line to the container prior to starting the pump.

Over the range of operating pressure, the normal case flow is 1-5% of the maximum pump flow. Vane pumps usually bypass more than piston-type pumps. If 10% of the maximum volume flows out of the case drain line, the pump is badly worn and should be repaired or changed.

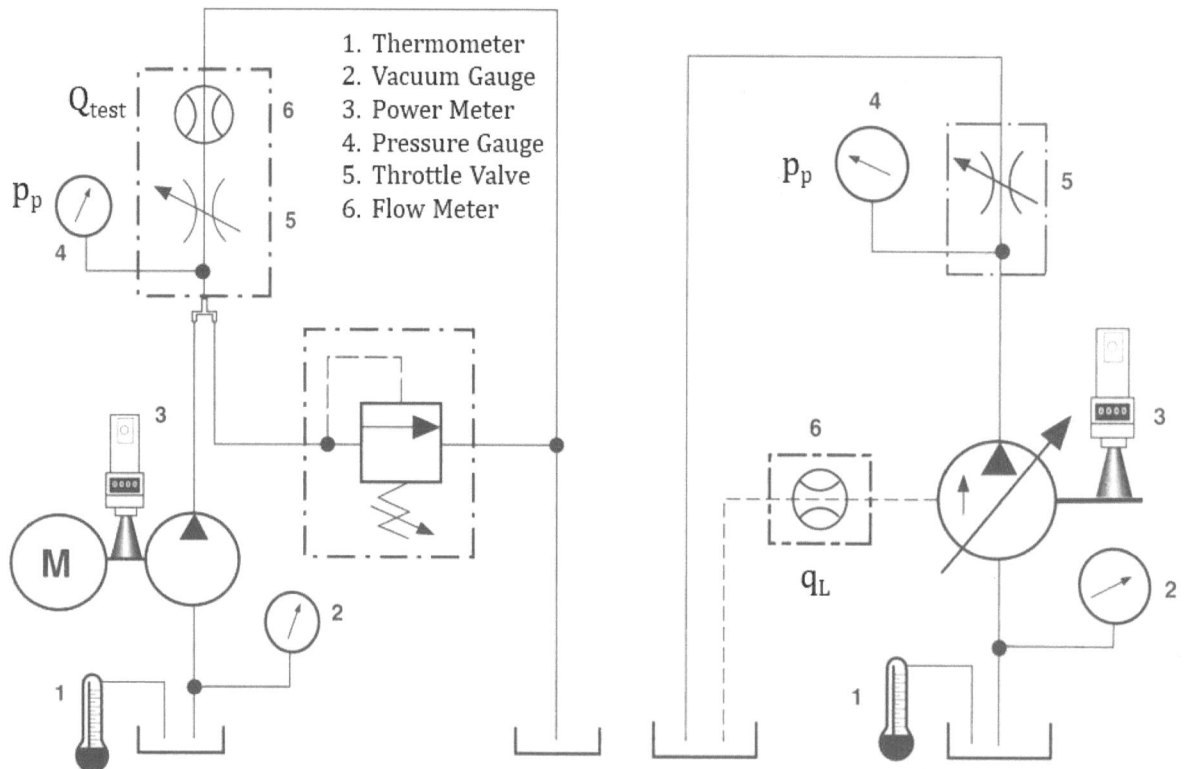

Fig. 4.33- Hydraulic Circuit for Pump Testing
(Courtesy of Fluid Power Training Institute – Rory S. McLaren)

4.4.2- Setting of Pressure Compensated Pump

Pressure compensated pumps are the most commonly used pump among the variable displacement pumps. As shown in Fig. 4.34, this pump works as a fixed displacement pump until the working pressure reaches the cracking (critical) pressure **(p_{CR})**. Any further increase of the pressure will cause the pump compensator (controller) to de-stroke the pump. The pump is fully de-stroked at maximum cutoff pressure **(p_{CO})**.

For the safety of the machine operation and the pump, a backup pressure relief valve is added and set a bit higher (usually 10 bar) than the pump compensator cutoff pressure. Setting such a pump in presence of the backup pressure relief valve is a bit challenging. The basic idea here is that the pressure relief valve should be set first because it is set higher than the pump compensator. If the pump compensator was set first, relief valve can't be set. Below are the steps of setting such a pump:

1. Turn OFF the power
2. Fully open the PRV
3. Fully close the pump compensator
4. Turn ON the power
5. Pressure now = tank pressure
6. Close the PRV gradually until **$p_p = p_{CO} + 10$ bar**
7. Open the pump compensator to lower pump pressure until **$p_p = p_{CO}$.**

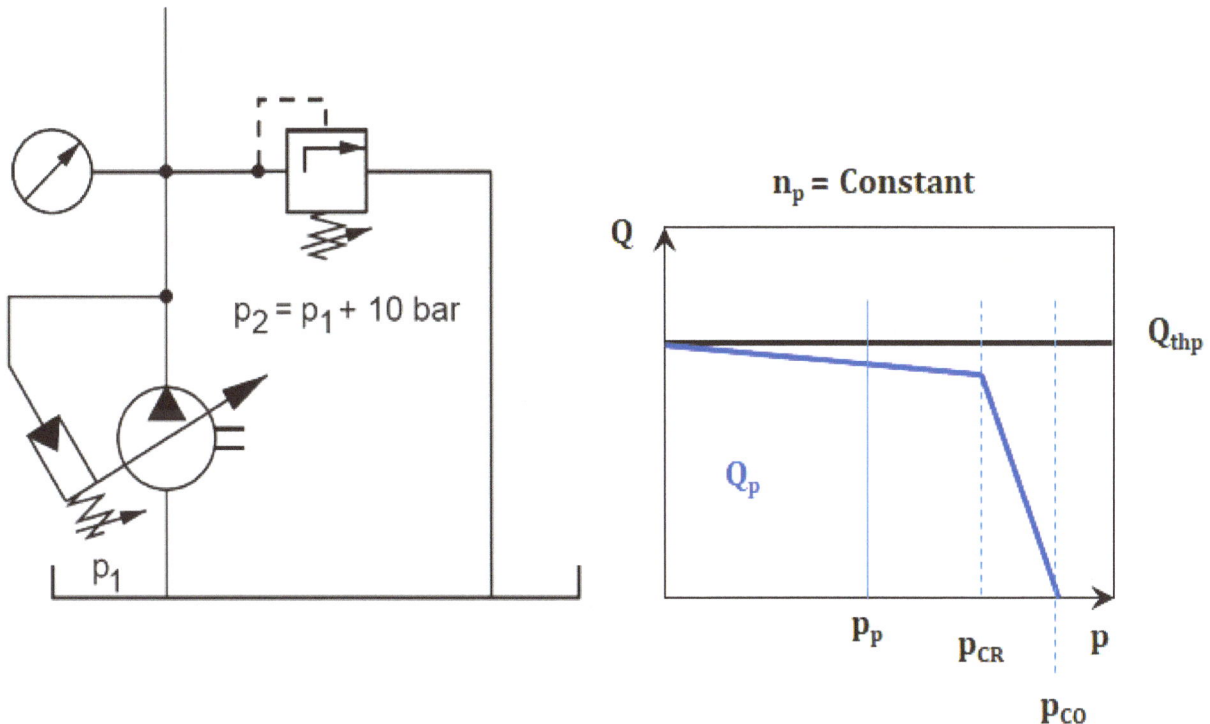

Fig. 4.34- Setting of a Pressure Compensated Pump

4.5- BP-Pumps-05-Transportation and Storage

4.5.1-Pump and Power Units Transportation

A pump could be of a significant weight and size. If not transported appropriately, it may lose its stability and thus be knocked over, fall or move in an uncontrolled way. It is highly recommended to review the instructions provided by the manufacturer. Figure 4.35 shows an example of given instructions for transportation a pump with *ring screw*. Figure 4.36 shows an example of given instructions for transportation a pump with *lifting strap*.

If no instructions are found, the following are guidelines for transporting a hydraulic unit:
- Ensure that no unauthorized persons are within the hazard zone.
- Check the weight and the center of gravity of the hydraulic power unit.
- Make certain that the lifting device has adequate lifting capacity.
- Only the intended locations and attachment points should be used for lifting the product.
- Place the product on a suitable surface rated to the weight of the unit or on the ground.
- A hydraulic power unit must never be attached to or lifted at the mounted components (piping, hoses, manifolds, electric motors, accumulators, etc.).

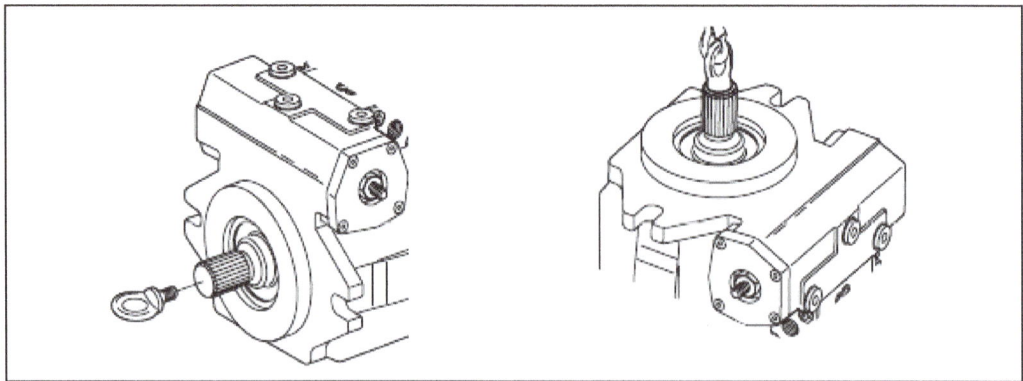

Transport with ring screw

The drive shaft can be used to transport the axial piston unit as long as only outward (pulling) axial forces occur. Thus, you can suspend the axial piston unit from the drive shaft.

▶ To do this, screw a ring screw completely into the thread on the drive shaft. The threaded sizes is stated in the installation drawing.

▶ Make sure that each ring screw can bear the total weight of the axial piston unit plus approx. 20%.

You can hoist the axial piston unit as shown with the ring screw screwed into the drive shaft without any risk of damage.

Fig. 4.35- Manufacturer's Instructions for Axial Piston Pump Transportation using Ring Screw (Courtesy of Bosch Rexroth)

Transport with lifting strap

► Place the lifting strap around the axial piston unit in such a way that it passes over neither the attachment parts (e.g. valves) nor such that the axial piston unit is hung from attachment parts

Risk of injury!

During transport with a lifting device, the axial piston unit can fall out of the lifting strap and cause injuries.

► Hold the axial piston unit with your hands to prevent it from falling out of the lifting strap.

► Use the widest possible lifting strap.

WARNING!

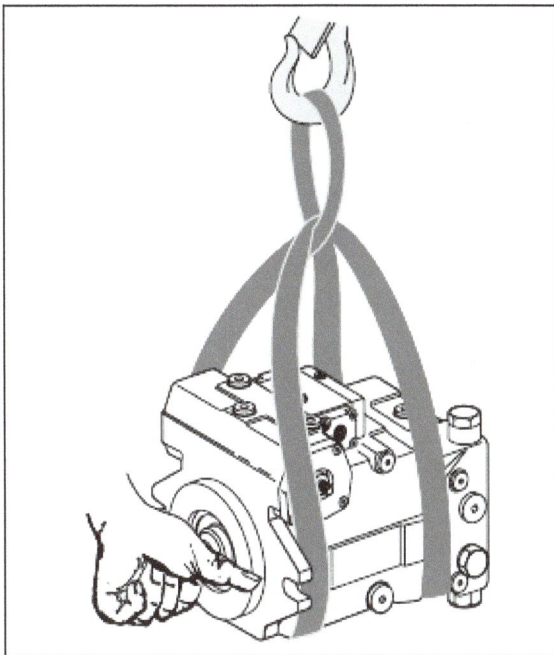

Fig. 4.36- Manufacturer's Instructions for Axial Piston Pump Transportation using Lifting Strap (Courtesy of Bosch Rexroth)

4.5.2-Pump and Power Units Storage

It is highly recommended to review the storage instructions from the manufacturer. Figure 4.37 shows an example of a manufacturer instructions for storage of a pump.

Storing the axial piston unit

Requirement

- The storage areas must be free from corrosive materials and gasses.
- The storage areas must be dry.
- Ideal storage temperature: +5 °C to +20 °C
- Minimum storage temperature: -50 °C.
- Maximum storage temperature: +60 °C.
- Avoid intense lights.
- Do not stack axial piston units and store them shock-proof.
- Do not store the axial piston unit on sensitive attachment parts, e.g. sensors.
- For other storage conditions, see Table 4.

▶ Check the axial piston unit monthly to ensure proper storage.

After delivery

The axial piston units are provided ex-works with corrosion protection packaging (corrosion protection film).

Listed in the following table are the maximum permissible storage times for an originally packed axial piston unit.

Table 4: Storage time with factory corrosion protection

Storage conditions	Standard corrosion protection	Long-term corrosion protection
Closed, dry room, uniform temperature between +5 °C and +20 °C. Undamaged and closed corrosion protection film.	Maximum 12 months	Maximum 24 months

Fig. 4.37- Manufacturer's Instructions for Storage of an Axial Piston Pump (Courtesy of Bosch Rexroth)

Table 4.4 shows an example of given instructions for storage of a power unit.

Storage Conditions	Packaging	Protective Agent	Storage Time (Months)	
			Test with the protective agent	Filling with the protective agent
Storage in dry rooms at constant temperature	For carriage overseas	Mineral oil	12	24
		Corrosion protection oil	12	24
	Not for carriage overseas	Mineral oil	9	24
		Corrosion protection oil	12	24
Outdoor storage (protect the product against damage and water ingress)	For carriage overseas	Mineral oil	6	12
		Corrosion protection oil	9	24
	Not for carriage overseas	Mineral oil	0	12
		Corrosion protection oil	6	24

**Table 4.4- Manufacturer's Instructions for Storage of a Power Unit
(Mcmillan Engineering Group)**

However, if no instructions were found, the following set of bullets explain general best practices of storing a pump and a power unit:

- **Pump Case:** Pump case is to be filled with hydraulic oil or a rust preventive oil, which is compatible with the rubber seals. Leave a small air space for the oil to expand if the pump should be exposed to heat.
- **Pump Ports:** Ports are to be plugged with suitable plugs or dust caps.
- **Pump Shaft:** The shaft and other external machined surfaces should be coated with grease.
- **Long term storage:** If the pump is going to be stored for more than a year, shaft seal should be replaced before putting it in service.

Chapter 5

Maintenance of Motors

Objectives

This chapter provides guidelines for **motors** selection, replacement, maintenance scheduling, installation, testing, storage and transportation. This chapter is supported by examples and figures granted by leading fluid power manufacturers.

Brief Contents

5.1-BP-Motors-01-Selection and Replacement
5.2-BP-Motors-02-Maintenance Scheduling
5.3-BP-Motors-03-Installation and Maintenance
5.4-BP-Motors-04-Standard Tests and Calibration
5.5-BP-Motors-05-Transportation and Storage

Chapter 5: Maintenance of Motors

The following set of best practices provide general guidelines and may not be applicable for all cases. They are not intended to replace the instructions given by the component manufacturer. It is strongly advisable to adhere to instructions provided by the manufacturer.

5.1- BP-Motors-01-Selection and Replacement

5.1.1- Selecting or Replacing Motors

The following best practices list provides guidelines for selecting a new or replacing an existing motor.

BP-Motors-01-Selection and Replacement:
- Review maximum/optimum operating pressure.
- Review min/maximum/optimum operating speed.
- Review maximum operating torque.
- Review maximum overall efficiency at the optimum operating conditions.
- Review size (**See Note 1**) and displacement control requirements.
- Review type of fluid.
- Review contamination tolerance.
- Review noise level.
- Review initial cost.
- Review approximate service life.
- Review availability and interchangeability.
- Review maintenance and spare parts.
- Review physical size and weight.

Note 1: if a motor is <u>undersized</u>, one or combination of the following could occur:
- Motor speed increases and may work inefficiently.
- Working pressure increases to maintain same external torque and may need to rset maximum system pressure.

5.1.2- Displacement Calculation of Legacy Motors

When replacing a legacy motor, usually no information is available about the motor. The main challenge in replacing such legacy motors is to find the motor size. Because pumps and motors are conceptually having the same mechanisms, then equations 4.1 through 4.8 are applicable for finding displacements of various motors mechanisms.

5.2- BP-Motors-02-Maintenance Scheduling

Unless otherwise stated by the components and systems manufacturer, Table 5.1 provides guidelines for *scheduling* preventive maintenance actions for hydraulic motors.

#	Preventive Maintenance Actions	Daily	Weekly	Monthly	Biannually	Annually
1	Clean around and outside surface	✔	✔	✔	✔	✔
2	Check for unusual sound	✔	✔	✔	✔	✔
3	Check temperature of the motor body	✔	✔	✔	✔	✔
4	Check tightness and leakage around hydraulic connections		✔	✔	✔	✔
5	Check electrical connections (if found)		✔	✔	✔	✔
7	Check for vibration and condition of dampers			✔	✔	✔
8	Standard tests and calibration					✔

Table 5.1- BP-Motors-02-Maintenance Scheduling

5.3- BP-Motors-03-Installation and Maintenance

Proper motor installation is an important process for the trouble-free operation of the system. The following best practices provide guidelines for *installation and maintenance* of motors. As previously stated, because pumps and motors are conceptually having the same mechanisms, installation and maintenance guidelines provided in Chapter 4 are applicable for motors. Regarding motor cavitation, system designers should consider using anti-cavitation check valves in the system design.

BP-Motors-03-Installation and Maintenance:
1. Install the motor to avoid Cavitation.
2. Identify Ports and Direction of Rotation.
3. Proper Shaft Alignment for Direct Drive.
4. Adequately Dampen Vibration.
5. Proper Priming.
6. Proper Case Drain.
7. Review Range of Rotational Speed.
8. Review Range of Working Pressure.
9. Review Range of Working Temperature.
10. Review Compatibility with Working Hydraulic Fluid.

As previously stated, it is highly recommended to review installation instructions given by the component manufacturer. Figures 5.1 and 5.2 provide examples of manufacturer's installation instructions.

Hydraulic motor - high speed
MAH 6.3 - MAH 12.5

Drain line

Max. pressure = 6 bar absolute.
Drain pressure must never exceed return pressure by more than 1 bar.

Installing the drain line
The drain line/motor must be positioned so that the motor cannot empty itself during standstill.

Temperature

Fluid temperature:
Min. +3°C to max. +50°C. at max. pressure
Min. +3°C to max. +60°C. at max. 100 bar

Ambient temperature:
Min. 0°C to max. 50°C.

In case of lower operating temperatures, please contact the Danfoss Sales Organization for Water Hydraulics.

Storage temperature:
Min. -40°C to max. +70°C.

Motor variants

MAH motors are optimized for operation in one direction and are therefore available in CW and CCw versions.

Filtration

The water supplied to the valve must be filtered: 10 μm absolute, β_{10}-value > 5000 filter is recommended.

For further information on filters, please contact the Danfoss sales department for water hydraulics.

**Fig. 5.1- Manufacturer's Instructions for Hydraulic Motor Installation
(Courtesy of Danfoss)**

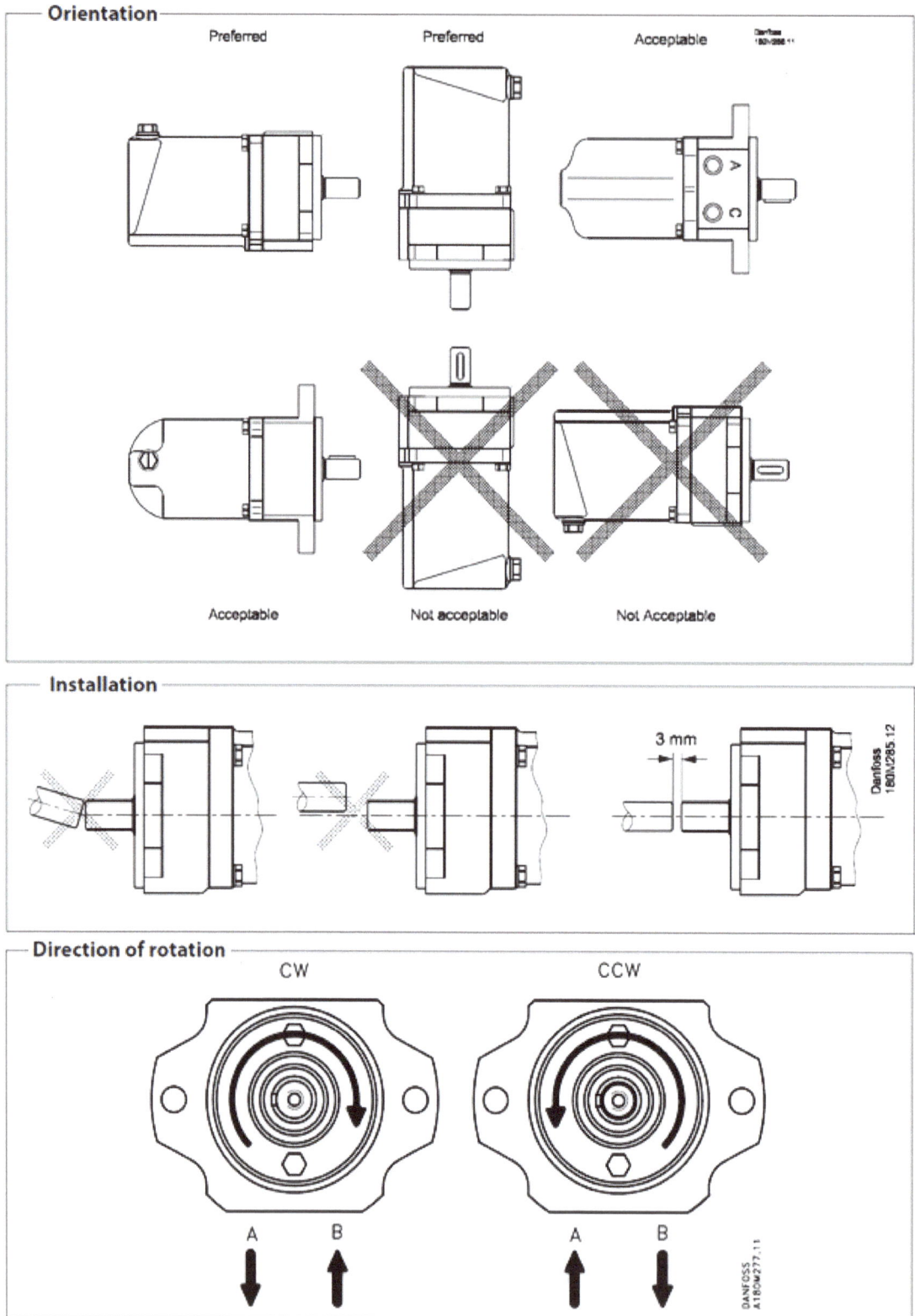

**Fig. 5.2- Manufacturer's Instructions for Hydraulic Motor Installation
(Courtesy of Danfoss)**

5.4- BP-Motors-04-Standard Tests and Calibration

Every new motor is factory tested to determine whether it meets acceptable performance standards. Motors are also tested after overhauling or major maintenance. Several standards have been developed for motor testing. **SAE-J746** specifies the requirements for pumps and motors test installations, test procedures under steady-state conditions and the presentation of test results. This test standard describes tests for determining characteristics of hydraulic positive displacement motors as used on construction and industrial machinery as referenced in SAE J1116. These characteristics are to be recorded on two data sheets: one at 49 °C (120 °F) and one at 82 °C (180 °F).

Hydraulic motor efficiency is a critical factor in the design of industrial machines and more important for off-highway machines because it affects the maximum vehicle payload and the top propulsion speed.

A positive displacement motor should run at a theoretical speed **(n_{th})** proportional to the motor size and inlet flow, no matter what the differential pressure across the motor is. As known, such a motor experiences internal leakage for the sake of self-lubrication. Internal leakage is affected by working pressure, driving speed, working temperature, and fluid viscosity. As a result, motor runs at a measured speed **(n_{test})** that is slower than the theoretical speed.

A positive displacement motor should develop a theoretical torque **(T_{th})** proportional to the motor size and differential pressure across the motor, no matter what the motor speed is. As known, such a motor experiences internal friction. Internal friction is affected by working pressure, rotational speed, working temperature, and fluid viscosity. As a result, motor develops a measured torque **(T_{test})** that is lower than the theoretical torque.

Plotting the test curves [Torques = f (pressure & flow) and Speed (pressure & flow)] give an idea about the volumetric efficiency and the mechanical efficiency where:

Volumetric Efficiency = n_{test} / n_{th}
Mechanical Efficiency = T_{test} / T_{th}
Overall Efficiency = Volumetric Efficiency x Mechanical Efficiency

For more detailed discussions about motor sizing and efficiency calculations, refer to Volume 1 of this series of textbooks "Introduction to Hydraulics for Industry Professionals".

Figure 5.3 shows a typical circuit diagram for hydraulic motor test stand. Unlike conventional test stands that use mechanical brakes or a hydraulic motor as a load device, Figure 5.4 shows an Energy-Saving test stand that was built by the Fluid Power Institute (FPI) at Milwaukee School of Engineering (MSOE). In this test stand, the loading device is a generator that generates part of the electrical power required to drive the electrical motor that drives the pump supplying power to the test motor.

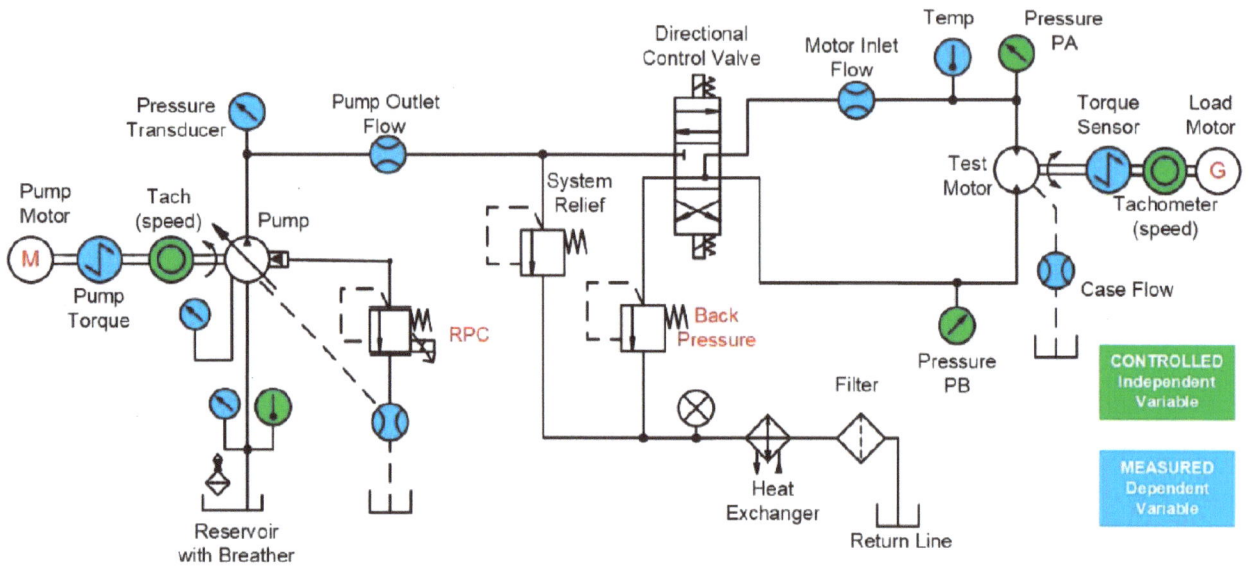

Fig. 5.3- Circuit Diagram for Hydraulic Motor Testing (Courtesy of Fluid Power Institute - MSOE)

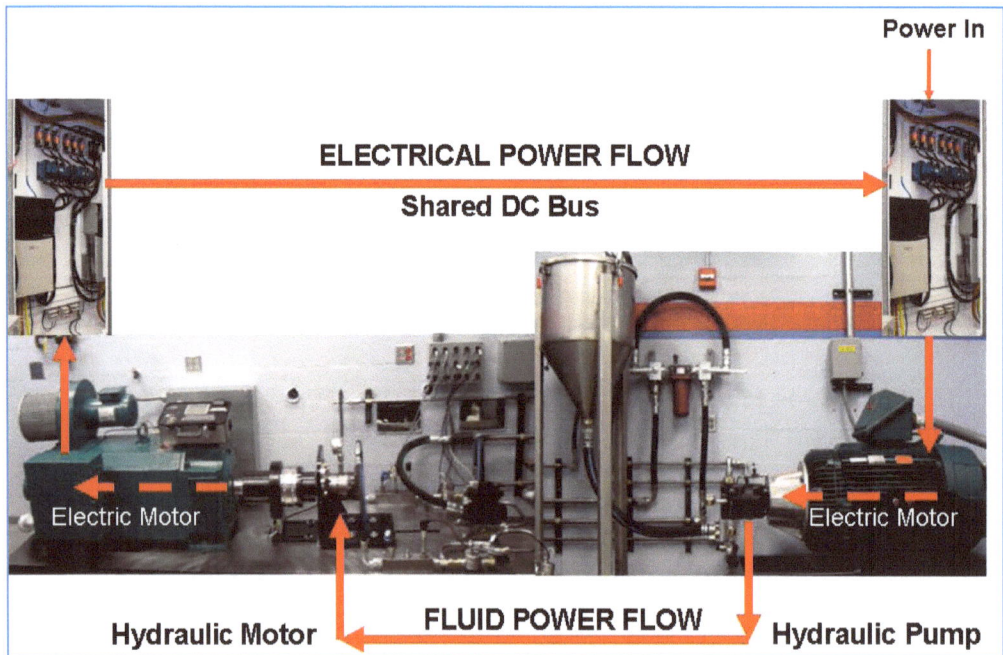

Fig. 5.4- Hydraulic Motor Energy-Saving Test Stand (Courtesy of Fluid Power Institute - MSOE)

5.5- BP-Motors-05-Transportation and Storage

Same guidelines for storage and transportation of hydraulic pumps are applicable for hydraulic motors. So, refer to section 4.5.2. in Chapter 4 of this book

Chapter 6

Maintenance of Cylinders

Objectives

This chapter provides guidelines for **cylinders** selection, replacement, maintenance scheduling, installation, testing, storage and transportation. This chapter is supported by examples and figures granted by leading fluid power manufacturers.

Brief Contents

6.1-BP-Cylinders-01-Selection and Replacement
6.2-BP-Cylinders-02-Maintenance Scheduling
6.3-BP-Cylinders-03-Installation and Maintenance
6.4-BP-Cylinders-04-Standard Tests and Calibration
6.5-BP-Cylinders-05-Transportation and Storage

Chapter 6: Maintenance of Cylinders

The following set of best practices provide general guidelines and may not be applicable for all cases. They are not intended to replace the instructions given by the component manufacturer. It is strongly advisable to adhere to instructions provided by the manufacturer.

6.1- BP-Cylinders-01-Selection and Replacement

The best practices list "BP-Cylinders-01-Replacement" provide guidelines for replacing an existing cylinder.

BP-Cylinders-01-Selection and Replacement:
- **Cylinder Size:** If the new *cylinder* has a different size than the old one, then speed and pressure of the new cylinder must be checked to determine if it will provide acceptable machine operation.
- **Cylinder Cushioning:** If the old cylinder is equipped with cushioning, the new one must also be equipped also with cushioning and the cushioning throttle valve should be reset at the time of installation.
- **Cylinder Installation Requirements:** Check if there is a need for changing the installation requirements such as mounting the cylinder with the machine body, engagement with the load, connection to transmission lines, etc.

6.2- BP-Cylinders-02-Maintenance Scheduling

Some hydraulic *cylinders* are maintenance-free after commissioning. However, after a new system has been commissioned, regular checks are necessary in order to determine whether the hydraulic cylinder functions perfectly.

Unless otherwise stated by components and systems manufacturer, Table 6.1 provides guidelines for *scheduling* preventive maintenance actions for hydraulic cylinders.

#	Preventive Maintenance Actions	Daily	Weekly	Monthly	Biannually	Annually
1	Clean the outer surface of the cylinder barrel + Clean around and the ports.	✔	✔	✔	✔	✔
2	Check if there is any sign of damage or even rubbing in the barrel paint.	✔	✔	✔	✔	✔
3	Check if there is any sign of leaking.		✔	✔	✔	✔
4	Check proper connection with the load.		✔	✔	✔	✔
5	Check proper connection to plumbing.			✔		
6	Check Surface temperature. Air bleeding.			✔	✔	✔
7	Greasing the cylinders connection points with the external loads.			✔	✔	✔
8	Cylinders maintenance and routine inspection from inside.				✔	✔

Table 6.1- BP-Cylinder-02-Maintenance Scheduling

6.3- BP-Cylinders-03-Installation and Maintenance

Proper cylinder installation and maintenance are important processes for trouble-free operation of the system. **ISO/TS 13725** provides detailed information about cylinder installations. However, the best practices list "**BP-Cylinders-03-Installation and Maintenance**" provides guidelines for *installation and maintenance* of hydraulic cylinders.

BP-Cylinders-03-Installation and Maintenance:
1. Proper Cylinder Disassembling.
2. Proper Cylinder Inspection and Maintenance.
3. Proper Seal Replacement and Installation.
4. Proper Cylinder Assembling
5. Proper Mounting on the Machine Structure.
6. Proper Alignment with the Load.
7. Proper Connection with Transmission Lines.
8. Proper Air Bleeding.
9. Propper Installation of External Limit Switches.
10. Protect End Caps Against Impact Load.
11. Protect Cylinder from External Hazard.
12. Protect Cylinder Rod from Corrosion.
13. Protect Air Chamber of a Single-Acting Cylinder from the Environment.
14. Review Range of Allowable and Maximum Working Conditions.

6.3.1- Proper Cylinder Disassembly

Practices of removing the cylinder from the machine is out of scope of this textbook. After removing the cylinder from the machine, the following guidelines describe disassembly process.

Know the Component: It is helpful to know the elements from which the cylinder consists of and the disassembly procedure. Although there are many types and sizes of hydraulic cylinders, they usually have many parts in common. Figure 6.1 shows the major parts of a hydraulic cylinder.

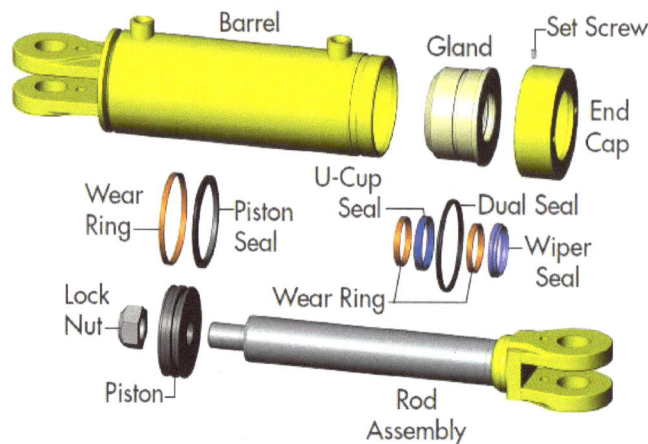

Fig. 6.1- Typical Construction of a Double Acting Cylinder (www. degelman.com)

Follow Proper Disassembly Procedure: It is assumed that such an important component is disassembled by a well-trained technician. However, the following simple procedure provides example for disassembling mill-type cylinder that has threaded end caps:
1. Loosen Set Screw and turn off end cap.
2. Carefully remove piston/rod/gland assemblies and place them on an appropriate fixture that won't damage the rod surface finish and plating.
3. Disassemble the piston from the rod assembly by removing lock nut.
4. DO NOT clamp rod by chrome surface.
5. Slide off gland assembly & end cap.
6. Remove seals and inspect all parts for damage.
7. Install new seals and replace damaged parts with new components.
8. Inspect the inside of the cylinder barrel, piston, rod and other polished parts for burrs and scratches.

Avoid Oil Spillage: There may be some hydraulic fluid left inside the cylinder. Therefore, it is a good idea to use a drip pan or other mean to make sure no *oil spillage* occurs during cylinder disassembly. Allow sufficient time to drain all fluid before pulling the cylinder rod out of the cylinder.

DO NOT Underestimate the Power: When *disassembling* a hydraulic cylinder for inspection, DO NOT underestimate the power. Figure 6.2 shows forbidden practices of disassembling a hydraulic cylinder using compressed air.

Proper Component Support: Disassembled cylinder rod shouldn't be placed on a worktable without proper supports such as V-Blocks shown in Fig. 6.3. That keeps the rod from rolling around and avoid sticking soft contaminants with the rod or damaging it by hard ones.

Fig. 6.2- Improper Cylinder Disassembling
(Courtesy Fluid Power Safety Institute)

Fig. 6.3- V-Blocks to Support Cylinder Rods
(www.ame.com)

6.3.2- Proper Cylinder Inspection and Maintenance

Cylinder Barrel Inspection: As shown in Fig. 6.4, a cylinder *barrel* should be at least be visually inspected from inside using flashlight to make sure there are no scratches. Such scratches make the cylinder operate sluggishly, lose power, or drift under load. Unless the barrel has harsh scratches that are unrepairable, one common way to remove minor scratches is to have the barrel honed. If needed, for expenses and cylinders used in critical applications, they must be checked for barrel straightness and out of round.

Fig. 6.4- Inspecting a Cylinder Barrel

Cylinder Rod Inspection: A bent rod in a hydraulic cylinder causes binding when moving, excessive rod bearing wear and seal deterioration. A damaged seal causes leakage preventing the cylinder from developing full power and ultimately premature failure. For that reason, *rod straightness* must routinely be inspected. As shown in Fig. 6.5, rod run-out is measured by a dial gauge and checked to see if it matches the allowable run-out. The allowable run-out should be stated by the cylinder manufacturer. However, 0.5 mm per linear meter (0.02 of an inch per every 20 inches) is generally considered an acceptable value. Straightening the rod is usually done using a hydraulic press.

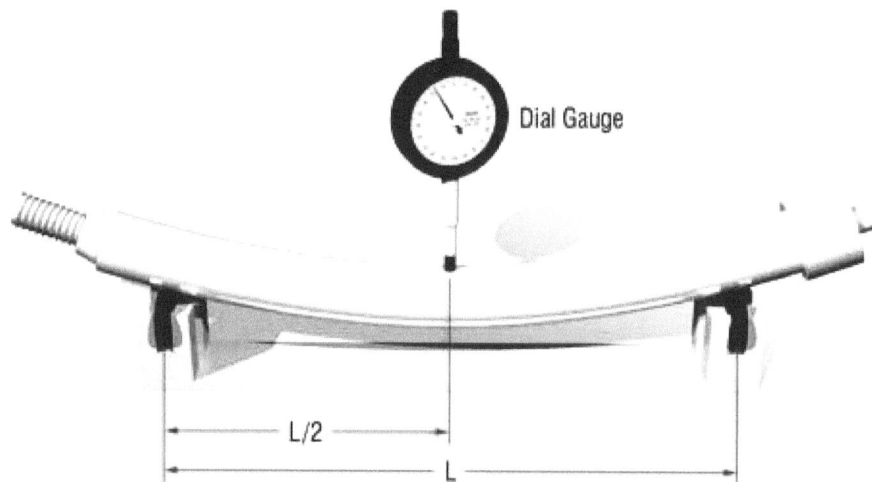

Fig. 6.5 – Checking Rod Straightness (www.machinerylubrication.com)

6.3.3- Proper Seal Replacement and Installation

Best Practices for Hydraulic Seals Replacement: If any of the cylinder *seals* need to be replaced, the following set of bullets must be considered list to make sure the new seal matches the old one:
- Seal Type.
- Seal Dimensions.
- Seal Lip Geometry.
- Seal Cross Section.
- Seal Material Based on Working Temperature.
- Seal Material Based on Working Pressure.
- Seal Material Based on Working Fluid.
- Seal Material Based on Hardness.
- Seal Material Based on General Properties.

Best Practices for Hydraulic Seals Installation: Hydraulic seal replacement is one of the most common cylinder maintenance actions. Proper installation of a hydraulic seal is an important process for the seal to perform reliably. The first advice is to review the seal installation instructions provided by the manufacturer. If not found, the following list provides guidelines for installing hydraulic seals.

1. **Figure 6.6, Properly Remove the Old Seal:** Carefully dismount the old seal without damaging the bores or shafts.
2. **Figure 6.7, Use Genuine Seals:** DO NOT use non-branded seals or seals that are not approved by the hydraulic component manufacturer.
3. **Figure 6.8, Never use a Pretensioned or Pre-Used Seal:** That is because used seals may be plastically deformed and have defects or geometrical shape changes that may not be seen by your naked eyes. In most cases where components must be disassembled for inspection, seals should be replaced.
4. **Figure 6.9, Inspect New Seals:** DO NOT use sharp objects to remove a seal out of its package. Before mounting the new seal, inspect it for damage on the circumference of the sealing lip or the outer diameter. DO NOT shorten the original tension spring if found. Double check the correct placement direction of the seal.
5. **Figure 6.10, Inspect Seal Groove:** It should be clean and free of damage or sharp edges.
6. **Figure 6.11. Inspect Assembly Tools:** They should be routinely inspected and calibrated. Before using them, make sure they are clean and free of sharp edges, scratches, and contamination.
7. **Figure 6.12, Cover Threads:** If the seal must be stretched over sharp edges, such as threaded parts or lead-in chamfers, these areas must be covered to prevent damaging the seals during the assembly process.
8. **Figure 6.13, Lubricate Seal Before Installation:** Lubricate the seal and the mounting surface before seal installation. That reduces the surface friction on the seal and makes it easy to install it. Do NOT use regular grease. Use the same hydraulic fluid or a predefined compatible seal lubricant. Make sue it is clean.
9. **Figures 6.14.A and 6.14.B, Use Adequate Installation Tools:** Use adequate tools for installing rod and piston seals to prevent damaging or twisting the seals.
10. **Figure 6.15, Compress the Seal in kidney Shape:** If needed or no assembly tools are found, compress the seal into a kidney shape. If the seal has a notch, DO NOT bend it from the position of the notch as this may cause overstretch or damage to the seal material.
11. **Figure 6.16, Check Proper Installation of the Seal:** make sure the seal is not tilted or twisted, and the seal axis coincides with the shaft or piston axis.
12. **Figure 6.17, Squeeze the Seal in its Groove:** In both static and dynamic applications, a certain amount of squeeze or compression is required upon installation to maintain contact with the sealing surfaces and prevent fluid leakage. This can be done by installation cones.

Fig. 6.6 - Properly Remove the Old Seal

Fig. 6.7 - Use Genuine Seals (Courtesy of Parker)

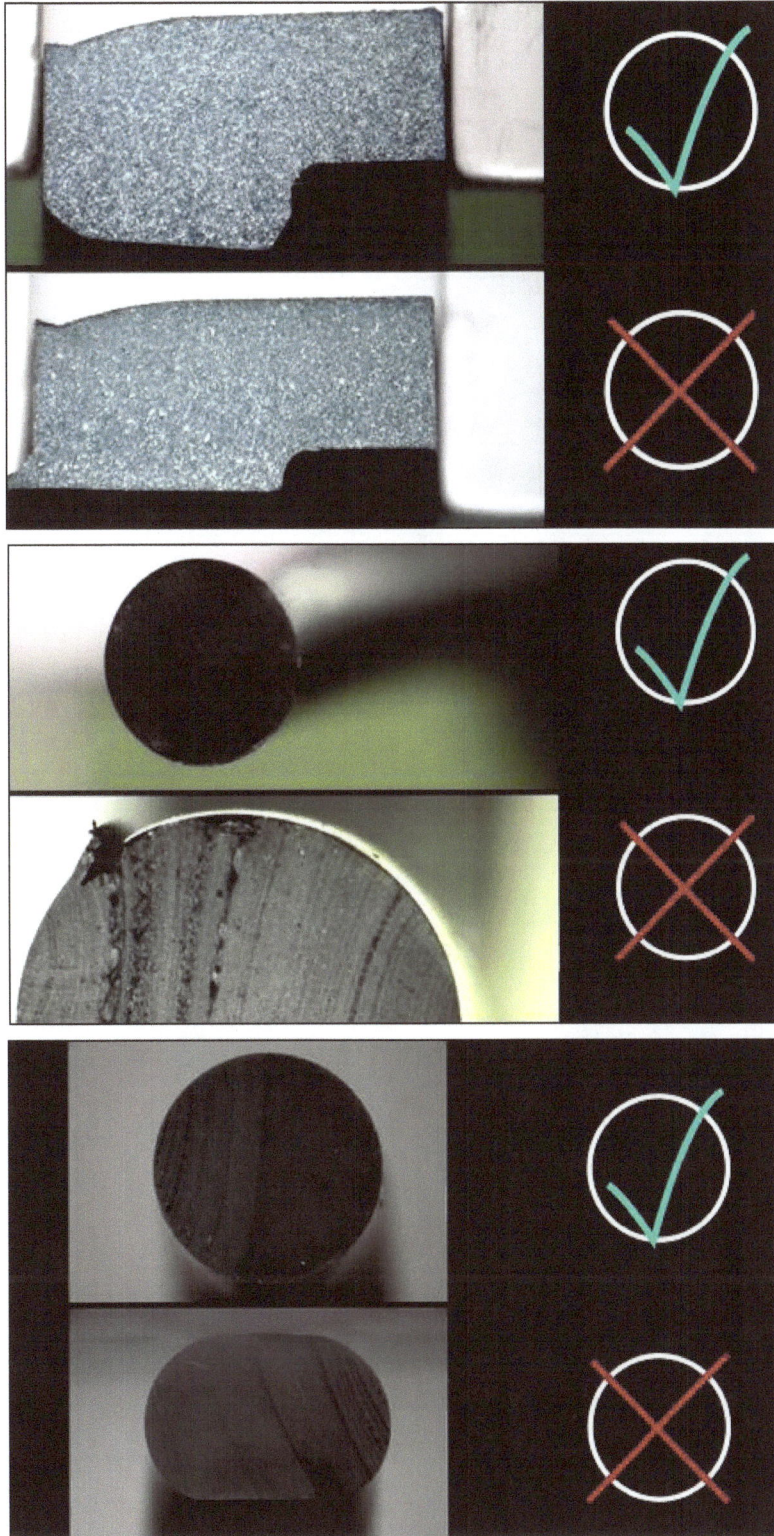

Fig. 6.8 - Never use a Pretensioned or Pre-Used Seal (Courtesy of Trelleborg)

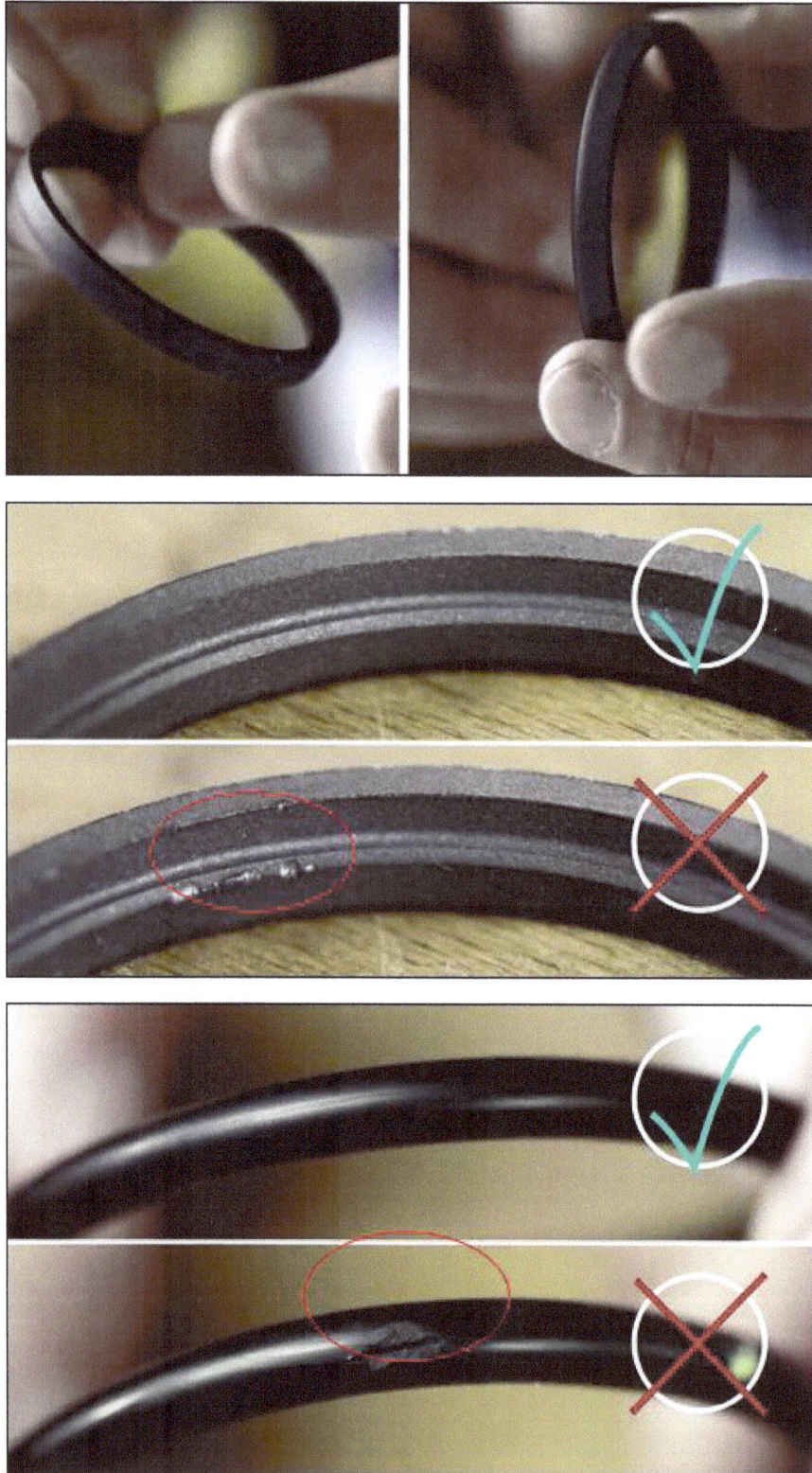

Fig. 6.9 - Inspect New Seals (Courtesy of Trelleborg)

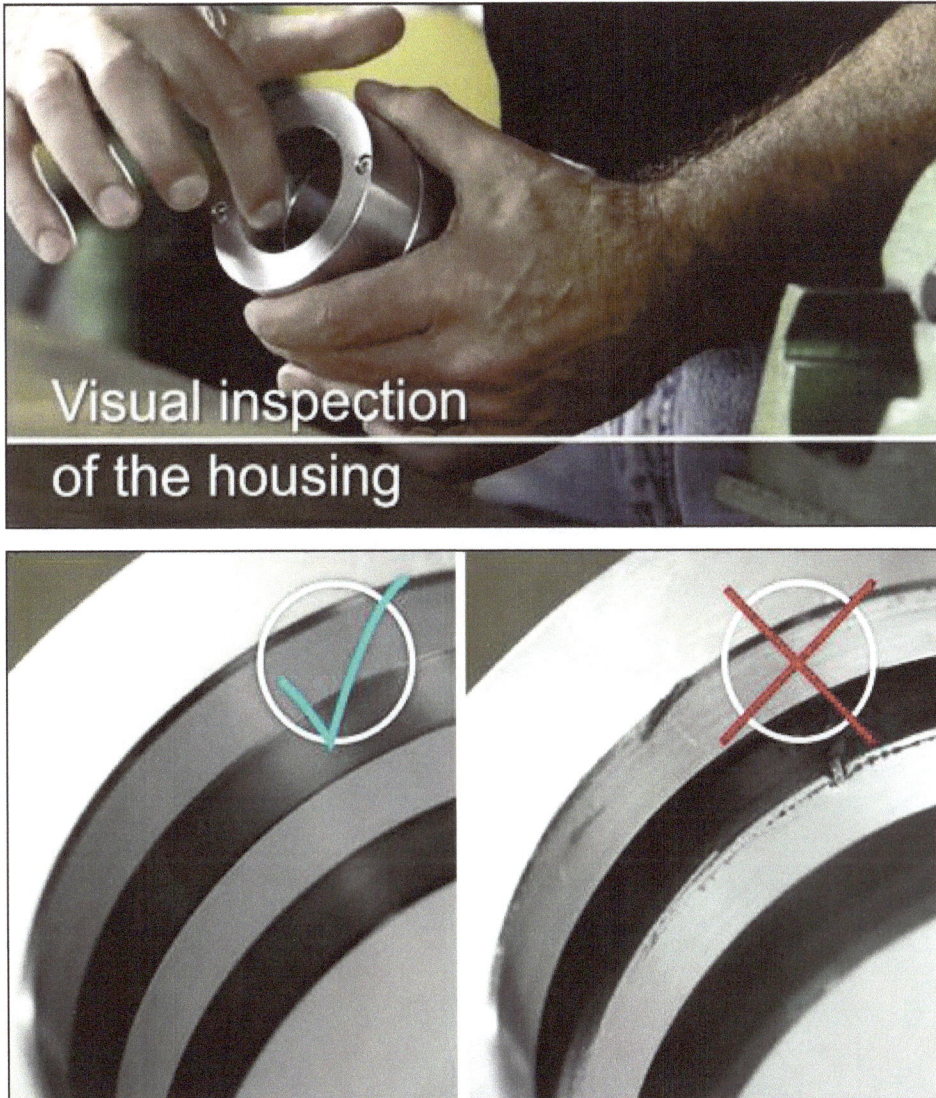

Fig. 6.10 - Inspect Seal Groove (Courtesy of Trelleborg)

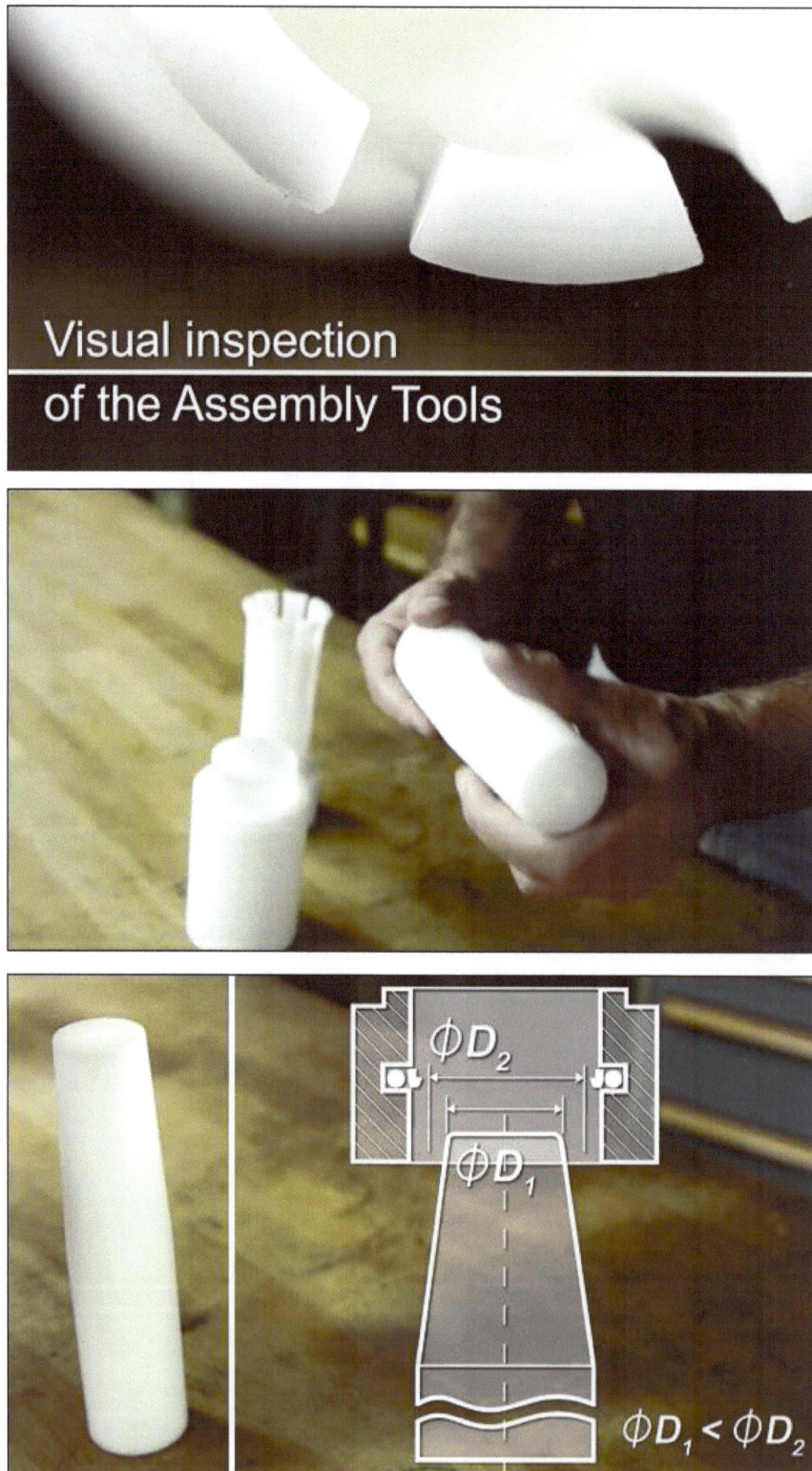

Fig. 6.11 - Inspect Assembly Tools (Courtesy of Trelleborg)

Fig. 6.12 - Cover Threads (Courtesy of Trelleborg)

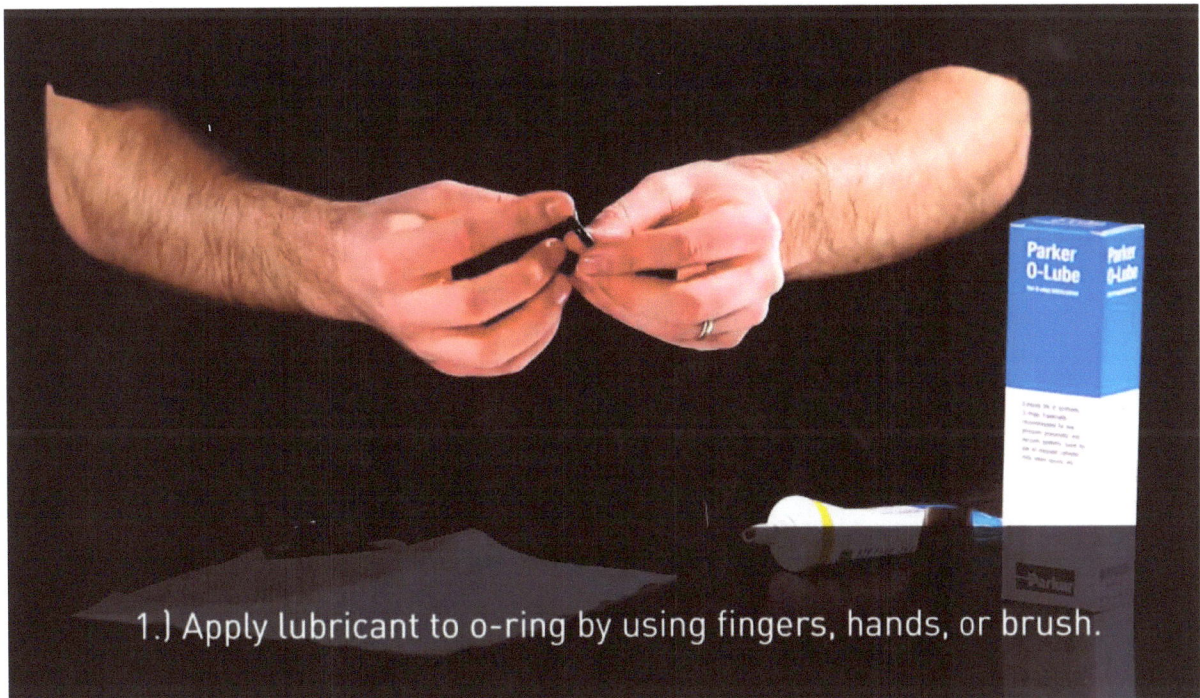

1.) Apply lubricant to o-ring by using fingers, hands, or brush.

Fig. 6.13 - Lubricate Seal Before Assembly (Courtesy of Parker)

Fig. 6.14.A - Use Adequate Installation Tools for Rod Seals (Courtesy of Trelleborg)

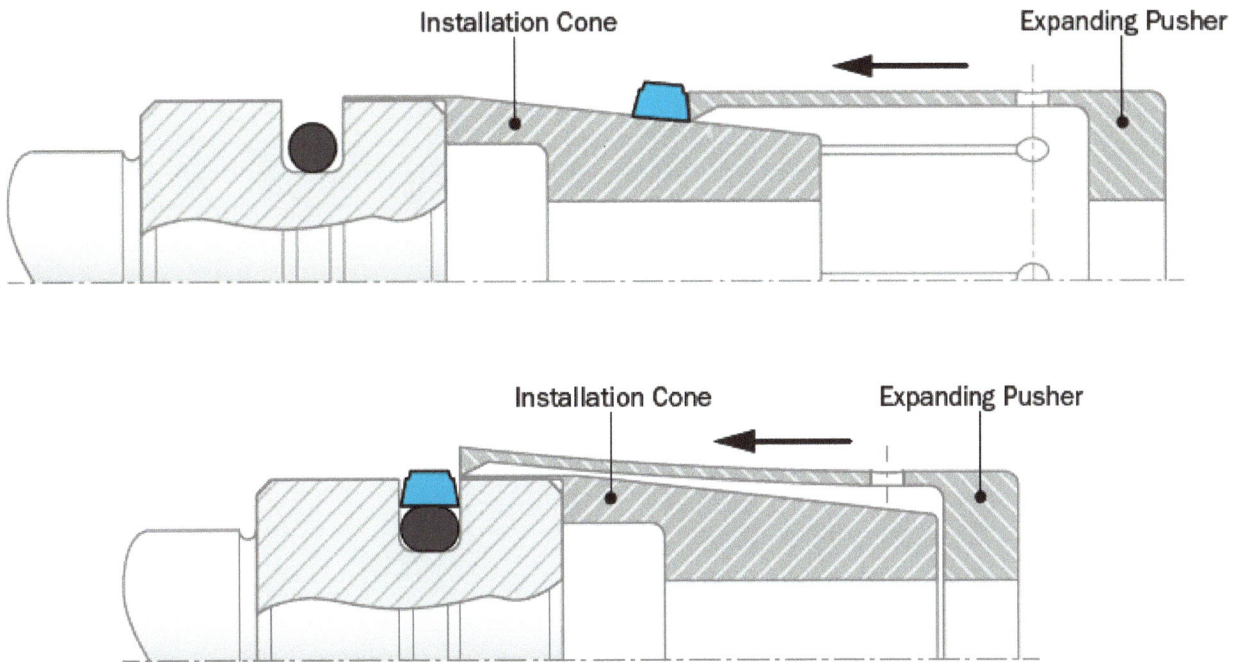

Fig. 6.14.B - Use Adequate Installation Tools for Piston Seals (Courtesy of Trelleborg)

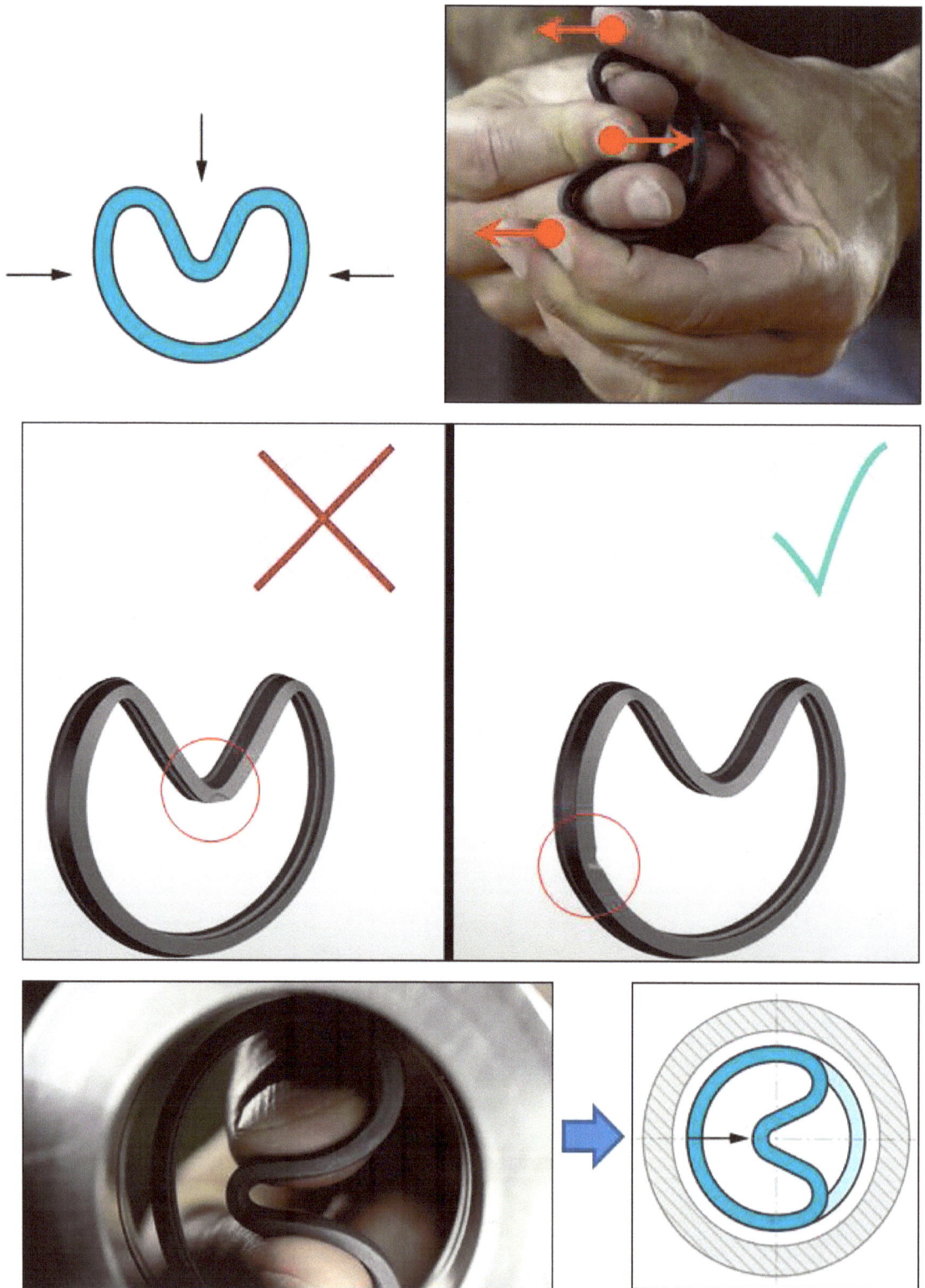

Fig. 6.15 - Compress the Seal in kidney Shape (Courtesy of Trelleborg)

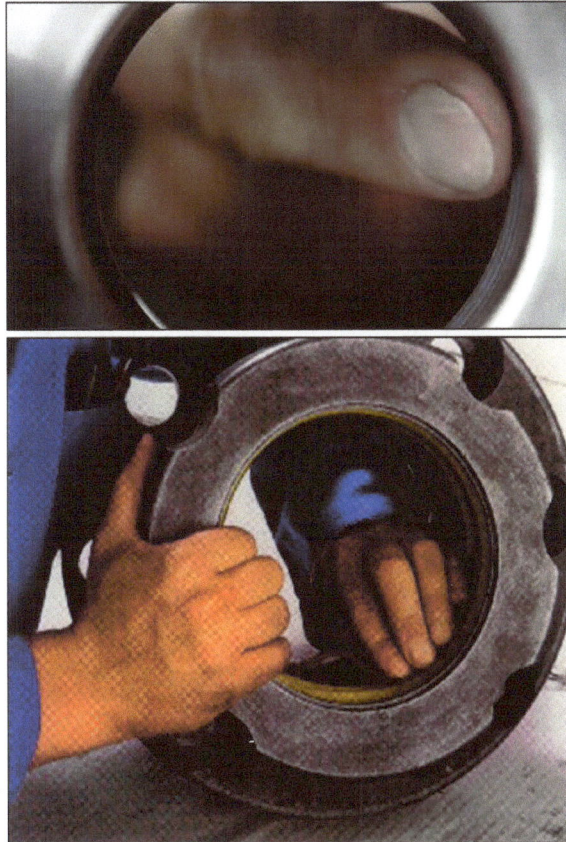

Fig. 6.16 - Check Proper Installation of the Seal

Fig. 6.17 - Squeeze the Seal in its Groove (Courtesy of Trelleborg)

6.3.4- Proper Cylinder Assembly

The following bullets provide guidelines for *assembling* the *cylinder*:

- **Cleaning:** Disassembled parts of the cylinder should be thoroughly cleaned using a solvent that is clean to standard and compatible with seals and parts.
- **Drying:** Cleaned parts should be dried by blowing with clean dry compressed air.
- **Lubricate:** Coat all parts with clean hydraulic fluid during assembly. Lubricating fluid should be same as the fluid that is used to fill the cylinder. No greases should be used.

6.3.5- Proper Mounting with the Machine Structure

This section provides guidelines for proper mounting of a cylinder with the machine structure.

- **Lifting and Transportation:** Regarding *lifting* and moving during installation of the hydraulic cylinder into the machine, the same rules apply as described in section 6.5 of this chapter.

- **Foot-Mounted Cylinders (Fig. 6.18):** As shown in the figure, the following points should be considered during such type of cylinder installation:
    - Supporting Surface: the surface of the machine on which the cylinder will be mounted must be straight, flat, and level without any degree of twisting so that any torsion of the hydraulic cylinder in the installed condition is avoided.
    - Pinning High Pressure Side: Foot-mounted cylinders can impose a shear force on the cylinder mounting bolts. Therefore, the foot leg at the high-pressure side must be pinned with fasteners that are adequate to carry the shear force.
    - Support Long Cylinder Barrel: Barrel of long stroke cylinders must be supported from the middle to avoid bending of the cylinder barrel under its own weight.

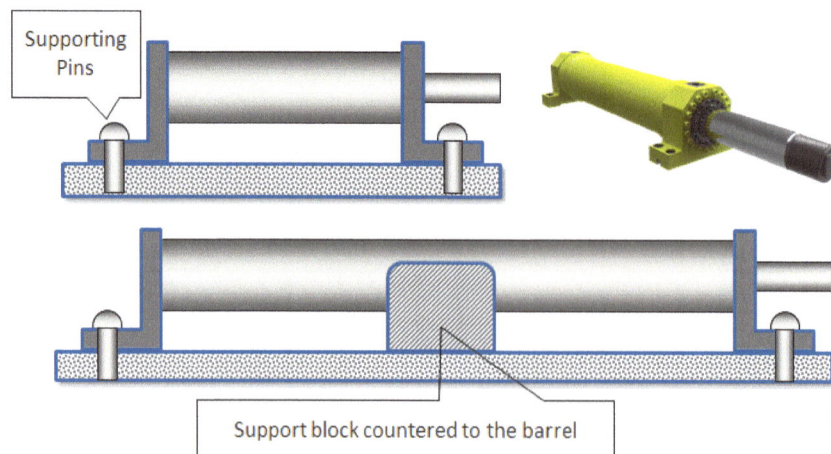

Fig. 6.18 – Foot-Mounted Cylinders

- **Flange-Mounted Cylinders (Fig. 6.19):** As shown in the figure, the following points should be considered during such type of cylinder installation:
  - Rear Flange Mount (left Side): When a rear (at the piston end) flange is used, buckling length equal double the cylinder stroke. That reduces the maximum allowable load.
  - Front Flange Mount (right Side): recommended for a cylinder that is subjected to a high compressive load.

Rear Flange Mount   Front Flange Mount

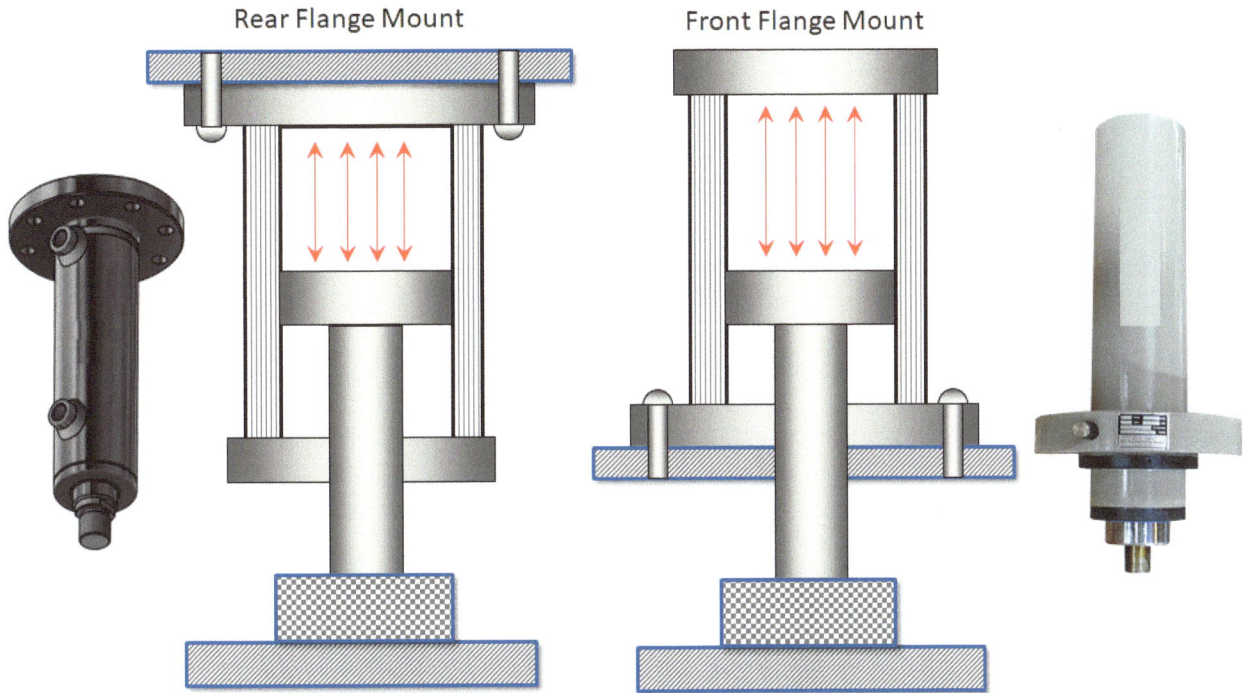

Fig. 6.19 – Flange-Mounted Cylinders

- **Eye-Mounted Clevis [Hinge-Mounted] Cylinders (Fig. 6.20):** As shown in the figure, the following points should be considered during such type of cylinder installation:
  - Cylinder Motion: Such a method is recommended for cylinders that might kinematically move with the load.
  - Cylinder Connections: Flexible hoses or swivel connections must be used to connect the cylinder to the control valves.

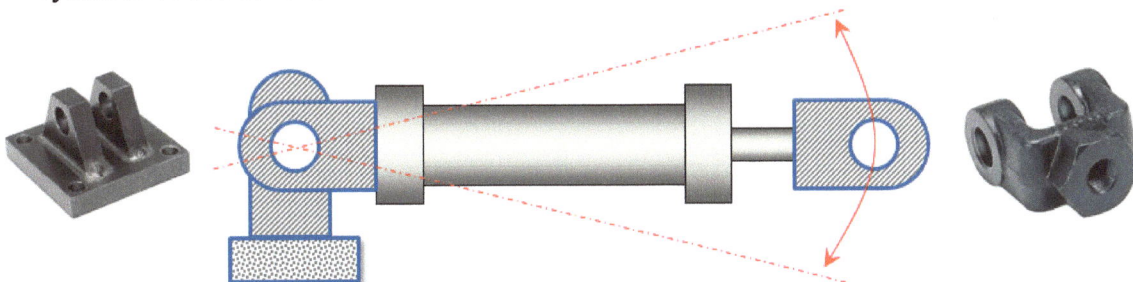

Fig. 6.20 – Eye-Mounted Cylinders

- **Trunnion-Mounted [Hinge-Mounted] Cylinders (Fig. 6.21):** As shown in the figure, the following points should be considered during such type of cylinder installation:
  - <u>Cylinder Motion:</u> Such a method is recommended for cylinders that might kinematically move with the load.
  - <u>Cylinder Connections:</u> Flexible hoses or swivel connections must be used to connect the cylinder to the control valves.
  - <u>Trunnion Bearing:</u> The trunnion bearings should be as close as possible to the cylinder barrel to reduce stresses on the trunnions.

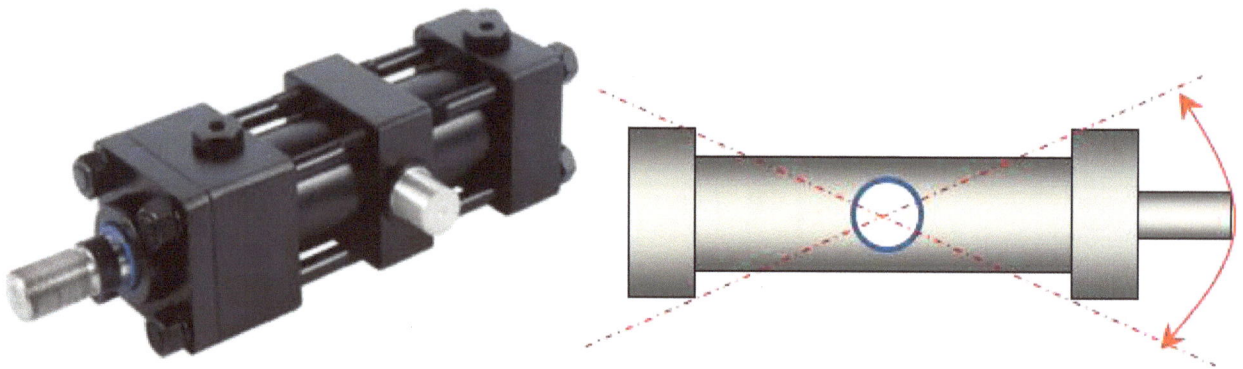

Fig. 6.21 – Trunnion-Mounted Cylinders

6.3.6- Proper Alignment with the Load

As shown in Fig. 6.22, perfect *alignment* between the load and the cylinder rod should be maintained to avoid lateral load, seals failure, external leakage, cylinder rod damage, and cylinder buckling. The following set of bullets provide best practices for cylinder rod alignment with the load:

- Decouple the load from the cylinder rod.
- Make sure that the eye at the rod connects easily with the eyes at the load without binding or transition fit.
- Make sure that the hinge pin is inserted without binding, forcing or hammering.
- Use shims, if needed, under the front and/or rear foot to help align the cylinder rod with the load.

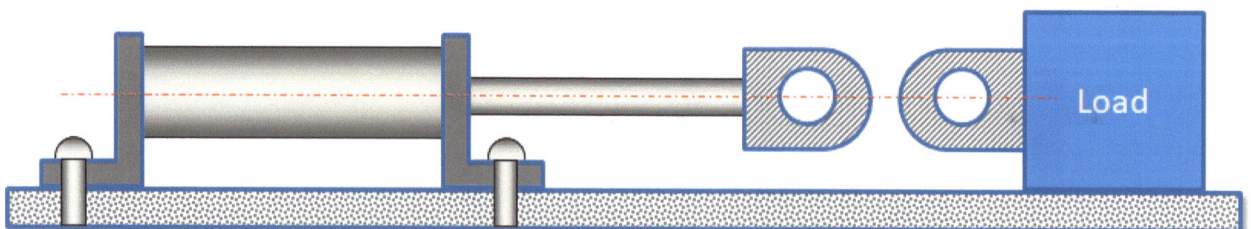

Fig. 6.22 – Cylinder Rod Alignment with the Load

6.3.7- Proper Air Bleeding

Why Air Bleeding:

Operating a cylinder with an air inside causes one or more of the following problems:
- Air in the cylinder is compressed and suddenly expands resulting in unexpected very fast movement of the piston and rod, damaging the cylinder end caps, and raising risk of bodily injuries.
- Actuators move erratically.
- Air compression can cause *dieseling* effect that damage the seals inside the cylinder.

Best Practices for Hydraulic Cylinders air Bleeding:

Air bleeding should be part of cylinder regular maintenance program and performed before installing a new hydraulic cylinder. It is highly advised to review air bleeding method instructed by the cylinder manufacturer. However, the following bullets are considered best practices when bleeding air from hydraulic cylinders.

- Make sure the hydraulic cylinder is not pressurized in any circumstances.
- Uncareful opening of the cylinder ports is a very hazardous situation and can cause not just property damage but sever bodily injury and even death to personnel in the vicinity of the cylinder. Therefore, before air bleeding, lock the machine elements to prevent load free falling when the cylinder ports are open.
- The bleeder valve should be pointing upwards when the hydraulic cylinder is positioned horizontally.
- Locate the air bleeding ports. As shown in Fig. 6.23, some cylinders are built with air bleed ports closed with screws on each end of the barrel. Do not confuse air bleed ports with cushioning adjustment screws at the end caps.

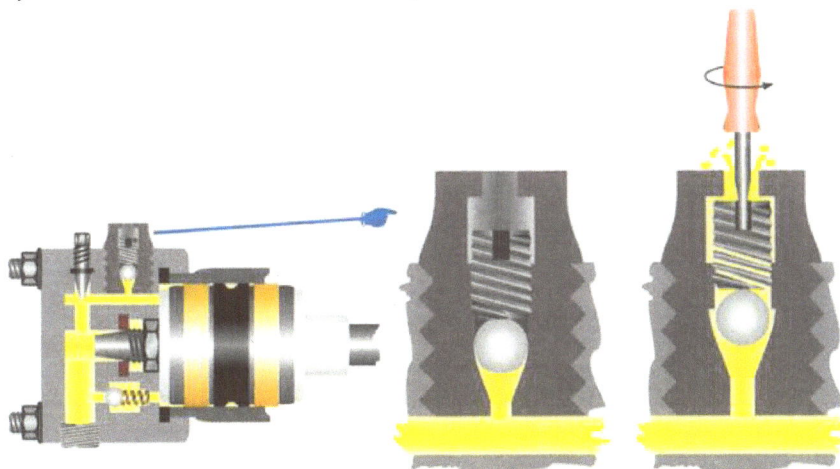

Fig. 6.23 – Operation of Air Bleeding Valve (Courtesy of Assofluid)

As shown in Fig. 6.24, *automatic air bleeds* are available in the marketplace. If they are used, it is essential to check them periodically following the manufacturer instructions. As shown in the figure, the valve consists of two parts with a setback function. The lower part is permanently assembled in the cylinder. In bleeding and filling, only the upper part of the valve is opened, in which an O-ring-seal is integrated. Through it neither air nor hydraulic fluid can penetrate through the opened thread. If there is a drop-in pressure, the valve closes automatically until pressure is applied again.

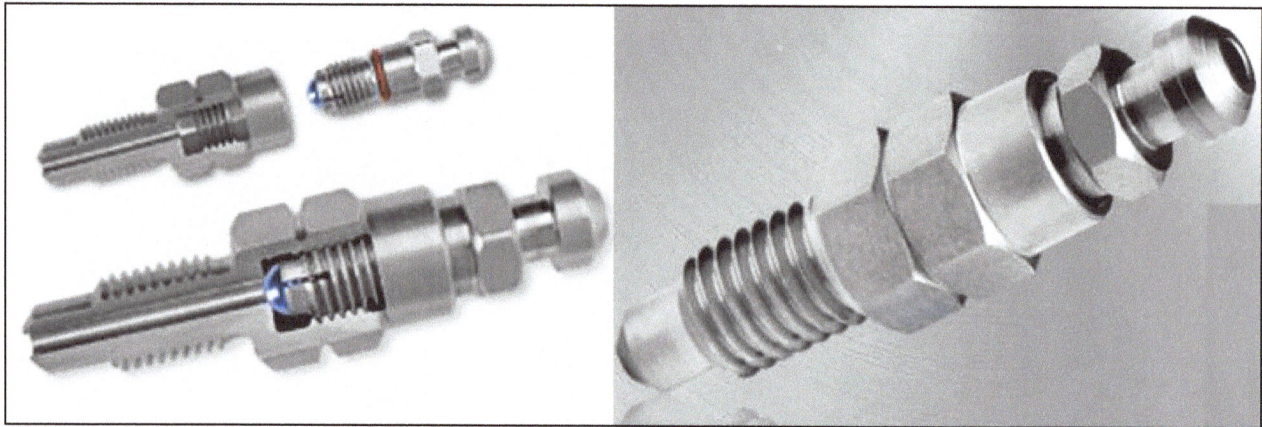

Fig. 6.24 – Automatic Air Bleeding Valve (www.stahlbus.com)

Examples of instructions provided by a manufacturer for air bleeding:

Bleeding Air from Double-Acting Type Hydraulic Cylinders
1. Check all hoses or pipes are connected properly.
2. Fully retract the cylinder by working fluid.
3. Open the air valve at the piston side of the hydraulic cylinder.
4. Set up the hydraulic system and start it up.
5. Extend the piston rod slowly with no pressure built up.
6. Keep extending the piston rod until there is only oil (no foam) come out of the air valve.
7. Shut down the system and close the air valve.
8. Depressurize the system.
9. Opening air bleeding valve at the rod side of the hydraulic cylinder.
10. Startup the hydraulic system.
11. Retract the hydraulic cylinder slowly with no pressure built up.
12. Keep retracting the cylinder until there is only oil (no foam) come out of the air valve.
13. Shut down the system and close the air valve.
14. Retest the system and cycle the hydraulic cylinder until it is running smoothly.

<u>Bleeding Air from Single-Acting or Lift Dump Truck Cylinders (Fig. 6.25)</u>
1. Raise the dump body and cylinder to full extension. Leave the dump body in this position for several minutes to allow air to rise to the top of the cylinder.
2. Lower the dump truck cylinder until the front of the dump body is approximately 2 feet off the chassis frame.
3. Hold the dump body mechanically in this position.
4. Crack the air bleeding valve open.
5. Wait until all trapped air has escaped from the valve and a full stream of hydraulic oil is escaping from the valve.
6. At this point, the cylinder is bled, and the bleeder valve can be closed.

Fig. 6.25 – Single-Acting Cylinder in Dump Trucks (www.hydrauliccylindersinc.com)

6.3.8- Proper Connection with Transmission Lines

Hydraulic lines can be connected directly to the cylinder ports or via a block mounted on the cylinder as shown in Fig. 6.26. They also can be connected to a valve that is directly mounted on the cylinder barrel.

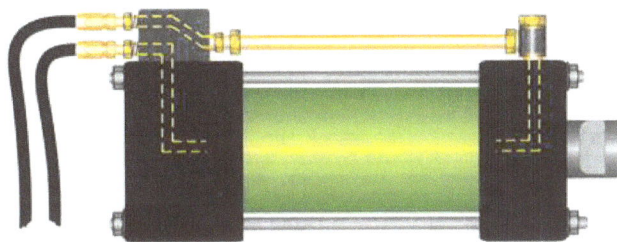

Fig. 6.26 – Hydraulic Lines Connection with the Cylinder (Courtesy of Assofluid)

6.3.9- Proper Installation of External Limit Switches

Mounting limit *switches* against the motion of cylinder rod should avoid damaging the limit switches by the cylinder overrunning. Figures 6.27 and 6.28 show best practices for installation of an electro-mechanical limit switch and a proximity switch against a cylinder rod; respectively.

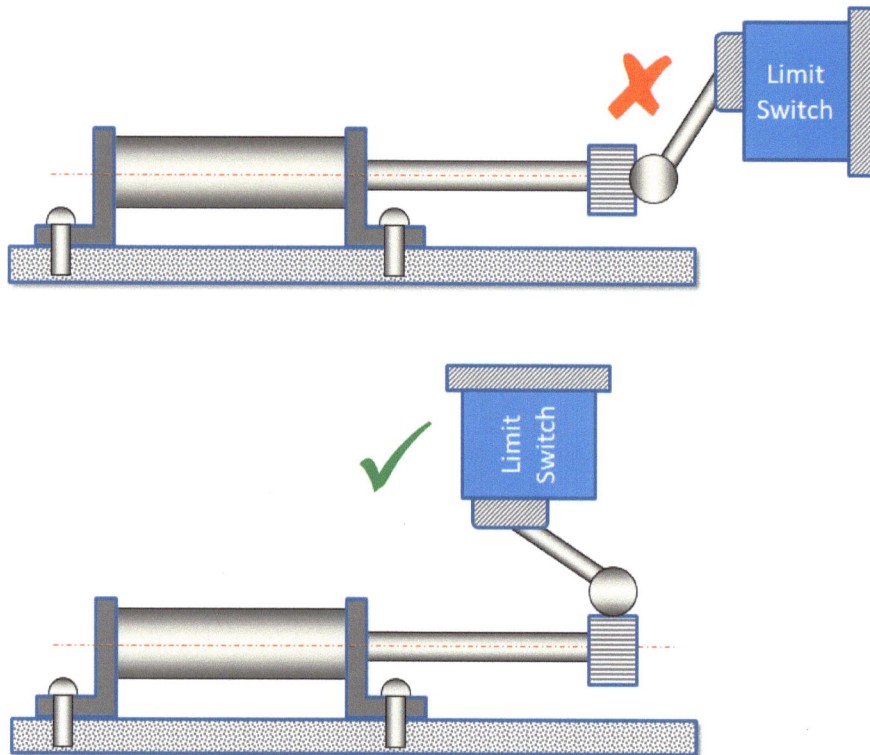

Fig. 6.27 – Best Practices for Electro-Mechanical Limit Switches Installation Against Cylinder Rod

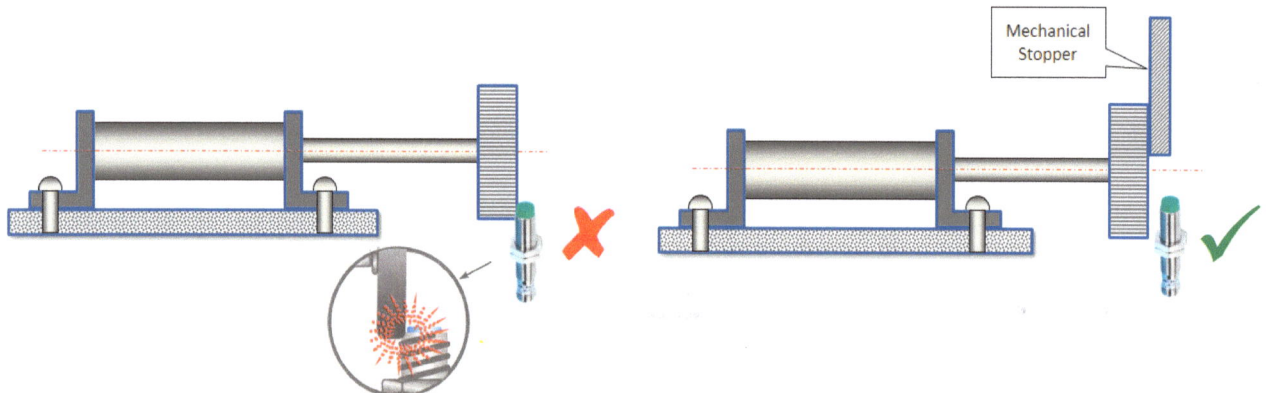

Fig. 6.28 – Best Practices for Proximity Switches Installation Against Cylinder Rod

6.3.10- Protect End Caps Against Impact Load

Standard cylinders are not designed to receive high impact on their end caps. If the cylinder piston continues to hit the two end caps frequently during the cylinder operation, the cylinder may be subject to fatigue failure causing sudden and serious damage. Therefore, in such cases some actions need to be taken to protect the cylinder as follows:

- For cylinder that moves at speeds below 10 m/s (4 in/s). usually no cushioning action is required.
- For cylinder that moves at speeds between 10-20 cm/s (4-8 in/s), built in cushioning system inside a cylinder is sufficient.
- For cylinder that moves at speed above 20 cm/s (8 in/s), built in cushioning is not sufficient and external cushioning or deceleration method is needed.
- One solution is to decelerate the cylinder hydraulically when the cylinder piston approaches the end caps.
- Another mechanical solution is to use an external braking system or a shock absorber as shown in Fig, 6.29.

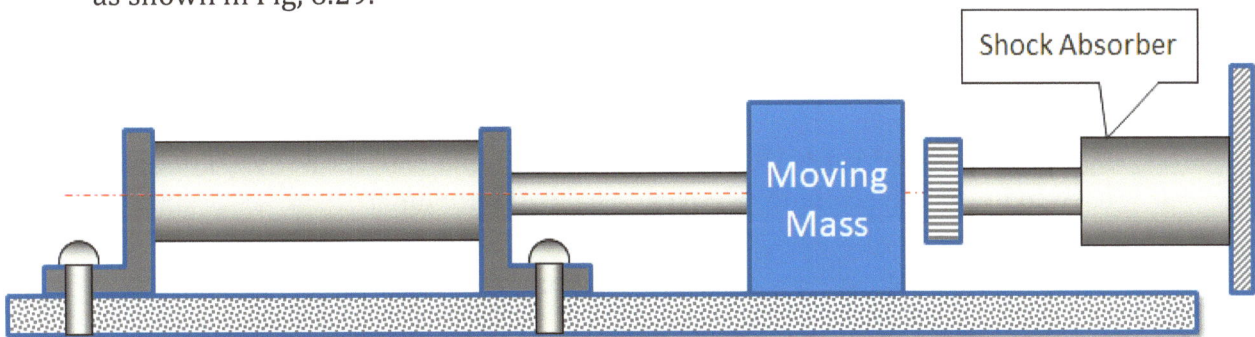

Fig. 6.29 – Protect End Caps Against Impact Load using Shock Absorber

6.3.11- Protect Cylinder from External Hazard

Protecting both the cylinder body and the cylinder rod from external hazard improves cylinder reliability and operating safely. The following set of bullets provide best practices for protecting the cylinder from external hazard:

- **Heat Sources:** Exposing cylinders to heat sources will lead to oil overheating, pressure intensification in closed oil volumes, and improper sealing performance. Therefore, as shown in Fig. 6.30, hydraulic cylinders must be *shielded* in outdoor applications where the cylinder is subjected to direct sunlight. In indoor applications, cylinder must be shielded or sufficiently separated from heat sources such as heaters, boilers, furnaces, etc.

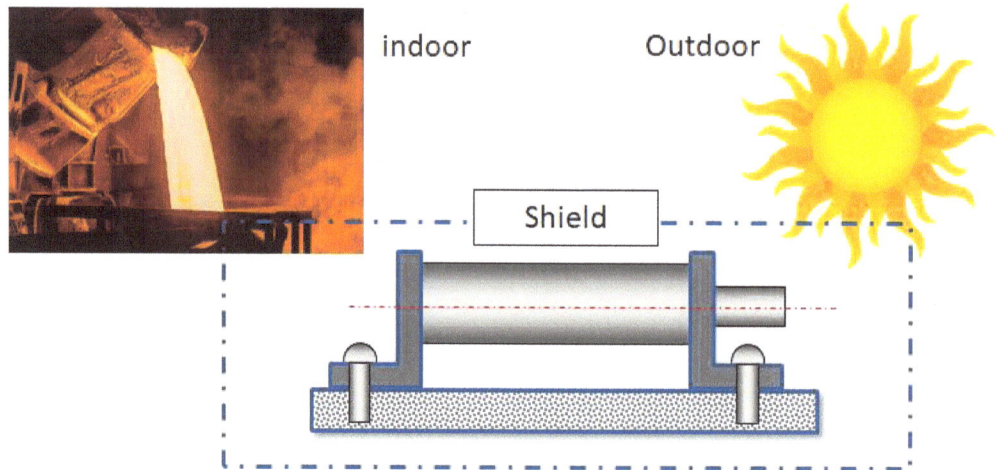

Fig. 6.30 – Shield Hydraulic Cylinders from Heat Source

- **External Hazardous:** Cylinders that are working in harsh conditions must also be protected. Figure 6.31 shows examples of hazardous conditions (weld splatter, fast drying chemicals, paint, excessive heat, abrasive contaminants etc.). As shown in the figure, a universal rubber bellow consists of 3 main parts Body, Joint and Collar. It is used to protect the cylinder from external contaminants.

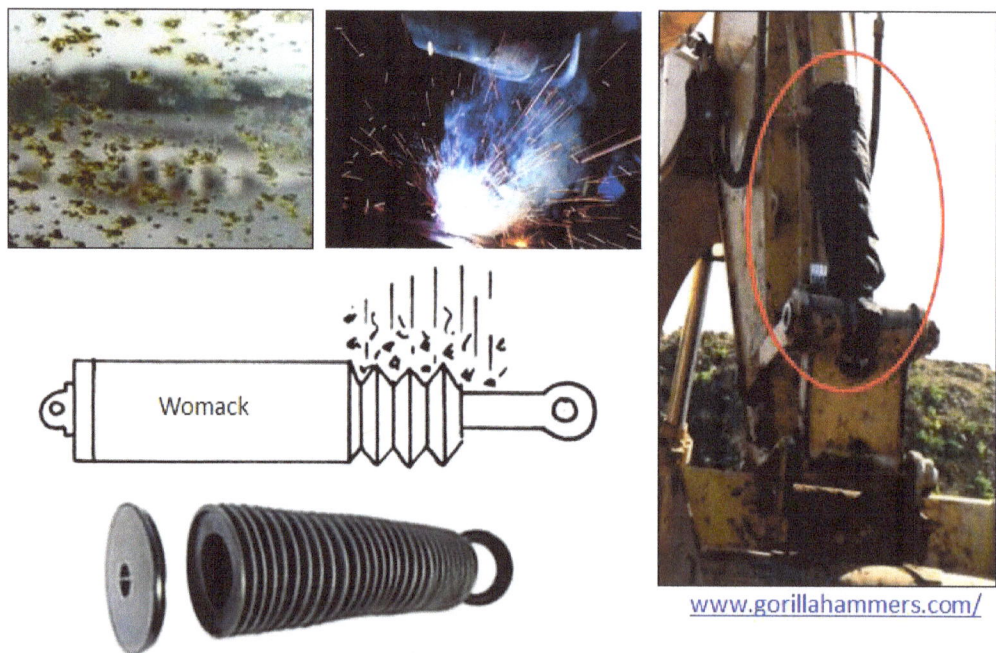

www.gorillahammers.com/

Fig. 6.31 – Shielding Hydraulic Cylinder Rods from Hazardous Conditions

6.3.12- Protect Cylinder Rod from Corrosion

Figure 6.32 shows a hydraulic cylinder that is used in excessively corrosive environment such as snow falling, sea water, or high humidity, or strongly fluctuating temperature. Corrosive conditions cause rod scratches and external leakage. In such cases, the following best practices have to be considered within the scope of the maintenance:

- Use fresh water to remove salt, sand, or any residues from the piston rod.
- Do NOT use steam cleaners or high-pressure water jets.
- Using industrial residue-free wipe, cover the cylinder rod with protection *oil film* of low viscosity. Apply the protection oil to the entire piston rod.
- Ensure compatibility of the oil used.
- Periodically check the protective oil film depends on how harsh the conditions are.

Hydraulic cylinder operating in a severe salt contaminated environment (www.systemseals.com).

Fig. 6.32 – Protecting Cylinder Rod from Corrosion

6.3.13- Protect Air Chamber of a Single-Acting Cylinder from the Environment

As shown in Fig. 6.33, a single acting cylinder may be initially retracted or initially extended with air chamber on the rod side or piston side; respectively. Such a cylinder is actuated by fluid power from one side only. The non-actuated side can't be kept closed. Otherwise, the air will continually be compressed and decompressed overheating the cylinder. So, the air chamber must breathe. As shown in the figure, air chamber must be protected by a high-quality *air breather* against dust and humidity. If not protected, dust and humidity will continue to build up inside the cylinder causing damage over the time. Air breather physical size and mesh size are usually specified by the system's designer to meet the general standard cleanliness class required for reliable system operation.

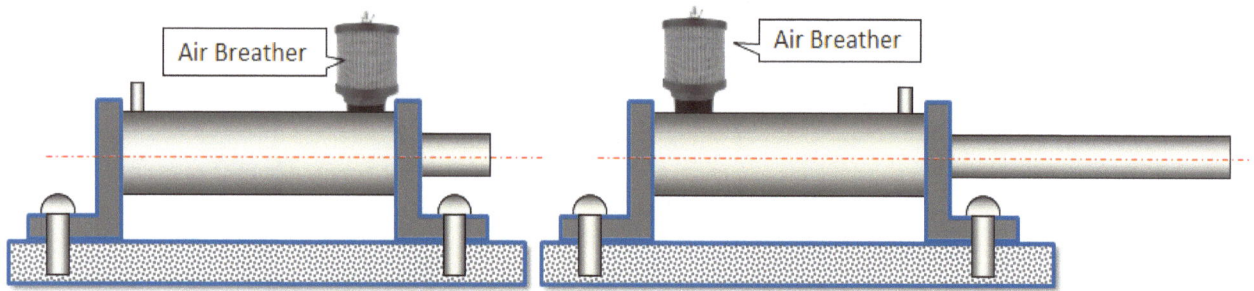

Fig. 6.33 – Air Breather to Protect Air Chamber of a Single-Acting Cylinder

6.3.14- Review Range of Allowable and Maximum Working Conditions

Maximum working conditions for a hydraulic cylinder (such as pressure, temperature, speed, load, etc.) are specified by a system manufacturer. During cylinder installation, it is good to be aware of these conditions and to make sure they won't be exceeded. Figure 6.34 shows a situation where the rod side of a differential cylinder is restricted. As a result, pressure could be intensified in in the rod side exceeding rated pressure limits.

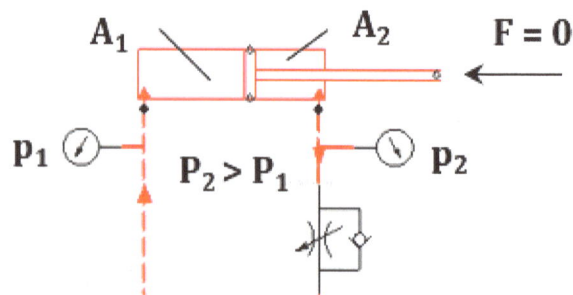

Fig. 6.34 – Pressure Intensification in a Double-Acting Differential Cylinder

6.4- BP-Cylinders -04-Standard Tests and Calibration

ISO10100-2001 standard specifies the requirements and procedure for cylinder testing. This test procedure is designed to check and determine the amount of cross-piston leakage and rod seal leakage. Seal leakage can be caused by damage or wear in the internal cylinder tube wall or the piston seal system. It will also determine the condition of the rod seal. It is recommended that the cylinder be tested in both the extend and retract positions.

Test Piston Leakage in Direction of Cylinder Retraction (Fig. 6.35):
- **Step 1:** Shut the prime mover off.
- **Step 2:** Lock out the electrical system or tag the keylock switch in accordance with local regulations.
- **Step 3:** Observe the system pressure gauge. Release any residual pressure trapped in the system by an accumulator, counterbalance or pilot-operated check valve, suspended load on an actuator, intensifier, or a pressurized reservoir.
- **Step 4:** Install flow meter (2) in series with the transmission line at the rod-end.
- **Step 5:** Install pressure gauge (2) in parallel with the connector at the rod-end.
- **Step 6:** Start the prime mover.
- **Step 7:** Allow the system to warm up to approximately 130 $^{\circ}$F. (54 $^{\circ}$C.).
- **Step 8:** Activate the directional control valve to retract the cylinder. When the rod is fully retracted, cylinder pressure increases to the value of the main pressure relief valve.
- **Step 9:** While holding this position, record on the pressure p_2, flow Q_2, and temperature T. If the cylinder isn't leaking past the piston seal in retract, Q_2 will be zero. Q_2 Will only measure leakage. Therefore, its range and resolution must be selected for low flow, i.e. below 4 lit/min (1 gpm). Rather than installing a flowmeter, it would be easier and cost effective is to disconnect the line on the cap end, and just collect the leakage in a bucket using a stopwatch to measure the flow.
- **Step 10:** Release the directional control valve and shut the prime mover off.
- **Step 11:** With the directional valve shifted to extend the cylinder, similarly, repeat step 1 through 10 to test the cylinder leakage in direction of cylinder extension considering the readings p_1, Q_1.

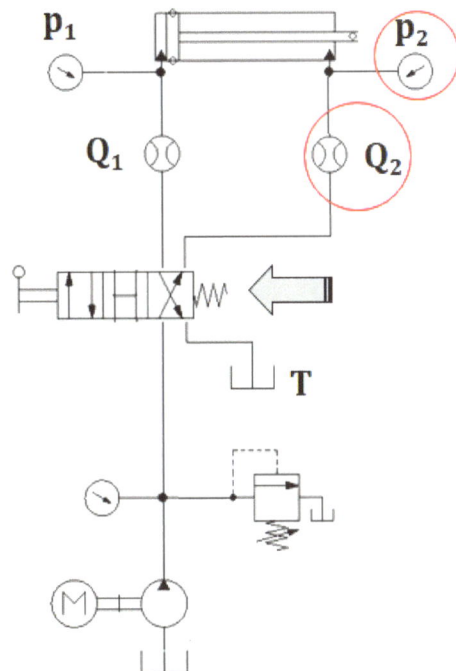

Fig. 6.35 – Hydraulic Cylinder Test Procedure

6.5- BP-Cylinders -05-Transportation and Storage

The best practices list **"BP-Cylinders-05-Transportation and Storage"** provide guidelines for transporting and storing hydraulic cylinders.

BP-Cylinders-05- Transportation and Storage (see Fig. 6.36):
- Pack the cylinders in plastic during shipping and transportation.
- Store cylinders indoor in a dry storage area.
- Store cylinders in fully retracted position.
- Store cylinders, particularly large ones, in a vertical position with the piston rod up. That avoids bending a cylinder barrel under its weight.
- Close cylinder ports with proper dust caps or steel plugs.
- Cover cylinder mounting ends and threaded rod ends with proper protecting covers.
- In case of storage of more than six months, the surface of the hydraulic cylinder must be coated or treated with a corrosion preventive that is compatible with cylinder paint, coating, and plating. Unprotected parts like fitting surfaces or mechanical interfaces must be protected with a corrosion preventative.
- For long storage, it is recommended to fill the rod side of the cylinder with protective oil that is compatible with internal seals and the fluid used in the specific application.
- Type of protective oil and interval of storage are usually specified by manufacturer.
- In cylinders that are filled with oil, pressure intensification due to temperature increase must be considered. Low flow thermal pressure relief valve can be used to depressurize the cylinder, or carefully and safely unplug the ports when needed.
- Pressure intensification due to temperature increase can be estimated by the formula:

$$\Delta P = \text{Bulk Modulus x thermal Expansion Coefficient x Temperature Difference}$$
$$\text{Example: } \Delta P \text{ (bar)} = 20,000 \text{ (bar)} \times 0.0005 \text{ (1/}^{\circ}\text{C)} \times 50 \text{ }^{\circ}\text{C} = 500 \text{ bar}$$

Fig. 6.36 – Best Practices for Storing Hydraulic Cylinders

Best Practices for Hydraulic Seals Storage:

Fundamental instructions on storage, cleaning and maintenance of elastomeric seal elements is described in international standards, such as: **DIN 7716 / BS 3F68: 1977, ISO 2230, or DIN 9088**. Properties of hydraulic sealing materials are affected by the storage environment. The following list provides guidelines for storing hydraulic seals.

Best Practices for Hydraulic Seals Storage:
1. **General Conditions:** Storage space should be kept cool, dry, dust free, and moderately ventilated.
2. **Humidity:** Optimum humidity is 40 to 65 percent, maximum 75%.
3. **Temperature:** Optimum temperature is 25 ºC (77 ºF), maximum 50 ºC (122 ºF). When taken from low temperatures, items should be warmed up to approximately 30ºC (86ºF) before they are used. Warming them up shouldn't be fast, an hour of time span is fine.
4. **Air:** Avoid exposure to direct and continuous stream of conditioned air.
5. **Heat:** Avoid exposure to direct heat source such as boilers or radiators.
6. **Light:** Avoid exposure to direct sunlight and ultraviolet light. Unless packed in opaque containers, it is advisable to cover windows with red or orange screens or coatings.
7. **Radiation:** Avoid exposure to Gamma radiation, otherwise seal compression set is severely affected.
8. **Ozone and Oxygen:** Avoid exposure to sources of ozone such as mercury vapor lamps, high-voltage electrical equipment, combustible gases, and organic vapors. Wrapping in airtight containers, or other suitable means should be used for vulcanized rubber items. Storage in containers that limit exposure to environmental conditions (e.g. sealed plastic bags) should be used for all materials when possible.
9. **Liquids:** Avoid exposure to vapors of gasoline, greases, acids, cleaning liquids (unless such liquids are part of the seals' design or manufacturer's packaging). No solvents, fuels, lubricants or cleaning agents, or similar products, are to be stored in the same area.
10. **Contact with Elastomers:** Avoid contact between seals made from dissimilar compounds. in such cases, each type should be individually packaged.
11. **Contact with Metals:** Avoid contact with certain metals that have degrading effects on some elastomers such as manganese, iron and particularly copper.
12. **Packaging:** Pack the seals in stress-free cases and DO NOT squeeze a hydraulic seal to accommodate it in a small storage area.
13. **Deformation:** DO NOT store seals on top of each other or place heavy objects on top of any stored seals. Where possible, rubber items should be stored in a relaxed position, free from tension or compression. Laying the item flat avoiding crushing keeps it free from strain and minimizes deformation.
14. **Hanging:** DO NOT hang or suspend seals on a hook in a vertical position as gravity will distort the seal over time.
15. **Cleaning:** Use cleaning fluid that is only specified by seal manufacturer. Organic solvents such as trichloroethylene, carbon tetrachloride, and petroleum are the most harmful agents. Soap, water, and methylated spirits are the least harmful, and all parts should be dried at room temperature before use.

16. **Stock Rotation (FIFO):** Stock the seals in rotation, i.e. First-In, First-Out manner (FIFO). This ensures that the next seal used in the rotation will be within its intended shelf life.

17. **Shelf Life:** Considerable storage life without detectable damage varies for different elastomers. In 1998, the Society of *Automotive Engineers* (SAE) issued an *Aerospace Recommended Practice* (ARP) for the storage time of elastomer seals and seal assemblies prior to installation. Table 6.2 shows approximate shelf life for standard elastomers under controlled storage conditions.

Compound Name	ASTM Polymer	Shelf Life
Aflas®	FEPM	Unlimited
Butyl Rubber, Isobutylene Isoprene	IIR	Unlimited
Chloroprene (Neoprene®)	CR	15 Years
Chlorosulphonated Polyethylene (Hypalon®)	CSM	15 Years
Epichlorohydrin (Hydrin®)	ECO	NA
Ethylene Acrylic (Vamac®)	AEM	15 Years
Ethelene Propylene, EPDM or LP	EP	Unlimited
Fluorocarbon (Viton®)	FKM	Unlimited
Fluorosilicone	FVMQ	Unlimited
Hydrogenated Nitrile, HNBR or HSN	HNBR	15 Years
Nitrile (BUNA-N or NBR)	NBR	15 Years
Perfluoroelastomer	FFKM	Unlimited
Polyacrylate	ACM	15 Years
Polyurethane (Polyester or Polyether)	AU/EU	5 Years
Silicone	Q,VMQ,PVMQ	Unlimited
Styrene Butadiene (Buna-S)	SBR	3 Years

Table 6.2 - Approximate Shelf Life for Standard Elastomers (Courtesy of MFP Seals)

Best Practices for Hydraulic Cylinders Lifting:

Improper lifting of hydraulic cylinder may result in minor-to-major damage to the cylinder depends on how heavy the cylinder is. Lifting instructions by manufacturer must e reviewed. However, the following are best practices for cylinders lifting.

- DO NOT Lift a cylinder in when the it is in vertical position and the cylinder rod faces the ground. This may result in unexpected cylinder extension during lifting.
- DO NOT hang the cylinder from the rod and the barrel simultaneously because this may cause rod seal damage. It rather recommended to hang the cylinder from the barrel only with the hanging attachments are equidistant from the cylinder's center of gravity.
- DO NOT lift the cylinder while it is filled with oil.

Chapter 7

Maintenance of Valves

Objectives

This chapter provides guidelines for **valves** selection, replacement, maintenance scheduling, installation, testing, storage and transportation. This chapter is supported by examples and figures granted by leading fluid power manufacturers.

Brief Contents

7.1-BP-Valves-01-Selection and Replacement
7.2-BP-Valves-02-Maintenance Scheduling
7.3-BP-Valves-03-Installation and Maintenance
7.4-BP-Valves-04-Standard Tests and Calibration
7.5-BP-Valves-05-Transportation and Storage

Chapter 7: Maintenance of Valves

The following set of best practices provide general guidelines and may not be applicable for all cases. They are not intended to replace the instructions given by the component manufacturer. It is strongly advisable to adhere to instructions provided by the manufacturer.

7.1- BP-Valves-01-Selection and Replacement

Valve types shall be selected to consider correct function, leakage control, maintenance and adjustment requirements, and resistance against environmental influence. This section is not intended to provide design solutions or valve sizing. For purposes of selection or replacing an existing valve, the following best practices shall be considered:

- **Valve Function:** The new valve must have identical symbol and part number. Many of the valves look the same or have similar (but not identical) symbols with minor differences. These minor differences may make a valve operate improperly and therefore they can't replace each other. As an example, Fig. 7.1 shows symbols for a counterbalance valve versus a sequence valve. But looking closely to these symbols, it is to be noticed that the spring chamber of the sequence valve is vented externally through the port Y. The sequence valve can replace the counterbalance valve but not the invers.

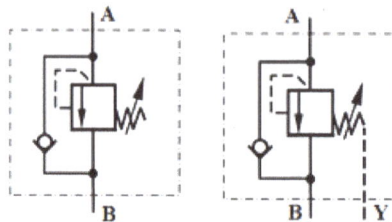

Fig. 7.1- Symbols for Counterbalance Valve (Left) and Sequence Valve (Right)

- **Directional Valve Transitional Conditions:** Even the symbol of a directional valve didn't give all information. The symbol is just an upper mask that hide below it more details! As shown in Fig. 7.2, a directional valve could have different transitional condition that needs to be the same when replacing a valve. Otherwise, system functional performance may be affected.

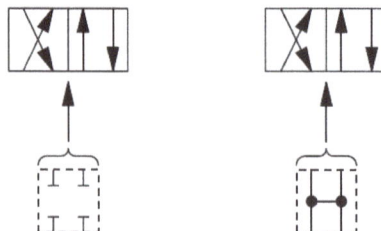

Fig. 7.2- Different Transitional Conditions for a Directional Valve

- **Valve Size:** Valves are primarily sized based on the flow rate. An Oversized valve has a very small active control zone so that it loses controllability. An undersized valve is always saturated and performs like a throttle valve generating heat.

- **Valve Operating Conditions:** Having identical symbols is not the only factor, but also valve size and maximum allowable working conditions must be identical. As an example, if a system that uses a switching valve in an On/Off control mode is upgraded to work in continuous control mode using a proportional valve. This is a different valve that may require upgrading the filtration system as well to meet the cleanliness requirement for a proportional valve.

- **Electrical Components on EH Valves:** Review electrical design parameters (voltage, current, power, switching time, cyclic rate, etc.)

7.2- BP-Valves-02-Maintenance Scheduling

Unless otherwise is stated by components and systems manufacturer, Table 7.1 provides guidelines for *scheduling* preventive maintenance actions for hydraulic valves.

#	Preventive Maintenance Actions	Daily	Weekly	Monthly	Biannually	Annually
1	Clean around and outside surface	✔	✔	✔	✔	✔
2	Check for proper connection with the hydraulic lines.		✔	✔	✔	✔
3	Check for proper connection with the electrical lines (if found)		✔	✔	✔	✔
4	Valve major maintenance and testing.				✔	✔

Table 7.1- BP-Valves-02-Maintenance Scheduling

7.3- BP-Valves-03-Installation and Maintenance

Hydraulic *valves*, particularly the *electro-hydraulic* ones, are complex and sensitive components. They are the components that perform the control functions. So, any faulty operation of a hydraulic valve will affect the overall performance of the system. Therefore, they should be kept well-maintained to assure proper system functionality. Standards **ISO/DIS 4411 and ISO 4411:2019** provide detailed information about valve installation and adjustment. However, the best practices list "**BP-Valves-03-Installation and Maintenance**" provides guidelines for *installation and maintenance* of hydraulic cylinders.

BP-Valves-03-Installation and Maintenance:
1. Proper Valve Disassembly.
2. Proper Valve Inspection and Maintenance.
3. Proper Valve Assembly.
4. Proper Hydraulic Connections
5. Proper Electrical Connections
6. Proper Valve Adjustment.

7.3.1- Proper Valve Disassembly

Hydraulic Valves are very sensitive to contamination. So, when disassembling a hydraulic valve for purposes of maintenance or testing, keep the surrounding area clean is highly recommended. As shown in Fig. 7.3, the following guidelines help secure clean housekeeping:

- **Cleaning Towels (1):** Towels for cleaning purposes must be industry specified and lint-free tissues or special material, e.g. 100% cotton.

- **Protective Plates and Dust Caps (2):** If a valve is shipped between departments or stored somewhere, it should be covered by suitable protective plates or plastic caps. Such protective elements are removed only immediately prior to installation.

Fig. 7.3- Proper Housekeeping when disassembling a Hydraulic Valve

7.3.2- Proper Valve Inspection and Maintenance

Clean Outside Surfaces of the Valve: Wherever an EH valve is used, particularly proportional and servo ones, it is highly advisable to keep the surrounding area clean and organized. It's worth it to hire a person dedicated to cleaning the outside surfaces of the valves and the surrounding area on daily basis. That helps prevent dirt from getting inside the valve or metallic chips from being attracted by the electric part of the valve. Therefore, one very important routine maintenance action is to keep the outside surfaces of the valve clean. An example of a dirty surrounding area is shown in Fig. 7.4.

Fig. 7.4- Improper Maintenance Practices Kept the Valve Dirty

Last chance Filter Inspection: As shown in Fig. 7.5, in servo valves and enhanced performance proportional valves, there must be a miniature filter installed usually in the inlet pressure and called *last chance filter*. It is mandatory to inspect and clean this filter during maintenance.

Fig. 7.5- Last Chance Filters in Servo Valves

7.3.3- Proper Valve Assembly

As shown in Fig. 7.6, when *mounting* valves, the following best practices shall be considered:

- **Gravity (1):** Horizontal is preferred to avoid spool drifting under its weight.
- **Line-Mounted Valves (2):** A line mounted valve must not use the connecting lines as support. Connecting lines must have independent support.
- **Actuator-Mounted Valves (3):** For a valve that is mounted on top of an actuator, avoid mounting a valve where the spool is in the direction of motion of the actuator. This avoid affecting the spool position by the vibration of the actuator.
- **Surface-Mounted Valves (4):**
  - Whether using a subplate (as in industrial applications) or directly on machine chasses (such as in mobile applications), the flatness and surfaces finish must be in accordance with the valve manufacturer's recommendations.
  - Mounting surface shall be away from possible mechanical stresses to avoid distortion.
  - Manifolds and subplates should be securely mounted.
- **Flange-Mounted Valves (5):** When welding a flange to a pipe, remove the flange seal until the flange has cooled down then replace the seal.
- **Mounting Torque:** Bolting torque mustn't exceed manufacturer specifications in order to avoid valve body distortion. Clearances inside valves are very small and over torqueing may result in spool seizure.
- **Space Surrounding:** the following should be considered:
  - Enough clearance for wrench and/or bolt access and electrical connections.
  - Access for removal, repair or adjustment.
- **Components Surrounding:** Consider minimizing the chance of damaging the valve by surrounding mechanical operating devices.
- **Electrical Connections:** Shall be in accordance with appropriate standards (e.g. IEC 60204-1 or manufacturer standard) and be designed with the suitable protection class (e.g. in accordance with IEC 60529).

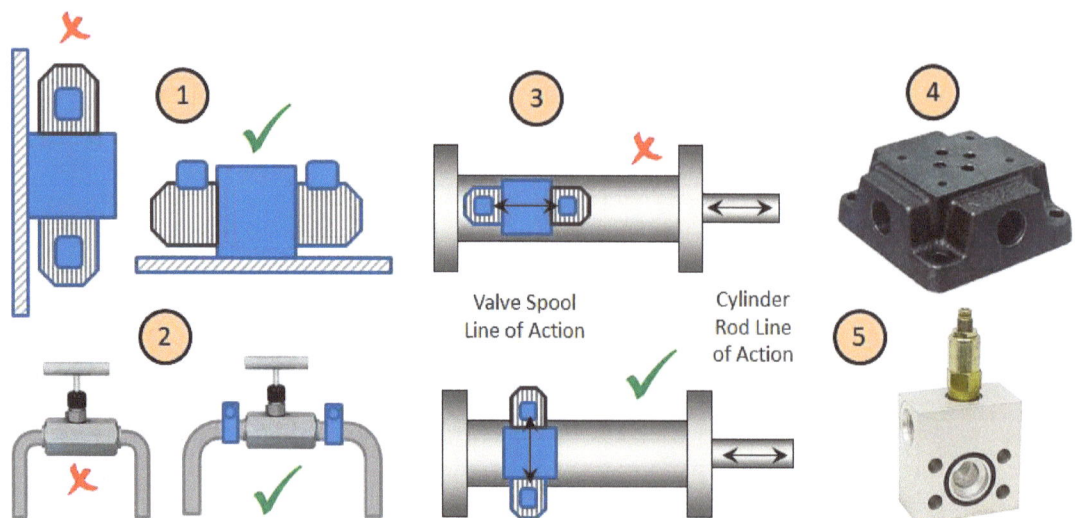

Fig. 7.6- Best Practices for Proper Valve Mounting

Follow Manufacturer's Instructions: In addition to the aforementioned general guidelines, it is highly advisable to review the manufacturer's instructions for specific valve installation. Referring to Fig. 7.7 The following example is excerpted from Hydraforce user's manual of a cartridge valve.

Step 1: Remove cartridge from packing. Inspect O-Ring to ensure there is no damage such as cuts or nicks. Check if the backup rings fit tightly within the O-ring groove.

Step 2: Immerse the hydraulic portion of the cartridge valve in clean compatible oil to lubricate the seals. Dry seals could cause the backup rings to spin out of the groove which damage or cut the seal.

Step 3: Insert the cartridge valve into the cavity and tighten by hand in a clockwise manner. You should be able to screw it in with little resistance up to the O-ring.

Step 4: Continue to screw in the cartridge with a torque wrench and tighten to the torque specified in the catalog. Tightening the valve above the specified torque value, this may cause the spool or poppet to stick.

Step 5: Install the waterproof O-ring on the cartridge hex if one is required.

Step 6: Install the coil with the lettering facing the hex nut. Install the coil nut and tighten the coil nut to specified torque.

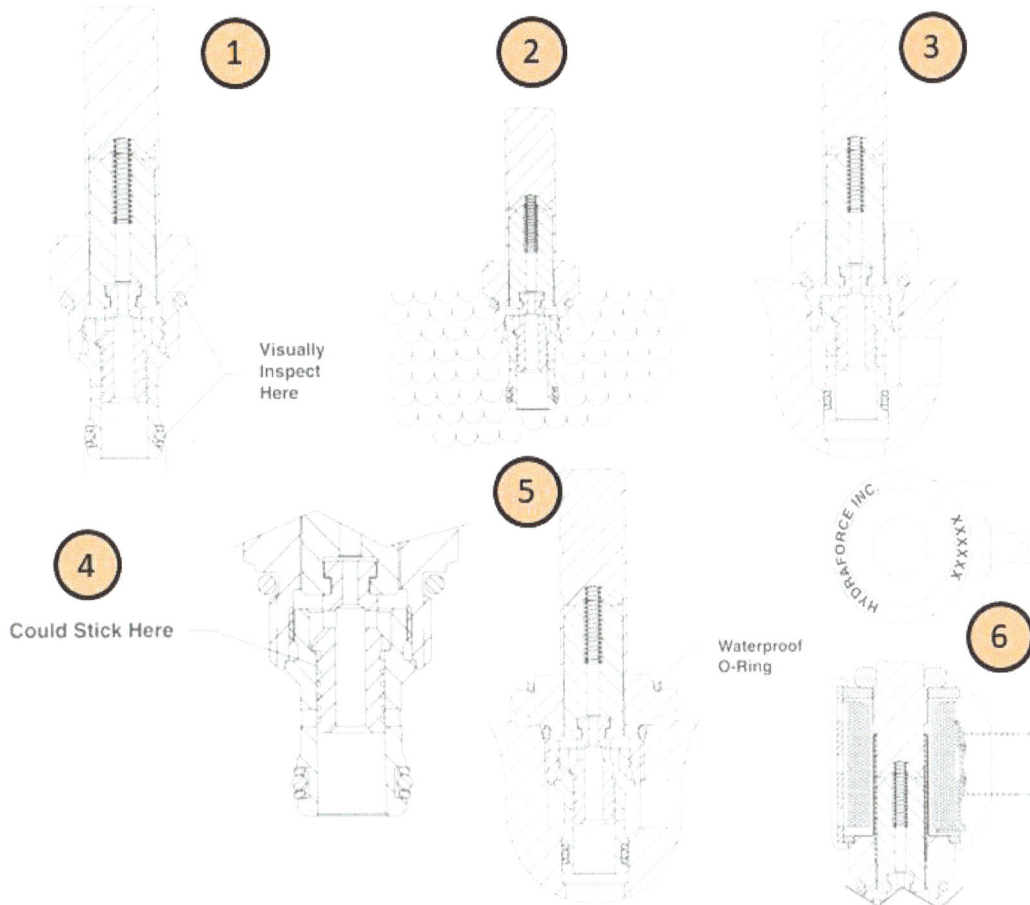

Fig. 7.7- Example of Manufacturer's Instructions for Valve Installation (Courtesy of Hydraforce)

7.3.4- Proper Hydraulic Connections

When connecting a hydraulic valve to the relevant hydraulic lines, a great care and attention should be considered. The following are examples of what should be considered:

Example 1 – Main Line Connection: As shown in Fig. 7.8, a four-way valve has the following main ports, **P** (Pressure), **T** (Tank), and **A** & **B** (Load Lines). If the load lines aren't connected to the right sides of the actuator, the actuator motion will be reversed. If the tank line isn't connected to the tank, the valve won't operate.

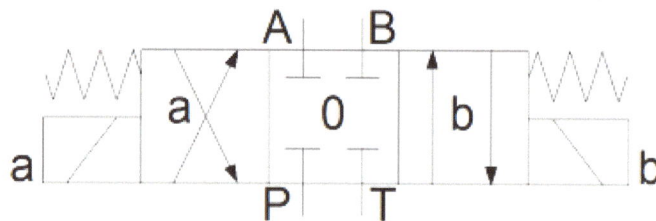

Fig. 7.8- Main Ports on a 4-way Directional Valve

Example 2 – Pilot Line Connection: As shown in Fig. 7.9, a pilot operated directional valve requires supply and drainage of pilot pressure. As explained thoroughly in volume 1 of this series of textbooks, depending whether the supply and the drain of the pilot pressure is internal or external, four configurations of the valve are found. If the supply and drain of pilot pressure are mistakenly connected, they may result in immediate valve malfunction.

For externally piloted valves, pressure filter without by-pass immediately before the valve in the control pressure line to port "X" (β10 > 75 to ISO OR cleanliness class 5 to NAS 1638). For cleanliness requirements of the oil used with EH valves, refer to Volume 2: Electro-Hydraulic Components and Systems Chapter 10).

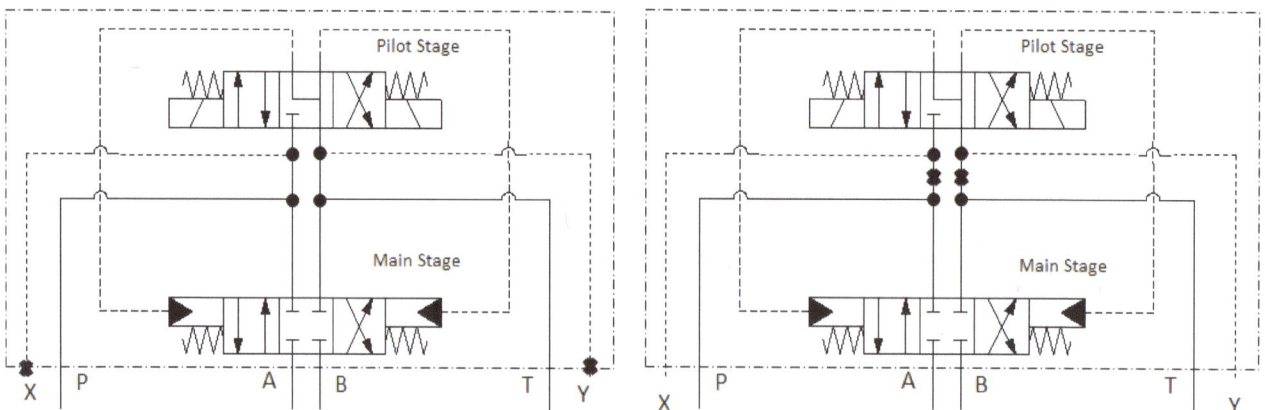

Fig. 7.9- Supply and Drain of Pilot Pressure in a Pilot-Operated DCV

7.3.5- Proper Electrical Connections

As shown in Fig. 7.10, the following bullets present best practices for connecting hydraulic valves with their relevant electrical connections:

- **Plug-in Connectors (1):** DO NOT connect a valve to flying wires to avoid short circuiting.
- **Plug with Built-in light Indicator (2):** Consider using plug-in connectors rather than permanent wiring to ease valve replacement. The plug must be securely screwed to avoid sparking and solenoid burning. Consider using such plugs to ease valve troubleshooting.
- **Keep Solenoids Covered (3):** Some legacy directional valves have removable solenoid covers. Make sure the solenoids are covered and protected from dust and humidity. Do NOT operate the valve while the solenoid is uncovered.
- **Match the Solenoids with the Actuator's Motion (4):** hydraulic lines A & B and solenoid electrical connections must match the actuator's motion; otherwise unexpected reverse motion could occur.
- **Electrical Cables:** Avoid running electrical cables within hydraulic plumbing or near sources of electrical noise such as electrical motors and variable frequency drivers (VFDs).

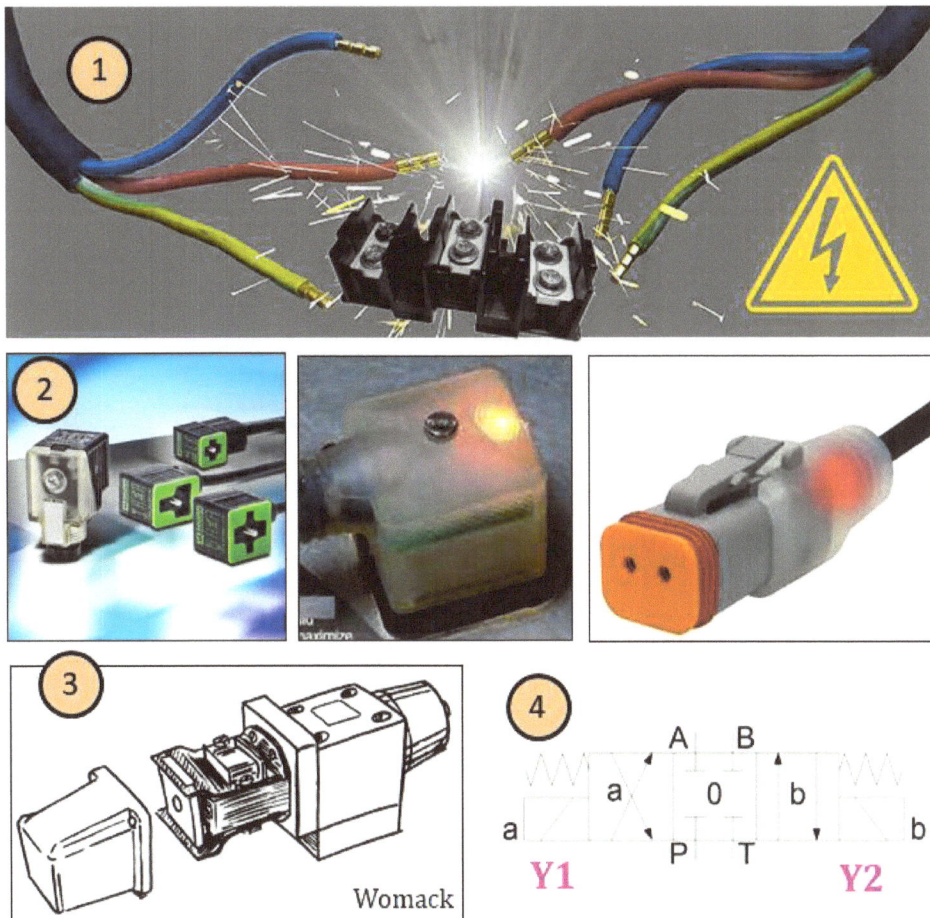

Fig. 7.10- Considerations for Proper Electrical Connection to Hydraulic Valves

7.3.6- Proper Valve Adjustment

Adjustment of hydraulic valves is a very critical process no matter the valve type is. Every valve adjustment affects system performance. Improper valve adjustment may result in potential hazard. Improper adjustment of flow and directional control valves may result in changing the system speed and/or sequence. Improper adjustment of pressure control valves may result in changing the load carrying capacity and potentially unsafe operation due to increase in system pressure. However, a common advice is DO NOT tamper with the relief valve unless it is permitted and done by experienced person.

When valves permit *adjustments* of one or more parameters, the following provisions should be incorporated, as appropriate:
- Means for securing the adjustment.
- Means for locking the adjustment, if required to prevent unauthorized change; or
- Means for preventing adjustment beyond a safe range.

Figure 7.11 shows an example from industry of given instructions for allowable adjustments on a servo valve.

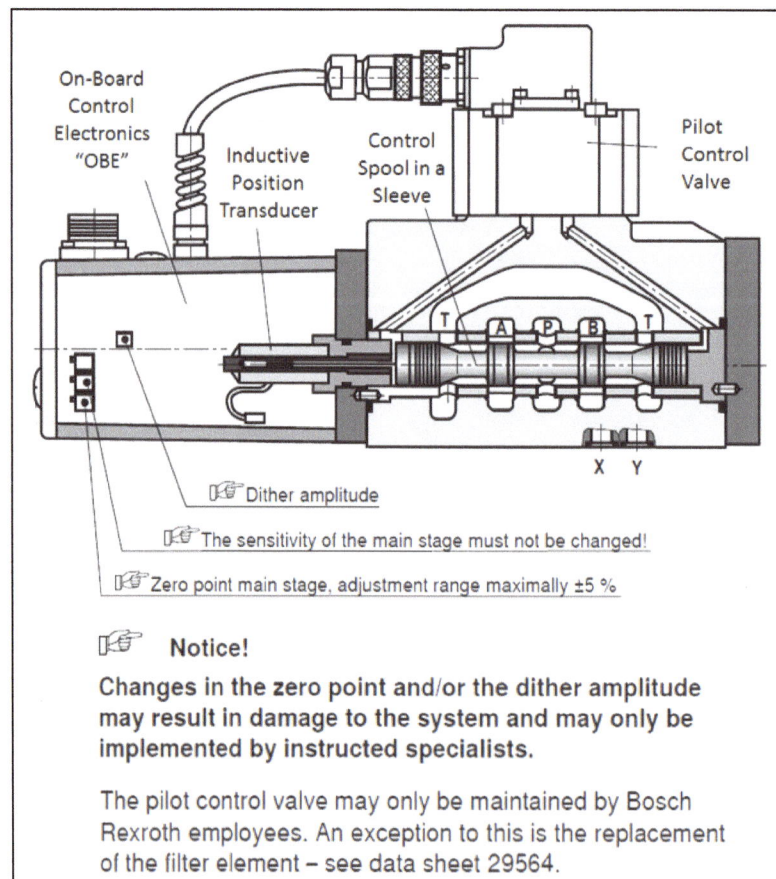

Fig. 7.11- Instructions for Servo Valve Adjustments (Courtesy of Bosch Rexroth)

7.4- BP-Valves-04-Standard Tests and Calibration

This section was graciously granted to this textbook as a Courtesy of "CFC Industrial Training".

CAUTION! The pressure and flow ratings of the diagnostic equipment which will be used to conduct these tests must be at least equal, if not greater than, the pressure and flow ratings of the valve being tested.

Test Conditions: To get comparable and unified test results, tests must be conducted at a predefined temperature (e.g. approximately 130 °F (54 °C.) using hydraulic fluid that has a predefined viscosity (e.g. 32 cSt.).

7.4.1- Flow Control Valve Test

Purposes of the Test: This test procedure will determine the amount of leakage across the ports of a *flow control valve* when it is in the "closed" position.

Test Procedure (Figure 7.12):
- Oil should be trapped between a hand pump (1) and the inlet port of the needle valve.
- Observe pressure gauge (2). The pressure should "hold" for a reasonable length of time.
- There is no general rule-of-thumb regarding leakage rate of needle valves. If it is necessary to determine accurate leakage rates, refer to the valve manufacturer's specifications.
- Open the pressure relief valve and release the pressure between the hand pump and the needle valve.

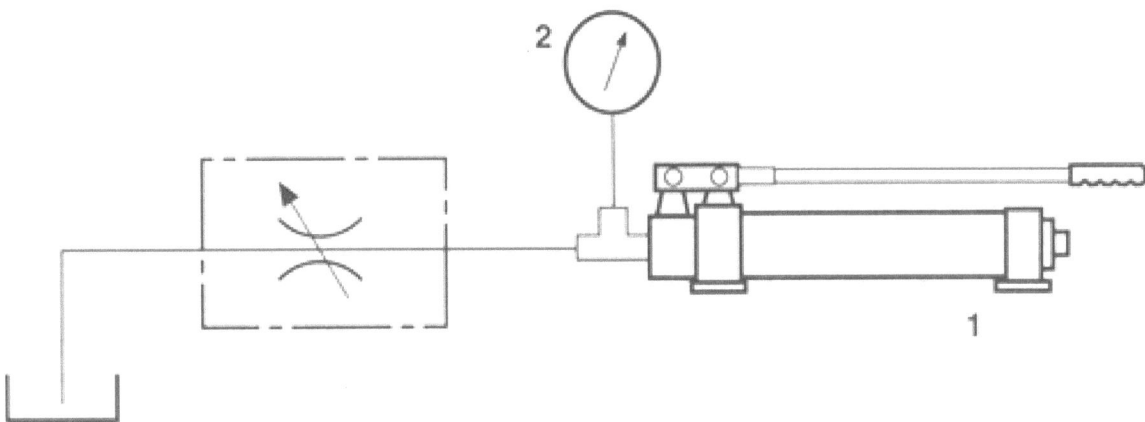

Fig. 7.12- Standard Test Circuit for Flow Control Valve (CFC Industrial Training)

7.4.2- Test for Pressure Relief Valves

Note: tests for pressure relief valves are also applicable for *counterbalance valves* and *sequence valves.*

Purposes of the Test: This test procedure is used to develop the characteristic curve of a direct-operated or a pilot-operated *pressure relief valve* and to verify that the valve is adjusted to manufacturer's specification.

Test Procedure (Figure 7.13):
❖ **A: check if the valve responds to pressure adjustment:**
  ▪ Keep the actuator in deadheaded "stalled" position.
  ▪ Keep the PRV fully opened.
  ▪ Keep the needle valve (3) fully closed (turn clockwise).
  ▪ Turn on the pump and observe pressure gauge (1), it should read tank line.
  ▪ Close the PRV gradually to maximum pressure → **p** increases accordingly.

❖ **B: develop the valve characteristic curve:**
  ▪ Keep the actuator in deadheaded "stalled" position.
  ▪ Keep the PRV in its required adjustment.
  ▪ Keep the needle valve (3) fully opened (turn counterclockwise).
  ▪ Turn on the pump and observe pressure gauge (1), it should read tank line.
  ▪ Gradually and incrementally, based on the pressure, close the needle valve (3).
  ▪ Plot the flow **"Q"** at the flow meter (4) versus the pressure "**p**" at the pressure gauge (1).

Test C: Purposes of the Test: This test procedure will determine the amount of leakage across the ports of a direct-operated or a pilot-operated pressure relief valve. Figure 7.x shows the test circuit diagram.

Test Procedure (Figure 7.14):
  ▪ Oil should be trapped between a hand pump (1) and the inlet port of the relief valve.
  ▪ Observe pressure gauge (2). The pressure should "hold" for a reasonable length of time.
  ▪ Open the pressure relief valve and release the pressure between the hand pump and the relief valve.

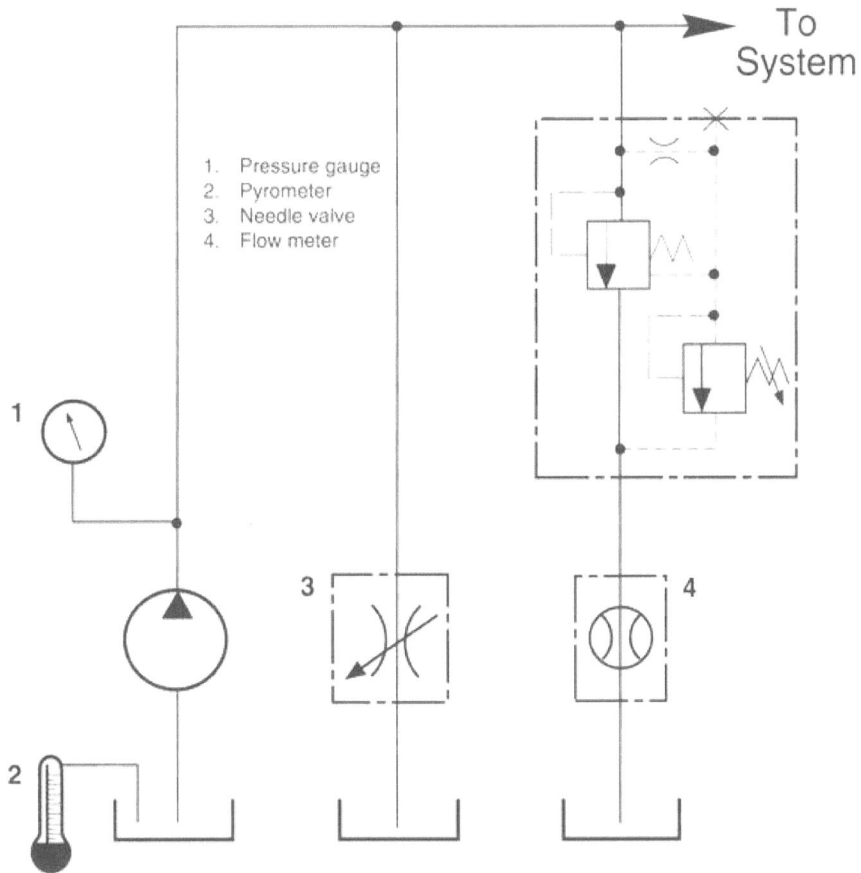

1. Pressure gauge
2. Pyrometer
3. Needle valve
4. Flow meter

Fig. 7.13- Circuit for Pressure Relief Valve Adjustment Test (CFC Industrial Training)

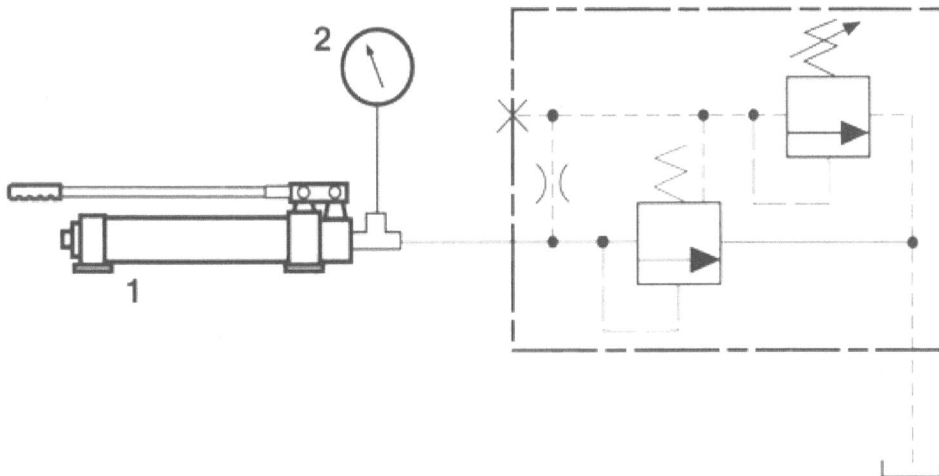

Fig. 7.14- Circuit for Pressure Relief Valve Leakage Test (CFC Industrial Training)

7.4.3- Test for Pressure Reducing Valves

Purposes of the Test: This test procedure will determine:
- Check the operation valve.
- The pressure in the external drain-line.
- If there is a restriction in the external drain-line.
- If there are pressure surges in the external drain-line.

Test Procedure (Figure 7.15):
- Open needle valve fully (turn counterclockwise).
- Turn on the pump.
- Gradually close the needle valve and observe the pressure readings at gauges 1,2 and 3.
- Gauge 3 supposed to read approximately constant reduced pressure (p_3) until the valve is in the inactive zone, where the differential pressure (p_2-p_3) is reduced below the value specified by the valve manufacturer.
- Gauge 2 shows if there is blockage or surge leakage in the spring chamber vent line.

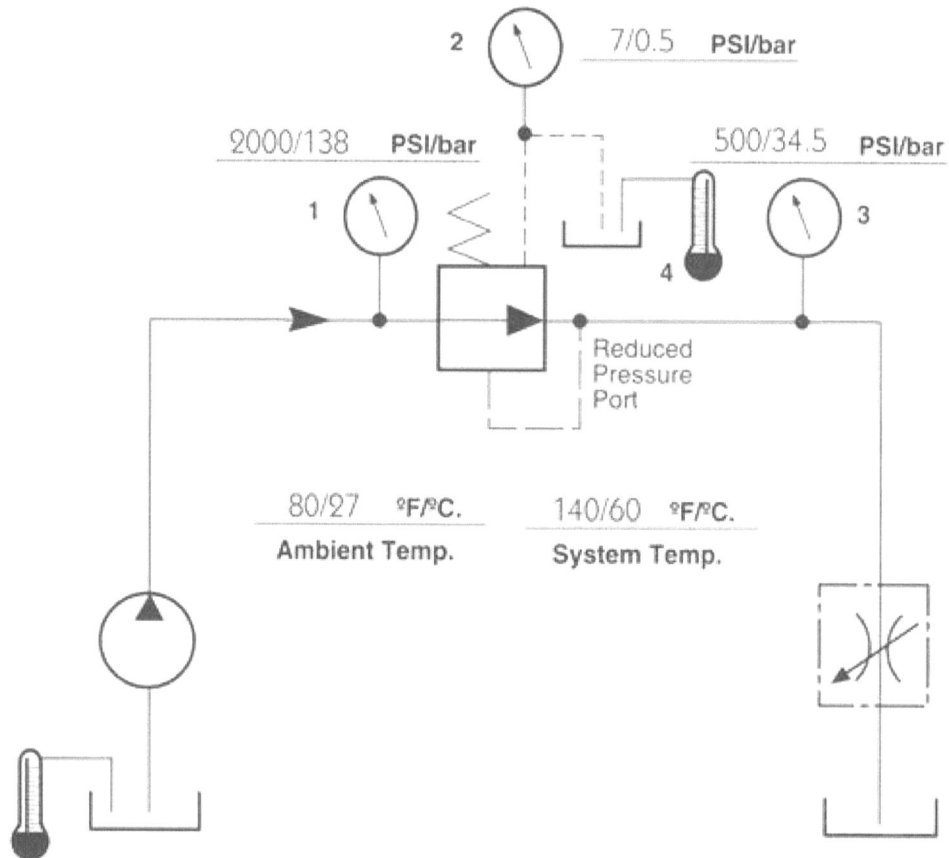

Fig. 7.15- Sample Test Worksheet for a Reducing Valve (CFC Industrial Training)

7.4.4- Test for Directional Control Valve

Purposes of the Test: This test procedure will determine the amount of leakage across the port "**P**" and the other ports in the valve.

Test Procedure (Figure 7.16):
- Block the port on the leakage path under the test. In this figure P-B.
- Pressurize this pass to predefined maximum pressure of the valve.
- Observe pressure at the gauge.
- The pressure should "hold" for a reasonable length of time.
- Open the pressure relief valve and release the pressure between the hand pump and the valve.

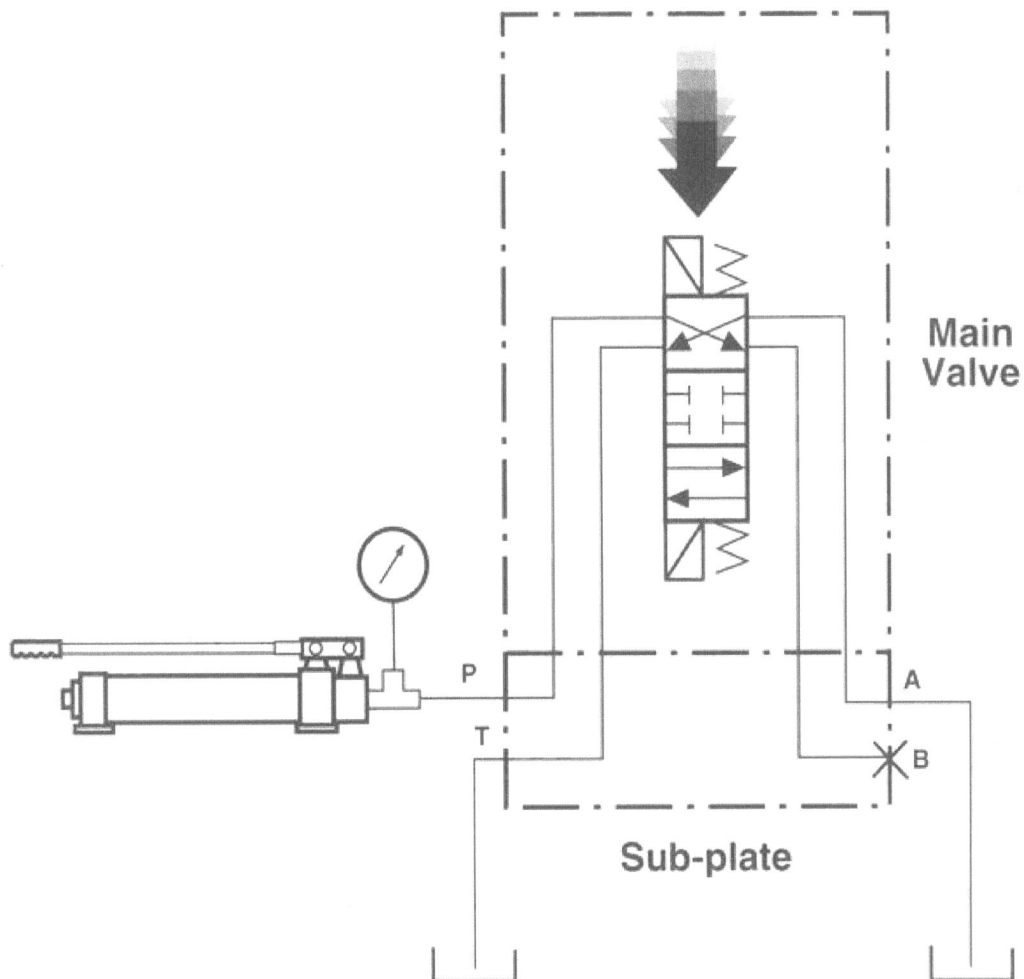

Fig. 7.16- Leakage Test for an Industrial Directional Valve (CFC Industrial Training)

7.4.5- Test for Proportional and Servo Valve

Regulating Standard (ISO 10770-1-2009): Describes methods for determining the performance characteristics of electrically modulated, hydraulic, four-port directional flow-control valves. This type of electrohydraulic valve controls the direction and flow in a hydraulic system.

Type of Tests: There are number of standard tests that can be conducted to check the performance of EH valves. These tests are conducted if required to diagnose some malfunctions of the valve. It can also be conducted frequently as part of preventive maintenance plans. The following are the standard tests:
  - Static Test: Flow Gain or Pressure Gain.
  - Static Test: Limiting Power.
  - Static Test: Hysteresis.
  - Static Test: Null leakage and Proof Pressure.
  - Dynamic Test: Step response.
  - Dynamic Test: Frequency response.
  - Dynamic Test: Fail-Safe Function.

Standard Test Conditions: Table 7.2 shows the standard test conditions enforced by the standard. These test conditions must be maintained to receive comparable results.

Parameter	Condition
Ambient temperature	20 °C ± 5 °C
Fluid cleanliness	Solid contaminant code number shall be stated in accordance with ISO 4406.
Fluid type	Commercially available mineral-based hydraulic fluid (i.e. L - HL in accordance with ISO 6743-4 or other fluid with which the valve is able to operate)
Fluid viscosity	32 cSt ± 8 cSt at valve inlet
Viscosity grade	Grade VG32 or VG46 in accordance with ISO 3448
Pressure drop	Test requirement ± 2.0 %
Return pressure	Shall conform to the manufacturer's recommendations

Table 7.2- Standard Test Conditions for Four-Ports EH Valve (Courtesy of NFPA)

Standard Test Circuit: Fig. 7.17 shows the standard test circuit that contains all the required components for testing, condition monitoring, and outputting the test results.

Caution: A valve MUST NOT be tested with electrical signal only before hydraulic power is supplied including pilot pressure to the first stage. Otherwise, the wire spring in a flapper nozzle stage may be severely stretched. Same is applicable for proportional valves in order to prevent spool friction without lubrication.

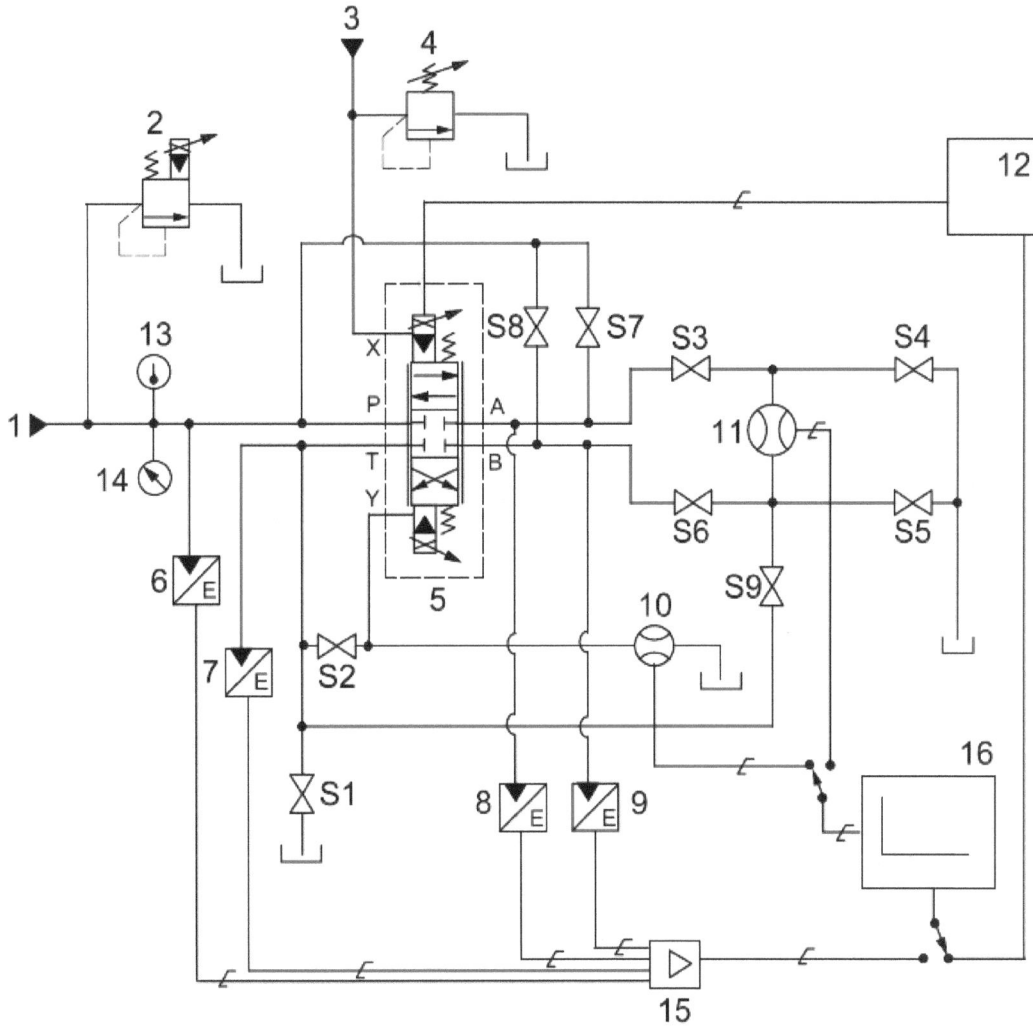

Fig. 7.17- Standard Test Circuit for Four-Ports EH Valve (Courtesy of NFPA)

#	Component	#	Component
1	Main Flow Source	14	Pressure Gauge
2	Main Relief Valve	15	Signal Conditioner
3	External Pilot Flow Source	16	Data Acquisition
4	External Pilot Relief Valve	S1 – S9	Shut-off Valves
5	Unit under Test	A, B	Load Ports
6 - 9	Pressure Transducer	P	Supply Pressure Port
10, 11	Flow Transducers	T	Tank Port
12	Signal Generator	X	Pilot Control Pressure Supply
13	Temperature Indicator	Y	Pilot Control Pressure Return

Test Accuracy: Instrumentation shall be accurate to within the limits shown in **Class B of ISO 9110-1:**

- Electrical resistance: ± 2 % of the actual measurement.
- Pressure: ± 1 % of the valve's rated pressure drop to achieve rated flow.
- Temperature: ± 2 % of the ambient temperature.
- Flow: ± 2,5 % of the valve's rated flow.
- Input signal: ± 1,5 % of the electrical input signal required to achieve the rated flow.

Commercial Servo and Proportional Valve Testers: the following couple of examples shows typical commercial proportional valve testers available in the market.

Example 1: Stationary Valve Tester: As shown in Fig. 7.18, the test rig is integrated with a hydraulic power supply. The features of this test rig in brief are as follows:
- Used for both manual and automatic tests.
- Used for both steady state and dynamic performance tests.
- Plug and Play test procedure, no high skills required.
- Suitable for various valve electrical input signals.

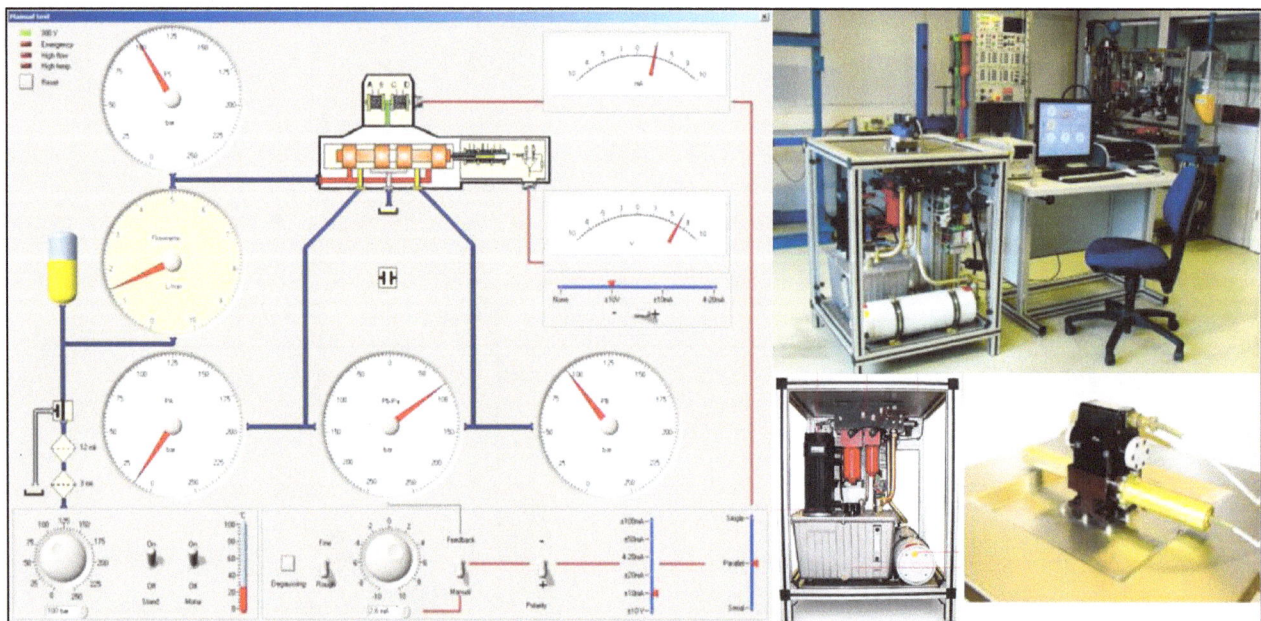

Fig. 7.18- Stationary Proportional and Servo Valve Tester (dietzautomation.com)

Example 2: Portable Valve Tester: As shown in Fig. 7.19, a portable servo valve tester is also available. It is enclosed in a protective case. It can be used for in field tests. It can be hooked to a data acquisition box for data monitoring, recording, and printing.

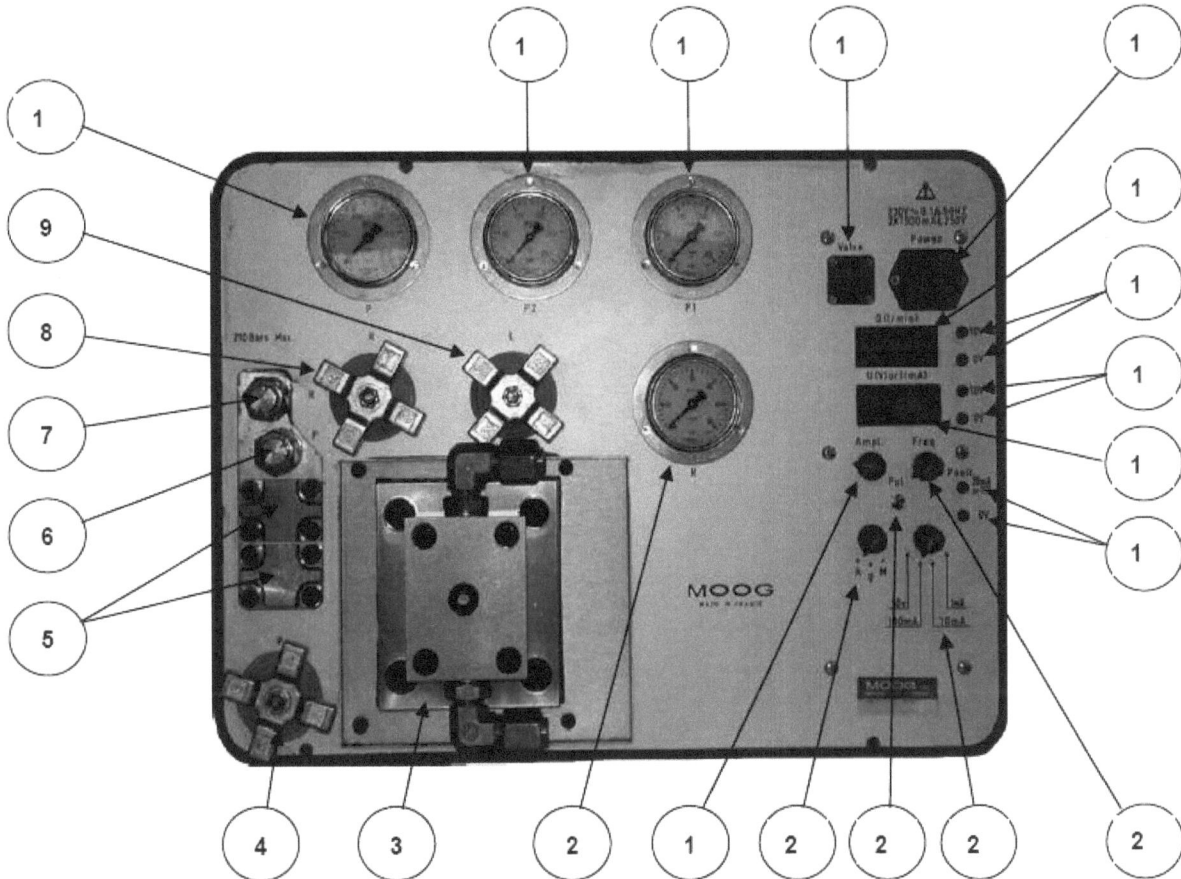

1	Current amplitude control	9	Loading valve	17	Multimeter monitor output
2	Return pressure gauge (P_T)	10	Supply pressure gauge (Pa)	18	Digital display multimeter
3	Manifold plate	11	Control pressure gauge on C2 (P2)	19	Spool position output
4	Main shut-off	12	Control pressure gauge on C1 (P1)	20	Frequency control adjustment
5	Two internal filters	13	Servovalve command Signal input connector	21	Scale current or voltage selector
6	Pressure fitting	14	Test box power supply VAC	22	Reverse polarity switch
7	Return fitting	15	Digital display flow meter	23	Manual/automatic mode selector
8	Return shut-off	16	Flow meter monitor output		

Fig. 7.19- Portable Servo Valve Analyzer Series F087-127 (Courtesy of Moog)

7.5- BP-Valves-05-Transportation and Storage

It is always advisable to refer to the manufacturer's specifications for longer term storage. However, the storage locations for valves, particularly EH ones, must be dry with low humidity and dirt-free. For storage longer than 3 months, fill the housing with compatible preservative oil and seal the valve with protective covers. The stored valves must be checked from time to time to make sure that they are kept in the right position.

Chapter 8

Maintenance of Accumulators

Objectives

This chapter provides guidelines for **accumulator's** selection, replacement, maintenance scheduling, installation, testing, storage and transportation. This chapter is supported by examples and figures granted by leading fluid power manufacturers.

Brief Contents

8.1-BP-Accumulators-01-Selection and Replacement
8.2-BP-Accumulators-02-Maintenance Scheduling
8.3-BP-Accumulators-03-Installation and Maintenance
8.4-BP-Accumulators-04-Standard Tests and Calibration
8.5-BP-Accumulators-05-Transportation and Storage

Chapter 8: Maintenance of Accumulators

The following set of best practices provide general guidelines and may not be applicable for all cases. They are not intended to replace the instructions given by the component manufacturer. It is strongly advisable to adhere to instructions provided by the manufacturer.

8.1- BP-Accumulators-01-Selection and Replacement

Accumulators are originally specified to work safely, store the required amount of energy (volume of oil under certain pressure), and to respond to the system needs on steady and dynamic modes. This section is not intended to provide design solutions or accumulator sizing. When replacing an existing accumulator, the following shall be considered:

Nominal Size and Initial Gas Pressure: Nominal size and initial gas pressure of the accumulator shouldn't be changed. Otherwise, accumulator may not be function correctly due to the change in the stored energy.

Information Marked on the Accumulators: Make sure the following information is marked permanently on the body of the vessel.
- Manufacturer's name, logo, serial number, and date of manufacture (month/year).
- Total shell volume (*Nominal Volume*), expressed in liters or gallons.
- Allowable operating temperature range expressed in ^0C or ^0F.
- Permissible maximum pressure expressed in megapascals or bar or psi.
- **Note 1:** The place and method of stamping shall not reduce strength of the shell. If space is not available to provide all this information on the accumulator, it shall be provided on tags that are permanently attached to the accumulators.

8.2- BP-Accumulators-02-Maintenance Scheduling

Unless otherwise is stated by components and systems manufacturer, Table 8.1 provides guidelines for *scheduling* preventive maintenance actions for hydraulic accumulators.

#	Preventive Maintenance Actions	Daily	Weekly	Monthly	Biannually	Annually
1	Check line connections			✔	✔	✔
2	Check mounting components			✔	✔	✔
3	Check gas precharge pressure	The gas precharge pressure should be checked at least once during the first week of operation. If there is no loss of gas precharge pressure, it should be rechecked in 3 to 4 months. Thereafter, it should be checked at least once a year.				

Table 8.1- BP-Accumulators-02-Maintenance Scheduling

8.3- BP-Accumulators -03-Installation and Maintenance

Gas loaded accumulators are very sensitive components in hydraulic systems because the issue of energy storage. Improper treatment of such components may lead to hazard of explosion. Therefore, they should be kept well-maintained to assure the proper and safe system functionality. What follows is a list of the ISO standards about accumulators:

- **ISO 5596** Hydraulic Fluid Power - Gas-loaded accumulators with separator Ranges of pressures and volumes and characteristic quantities.
- **ISO 10945** Hydraulic Fluid Power - Gas-loaded accumulators - Dimensions of gas ports.
- **ISO 10946** Hydraulic Fluid Power - Gas-loaded accumulators with separator selection of preferred hydraulic ports.
- Standard **NFPA/T3.4.7 R2-2000 (R2014)** provides detailed information about accumulator's installation and adjustment.

However, the best practices list "**BP-Accumulators-03-Installation and Maintenance**" provides guidelines for *installation and maintenance* of hydraulic cylinders.

BP-Accumulators-03-Installation and Maintenance:
1. Review Disassembly Instructions.
2. Review Inspection Instructions.
3. Review Assembly Instructions.
4. Review Charging Instructions.

8.3.1- Disassembly Instructions

The following set of bullets provides general guidelines for accumulator disassembly:

- **Safety:** Accumulators are inherently dangerous due to high pressure gases and fluids.

- **DO NOT ATTEMPT** to maintain these systems unless:
  o Adequately trained.
  o Wear appropriate safety equipment.
  o Read and understand all instructions provided by the manufacturer.

- **Discharge Fluid:** The free release of the fluid based on how fast the gas expands results if flow surge. Therefore, gas-loaded-accumulator discharge rates shall be related to the manufacturer's rating.

- **Discharge Gas:** DO NOT attempt to open the accumulator without completely discharging the gas. Attach the proper charging and gauging unit and completely relieve the gas precharge (referee to Figure 1.76 in Chapter 1).

Case Study for Accumulator Disassembling: Figure 8.1 shows an example of a bladder accumulator. The listed items in this figure are used in the following accumulator disassembly example.

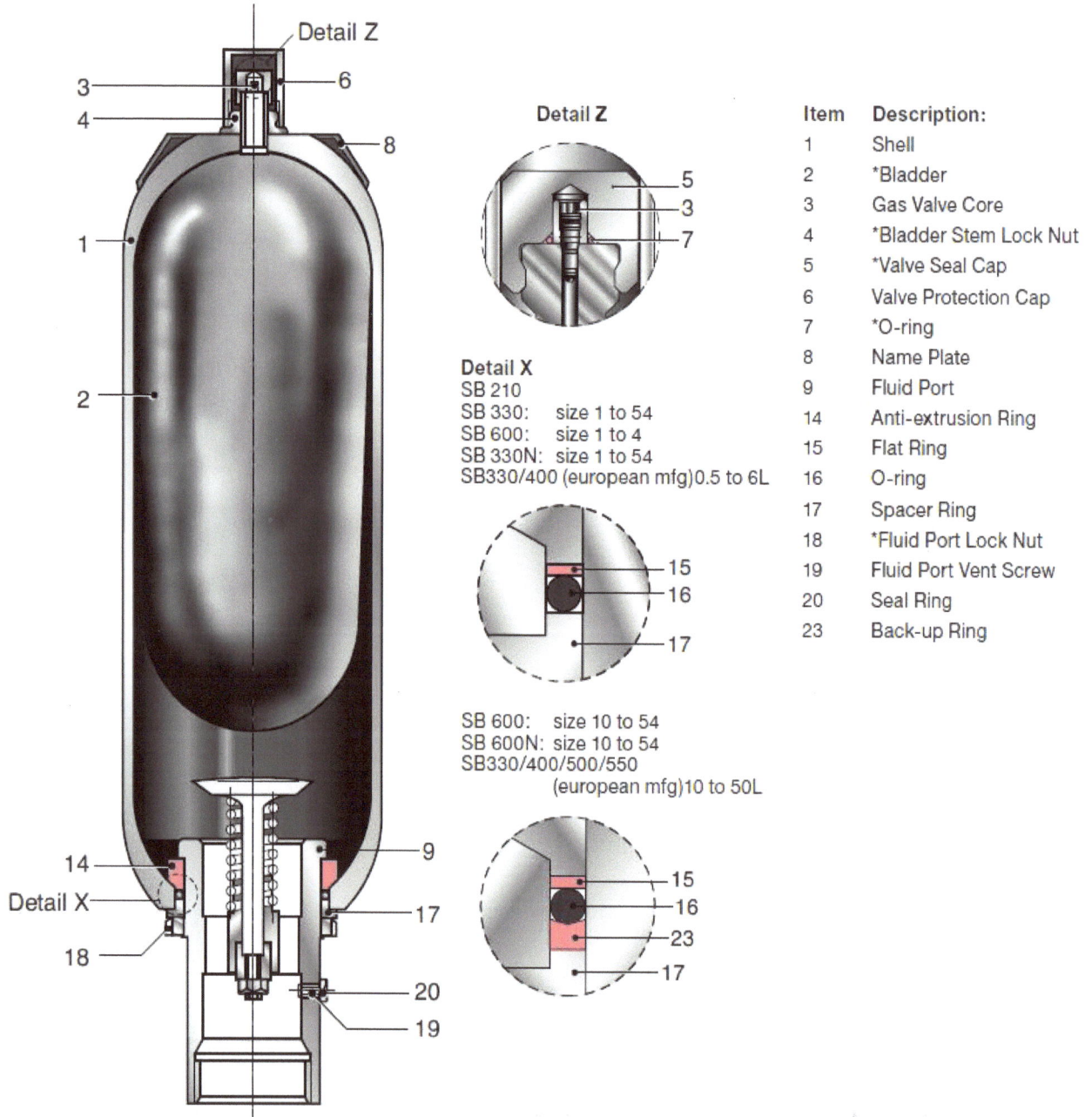

Item	Description:
1	Shell
2	*Bladder
3	Gas Valve Core
4	*Bladder Stem Lock Nut
5	*Valve Seal Cap
6	Valve Protection Cap
7	*O-ring
8	Name Plate
9	Fluid Port
14	Anti-extrusion Ring
15	Flat Ring
16	O-ring
17	Spacer Ring
18	*Fluid Port Lock Nut
19	Fluid Port Vent Screw
20	Seal Ring
23	Back-up Ring

Detail X
SB 210
SB 330: size 1 to 54
SB 600: size 1 to 4
SB 330N: size 1 to 54
SB330/400 (european mfg)0.5 to 6L

SB 600: size 10 to 54
SB 600N: size 10 to 54
SB330/400/500/550
(european mfg)10 to 50L

Fig. 8.1- Bladder Accumulator (Courtesy of Hydac)

Figure 8.2 shows and example of a given instructions for disassembling the bladder accumulator that is shown in the previous figure.

A: After removal from the system, place the accumulator in a vice or secure it to a workbench. Remove **valve protection cap** *(item 6)* and unscrew **valve seal cap** *(item 5)*. Attach the proper HYDAC Charging and Gauging Unit and completely relieve the gas precharge *(refer to HYDAC Charging and Gauging brochure #02068202)*. Remove **gas valve core** *(item 3)* by using the gas valve core tool.

B: Unscrew **vent screw** *(item 19)* and remove **seal ring** *(item 20)*. Unscrew **lock nut** *(item 18)* by using spanner wrench. Remove **spacer ring** *(item 17)*. If necessary, tap spacer ring with a plastic hammer to loosen.

C: Loosen **fluid port** *(item 9)* and push it into the shell. Remove **back-up ring**, *(item 23)* where applicable, **O-ring** *(item 16)* and **flat ring** *(item 15)* from fluid port.

D: Pull **anti-extrusion ring** *(item 14)* off fluid port and remove it through fluid side opening by folding it in half.

E: Remove **fluid port** *(item 9)*.

F: Remove **bladder stem lock nut** *(item 4)* and **name plate** *(item 8)* from the gas side. Remove **bladder** *(item 2)* from fluid side. It may be necessary to fold the bladder lengthwise to remove it.

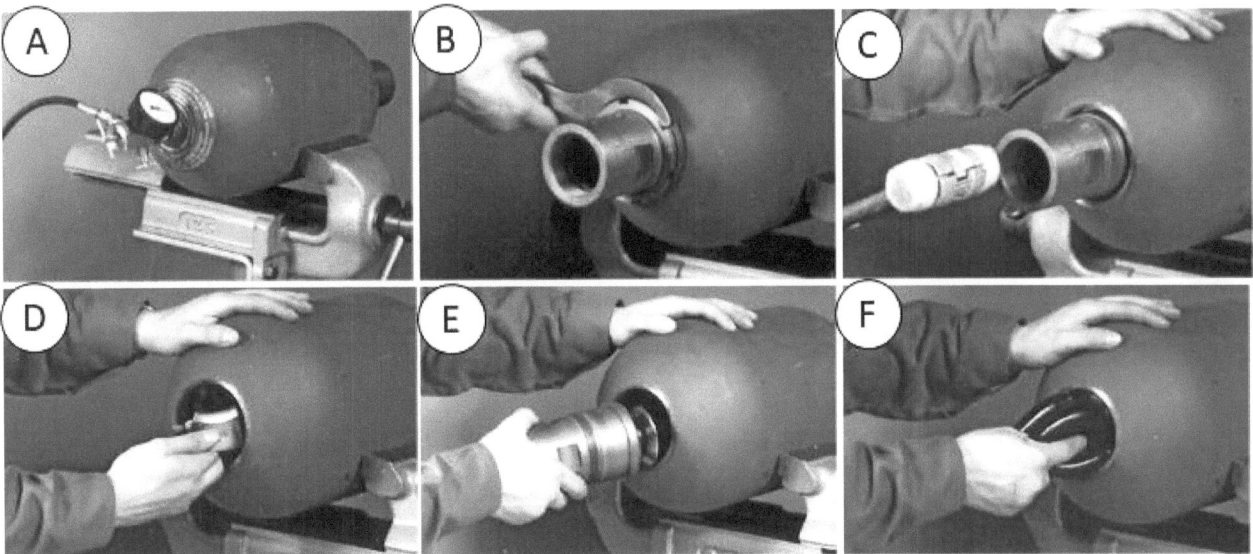

Fig. 8.2- Bladder Accumulator Disassembling Instructions (Courtesy of Hydac)

8.3.2- Inspection Instructions

After disassembling an accumulator, it should be inspected for any sort of defects or damages. The following are inspection guidelines:

- **Shell:**
  - Inspect shell interior to ensure it is free of debris, rough spots, or damage marks.
  - Exterior for any sign of damage such as scratches.
  - If any interior or exterior sign of damage is found, contact the supplier for proper repair or replacement instructions.

- **Bladder:**
  - The bladder must be visually inspected for lateral grooves and deep chafe marks.
  - Fill the bladder with nitrogen or compressed air to its natural shape and inspect for leakage.
  - If leakage occurs, first check the gas valve core and replace it if necessary.
  - If leakage still occurs, then the bladder must be replaced.
  - If any are found, the bladder should be replaced.
  - **Note:** Bladders can't be repaired.

- **Fluid Port:**
  - Visually inspect poppet, threads, sealing surfaces, seats, and anti-extrusion ring for any sort of damage.
  - If any damage is found, the fluid port should be replaced.

- **Seals:**
  - New seals should always be used whenever reassembling any bladder accumulator.

- **Other Parts:**
  - Visually inspect for damage and replace if necessary.

8.3.3- Assembly Instructions

Assembling: usually assembling is the reverse operation of disassembling.

Safety Block: An accumulator isn't a line mounted type of component. As shown in Fig. 8.3, It must be mounted on a safety base that contains the following components:

1. A pressure relief valve to adjust maximum charging pressure for the accumulator.
2. A pressure gauge to read the pressure.
3. An isolation valve to isolate the accumulator from the circuit if needed. It could be lockable if the accumulator function is critical.
4. A manual discharge valve to discharge the accumulator manually if needed.
5. Optionally, a normally open solenoid-operated 2/2 poppet type directional valve for automatic discharging the accumulator in case of emergency or machine shutdown.
6. Gas valve must be guarded by a sealing cap. Otherwise, if it is accidentally broken due to any reason, accumulator may act as a rocket.
7. If fittings are needed to connect the accumulator to the system, originally specified fittings must be used and never use commercial fittings from off-the-shelf.
8. Gas-loaded accumulators shall not be modified by machining, welding or any other means.

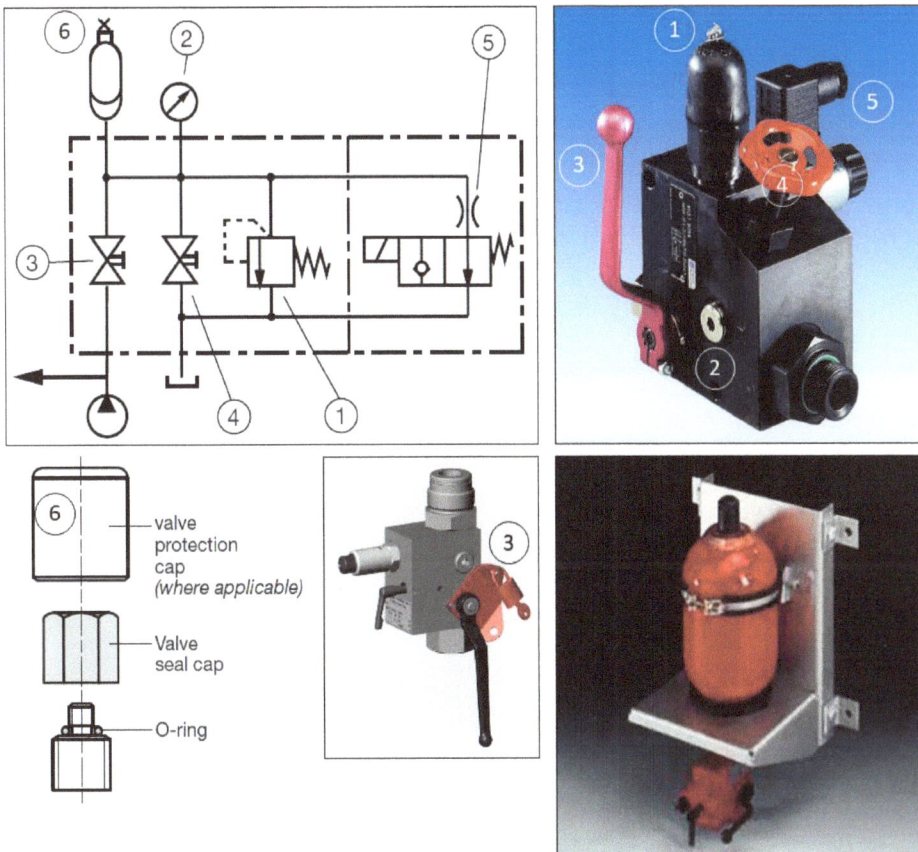

Fig. 8.3- Safety Base (Manifold) for an Accumulator (Courtesy of Hydac)

Mounting Guidelines: Referring to Fig. 8.4, the following guidelines should be considered when mounting an accumulator.

- **Mounting Position:** The best and safest mounting position for any accumulator is vertical with the fluid port downwards (i.e. gas valve upwards). Reason of that are:
  - o **Bladder Accumulators:** Horizontal orientation can result in increased bladder bag wear due to friction with the internal surface of the pressure shell, along with any system contaminants which collect in the shell. Horizontal mounting tends to damage the bladder, particularly for accumulators with cyclic charge/discharge. The extent of the damage is proportional to the cycle rate, compression ratio, and storage capacity. Distortion of the bladder can cause fluid to trap in cavities and has no way to get out.
  - o **Piston Accumulators:** horizontal mounting causes uneven and accelerated seal wear.
  - o **All Acc.:** In case of the gas valve failure, the accumulator will try move downward to the ground rather than moving horizontally acting as a rocket potentially causing sever damage or injury to surroundings.

- **Mounting Location:** Accumulators should be located in a place away from movement of other components with the least possibility of physical damage.

- **Space Above the Accumulator:** Maintain a space of at least 20 cm above the gas valve for testing and recharging services.

- **Supports:** Gas-loaded accumulators and any associated pressurized components shall be supported in accordance with the instructions of the accumulator supplier. However, accumulators must be well clamped. It is absolutely forbidden welding of supports or machining on the accumulator shell.

- **Heat Shield:** Accumulators should not be exposed to direct sunlight. Install a sunshade. Protect them with a baffle from other radiant heat sources such as die casting machines, furnaces, steam boilers, etc.

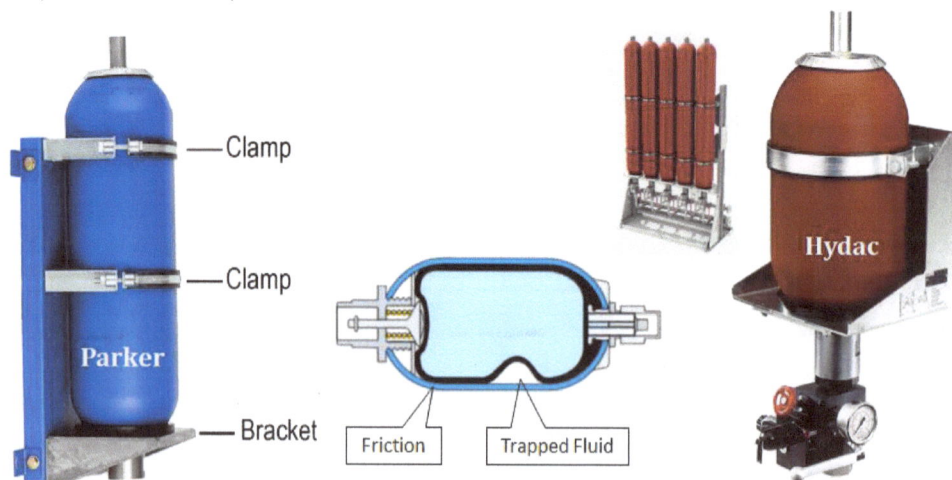

Fig. 8.4- Proper Mounting of Accumulators

8.3.4- Charging Instructions

The following set of bullets provide guidelines for charging accumulators:

- **Gas Type:** Never use Oxygen or air because of possibility of generating fire in case of accumulator accidental explosion. Charge only with *Nitrogen* (N_2) with highest purity (class 4) of purity level 99.99% by volume.

- **Initial Gas Pressure:** All hydraulic accumulators require charging before first use Initial gas pressure (also called *Pre-charge Pressure*) of an accumulator should be slightly less than the minimum working pressure of the hydraulic system. The reason is to make sure the separator (piston, bladder, or diaphragm) will always above and doesn't hit the bottom part of the accumulator every time the accumulator ejects all the fluid out. Unless otherwise stated by the manufacturer, consider the rule of thumb for initial gas pressure is to be 100 psi (7 bar) below minimum system working pressure for piston accumulators, and 90% of the minimum system pressure for bladder accumulators. As an example of manufacture's instruction for initial gas pressure, Hydac accumulators are pre-charged based on the application regardless the type of accumulator as follows:
  - For Energy Storage Applications   **$P_0 = 0.9\, P_1$.**
  - For Shock Absorption Applications   **$P_0 = 0.9\, P_m$.**
  - For Pulsation Damping Applications   **$P_0 = 0.9\, P_m$.**
  - Where **P_1** and **P_m** are the minimum and median system pressure; respectively.

- **Maximum Gas Pressure:** Maximum accumulator gas pressure can't exceed the maximum system pressure. Otherwise, the relief valve will be open during the charging process. It can be adjusted by a separate PRV mounted on the accumulator base. The second important consideration is that the maximum gas pressure should respect the allowable *Compression Ratio* for each accumulator type. Unless otherwise being stated by the manufacturer, the following are recommended compression ratios: (8-9) for the piston accumulators, (4-5) for bladder accumulators, and (1.5-2) for diaphragm accumulators.

- **Temperature Effect:** To ensure that the recommended initial gas pressure is maintained, even at relatively low or high operating temperatures, the initial gas pressure should be adjusted for temperature as follows:
  - **Fahrenheit:** $P_{0,T0} = P_{0,T2} \times (T_0 + 460)/ (T_2 + 460)$
  - **Celsius:** $P_{0,T0} = P_{0,T2} \times (T_0 + 273)/ (T_2 + 273)$
  - Where:
  - **T_0** = Precharge temperature.
  - **T_2** = Maximum operating temperature.
  - **$P_{0,T0}$** = Gas precharge Pressure at precharge temperature.
  - **$P_{0,T2}$** = gas precharge pressure at maximum operating temperature.

- **Charging Device:** Figure 8.5 shows an example of a *charging* and a *gauging* kit that is used to service hydraulic accumulators. Charging kits are used to fill accumulators with nitrogen. Gauging kits test the precharge pressure to ensure it is within a safe range. The gauging unit is available with male or female cap nut.

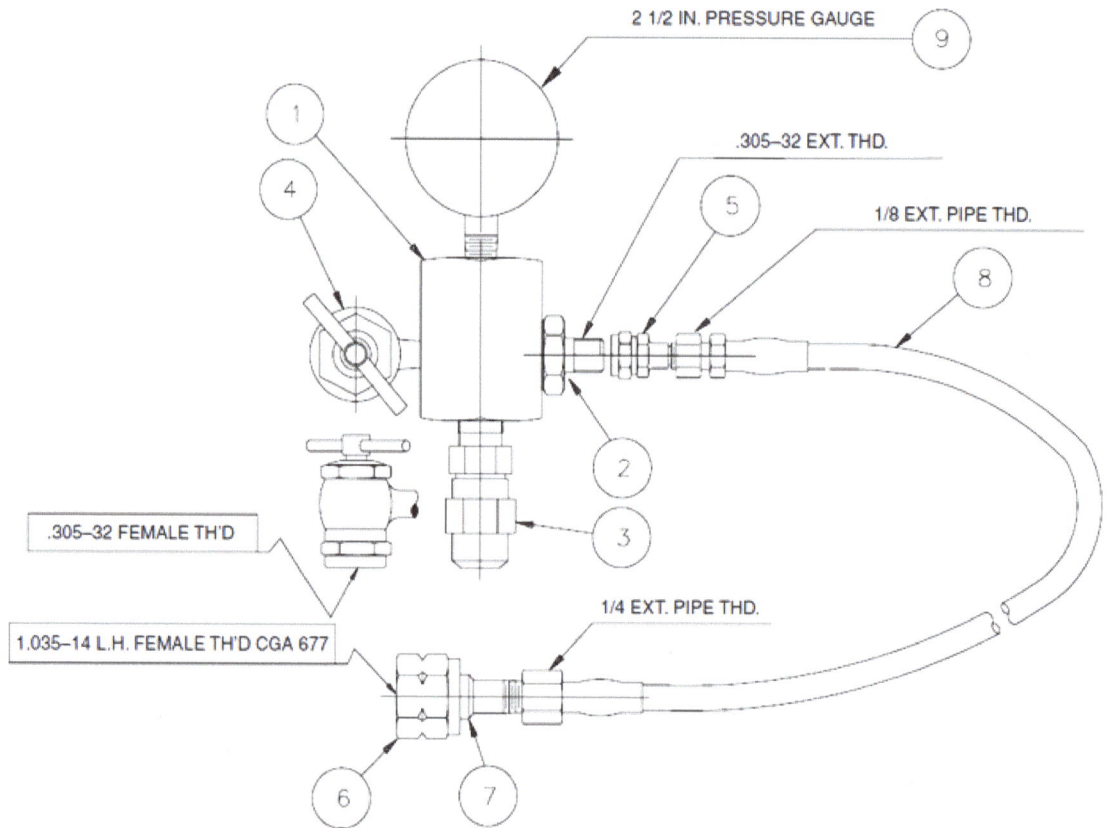

1	Adapter Cross
2	Gas Valve Assembly
3	Bleed Valve
4	Gas Chuck
5	Valve Connector
6	Nitrogen Bottle Nut with LH Connection
7	Gas Tank Nipple (CGA 677)
8	Charging Hose (10 ft.)
9	2.5" Pressure Gauge

Fig. 8.5- Accumulator Charging Kit (Courtesy of Parker)

Figure 8.6 shows a servicing kit from other manufacturer.

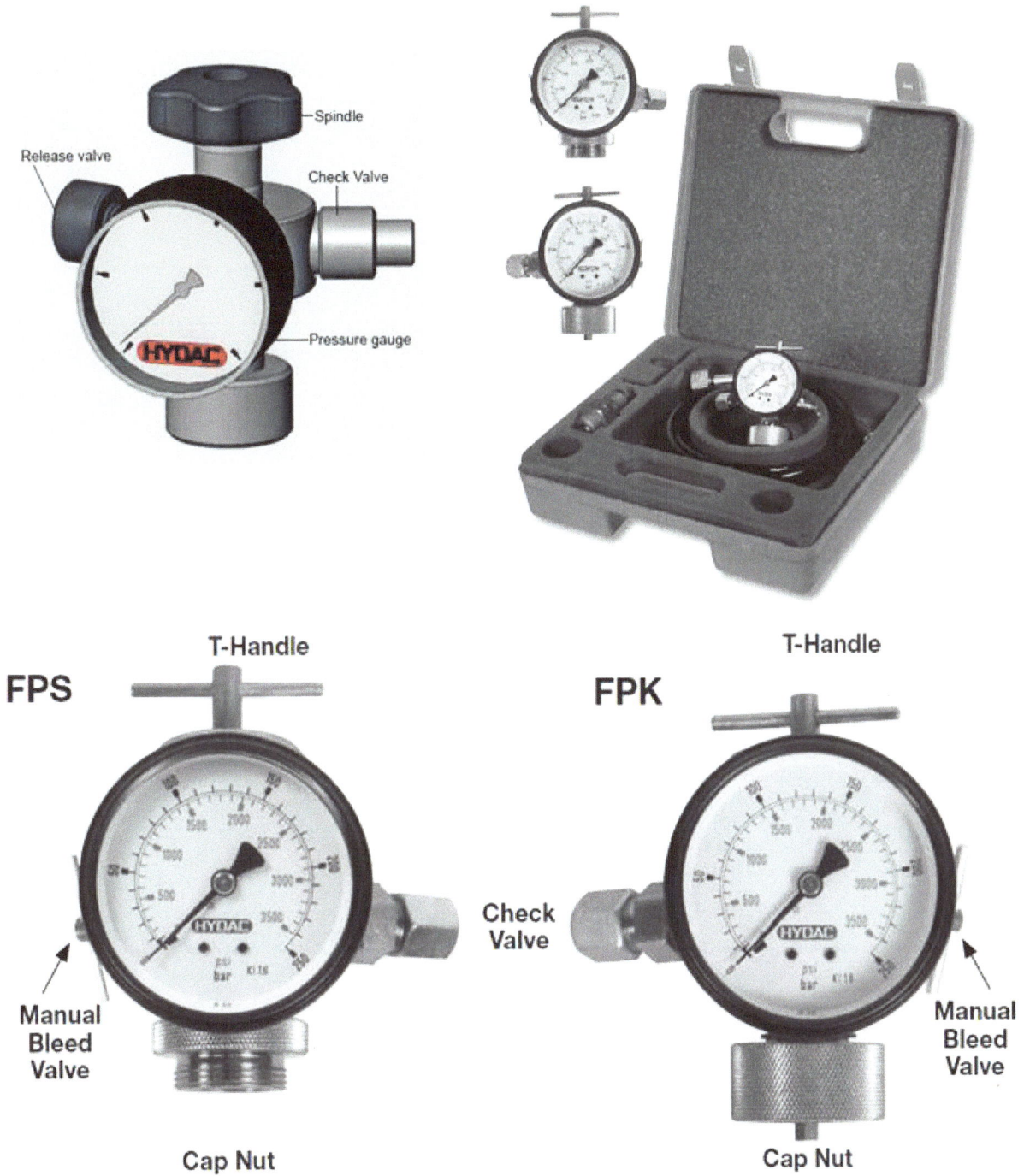

Fig. 8.6- Accumulator Charging and Gauging Kit (Courtesy of Hydac)

- **Charging Process:** Referring to Fig. 8.7, the following step-by-step process should be considered when charging an accumulator.

Connection to Accumulator:
o   Prior to gas pre-charging process, verify that the oil side of the accumulator is at zero pressure.
o   Remove the sealing cap of the accumulator's gas valve.
o   Make sure the T-Handle (6) on the charging unit is fully closed.
o   Connect the charging kit (1) to the accumulator.
o   Open the knop (4) of the charging kit and measure the current precharge pressure using the pressure gauge (3). If the pressure is higher than specified, press and release the release button (2) and lower precharge pressure to specified value. If pressure is below specified value, then proceed to the following step.

Connection to Nitrogen Cylinder:
o   Make sure that the shut-off valve of the nitrogen bottle (5) is fully closed.
o   Connect the other end of the charging assembly to the nitrogen bottle.

Charging the Accumulator:
o   Gradually and slowly open the shut-off valve of the nitrogen bottle (5).
o   Adjust the T Handle on the pressure regulator (6) to desired precharge value.
o   Allow gas to flow slowly into the accumulator. The first 20-25 psi should take 2-3 minutes. Then adjust the charge rate to be one minutes for every increase of 25 psi.
o   For every 25 psi, close the shutoff valve (5), give 5 seconds for the pressure to stabilize, and check the charge pressure at the pressure gauge.
o   Repeat the previous two steps until the accumulator is charged to the assigned initial gas pressure.
o   If needed, reduce excess charge pressure by carefully open the bleeding release valve (2) to get the precharge pressure reduced.
o   Wait for proper time until the accumulator is cooled to the room temperature and the initial gas pressure to reach equilibrium. When adjusting an existing gas initial gas pressure allow 5 to 10 minutes for the initial gas pressure to reach equilibrium. When charging for the first time or after performing maintenance work, allow 20 to 30 minutes for the initial gas pressure to reach equilibrium.
o   Close the shut-off valve (5) of the nitrogen cylinder and the knob (4).
o   Disconnect the charge assembly from the nitrogen bottle then from the accumulator.

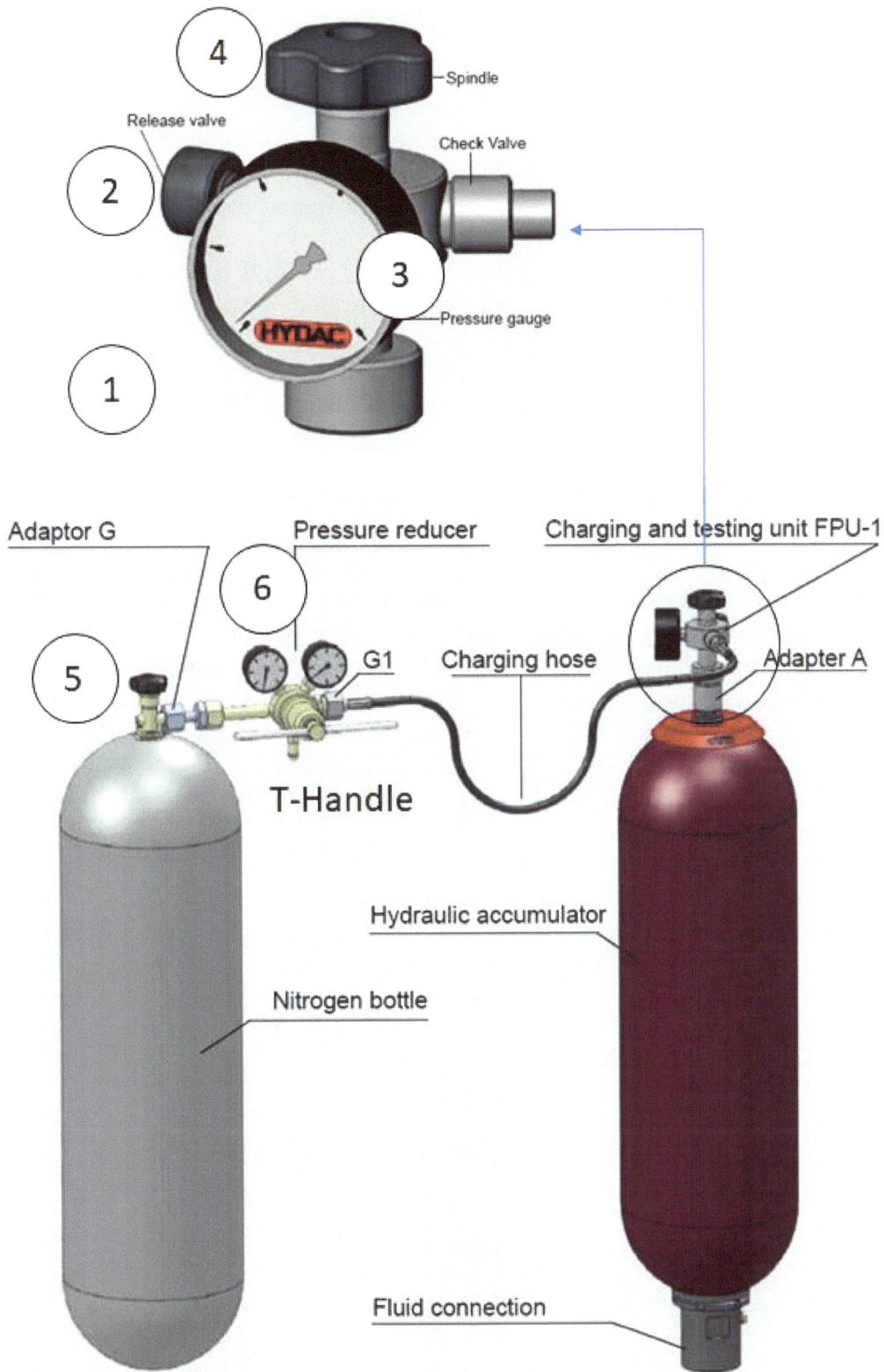

Fig. 8.7- Accumulator Charging Process (Courtesy of Hydac)

8.4- BP-Accumulators -04-Standard Tests and Calibration

Accumulators are, by default, tested by the manufacturer in different ways against fatigue life and explosion, etc. It is not the responsibility of end users to perform such tests. However, a test certificate should be provided by the manufacturer and must be kept in a safe place. The following set of bullets provide guidelines for testing an accumulator:

Testing Precharge Pressure: When the hydraulic system is turned off, a charging rig with a gauge can be used to check the pre-charge level as previously discussed.

Testing Working Temperature: Referring to Fig. 8.8, To verify that the accumulator is operating properly, the side of the shell can be checked with a temperature gun or infrared camera as follows:

- Normally, the bottom half or two-thirds (in the fluid chamber) should be hotter than the top half (the gas chamber).
- If heat is only concentrated at the very bottom part of the accumulator (in the fluid chamber), this means that the accumulator may be overcharged. The reason is that the fluid chamber in an overcharged accumulator contains less amount of oil volume because the differential pressure between the maximum system pressure and the precharge pressure is small.
- If heat goes all the way to the top, this means that the accumulator is undercharged. The reason is that the undercharged accumulator contains more oil volume because the differential pressure between the maximum system pressure and the precharge pressure is large.
- If no heat is measured, that may b because:
  o Automatic discharge device is stuck open.
  o The precharge pressure may be above the pump compensator setting (in a pressure compensated pumps) or above the relief valve setting (in fixed displacement valves).

Fig. 8.8- Accumulator Temperature Test

8.5- BP-Accumulators -05-Transportation and Storage

The following set of bullets provides guidelines for shipping and transportation of accumulators:

- **Transportation:**
  o Accumulators should be gas and fluid discharged.
  o If charged, gas precharge should not exceeds 150 psi (10 bar) during its shipping and transportation, and proper labeling should be used to indicate the charging pressure.
  o Keep the isolation valve closed.
  o Review instructions for lifting. However, a gas valve or fluid port on an accumulator shouldn't be considered as liftings points in any case.

- **Storage:**
  o In order to prepare an accumulator for proper storage for future use, the accumulator must be discharged from both gas and oil before storage.
  o Piston accumulators are recommended to be stored in vertical position to prevent the seals from developing a set (flat spot) on the side that the piston weight is exerted.
  o Accumulators should be stored in a cool, dry place away from sun, ultraviolet and fluorescent lights as well as electrical equipment. Direct sunlight or fluorescent light can cause the seals to dry out.
  o The ideal temperature for storage is 70°F.

Chapter 9

Maintenance of Reservoirs

Objectives

This chapter provides guidelines for **reservoirs** selection, replacement, maintenance scheduling, installation, testing, storage and transportation. This chapter is supported by examples and figures granted by leading fluid power manufacturers.

The following topics are discussed in Chapter 2 in Volume 4 "Hydraulic Fluids Conditioning" of this series of textbooks:
- Contribution of Hydraulic Reservoirs
- Configurations of Hydraulic Reservoirs
- Construction of Hydraulic Reservoirs
- Design of Hydraulic Reservoirs
- Hydraulic Reservoir Design Case Study

The following topics are discussed in Chapter 9 in Volume 6 "Troubleshooting and Failure Analysis" of this series of textbooks:
- Hydraulic Reservoirs Inspection
- Hydraulic Reservoirs Troubleshooting
- Hydraulic Reservoirs Failure Analysis

Brief Contents

9.1- BP-Reservoirs-01-Selection and Replacement
9.2- BP-Reservoirs-02-Maintenance Scheduling
9.3- BP-Reservoirs-03-Installation and Maintenance

Chapter 9: Maintenance of Reservoirs

The following set of best practices provide general guidelines and may not be applicable for all cases. They are not intended to replace the instructions given by the component manufacturer. It is strongly advisable to adhere to instructions provided by the manufacturer.

9.1- BP-Reservoirs-01-Selection and Replacement

Hydraulic reservoirs are selected and designed based on application, size of pumps, intended heat dissipation and many other factors that are discussed in Volume 4.

The following set of bullets briefs the best practices for replacing an existed hydraulic reservoir with other one:

- New reservoir should be able to host same amount of oil volume with the same free space above the oil surface as the old one.
- New reservoir should have the same surface area as the old one to maintain the same surface cooling capacity.
- DO NOT change the placement of the suction and the return lines.
- DO NOT change the distribution of oil heaters inside the reservoir.
- Maintain the same electrical connections with the reservoir (fluid level, fluid temperature, etc.)

9.2- BP-Reservoirs-02-Maintenance Scheduling

Unless otherwise is stated by components and systems manufacturer, Table 9.1 provides guidelines for *scheduling* preventive maintenance actions for hydraulic reservoirs.

#	Preventive Maintenance Actions	Daily	Weekly	Monthly	Biannually	Annually
1	Check and repair any source of external leakage	✔	✔	✔	✔	✔
2	Check oil level in the tank and make up oil of needed	✔	✔	✔	✔	✔
3	Clean around and the outer surface including the sight glasses		✔	✔	✔	✔
4	Check proper connection to hydraulic lines.			✔	✔	✔
5	Check electrical connections			✔	✔	✔
6	Check the state of the breather filter and the presence of the filler cap **(Note 1)**.			✔	✔	✔
7	Wash the suction strainers				✔	✔
8	Reservoir Major Maintenance **(Note 2).**: ▪ Drain the oil ▪ Clean inside the reservoir ▪ Repaint with compatible paint ▪ Change the cover gasket ▪ Refill the reservoir ▪ Clean/Replace air filter ▪ Clean the strainer				✔	✔

Table 9.1- BP-Reservoirs-02-Maintenance Scheduling

Note 1: Most breathers are consumable elements. So, they must be changed periodically. As it has been stated previously, the frequency of changing the breather is based on application and ambient conditions. The rule of thumb is to change the filter every three months in dirty environments, such as a foundries, and every six months in cleaner environments.

Note 2: The best time to check a reservoir is during the routine shutdown time of the equipment.

9.3- BP-Reservoirs-03-Installation and Maintenance

Well-designed reservoirs can cause problems if not installed properly. The following set of bullets provides guidelines about the best practices for hydraulic reservoirs maintenance and installation.

Clean Outside the Reservoir: A hydraulic reservoir has the prime function of hosting the required oil volume. However, it has other important functions such as dissipating heat through the side surfaces. If the reservoir is not periodically cleaned, its ability to dissipate heat is reduced over time and it will act as a heat sink. Accumulating heat inside the reservoir stimulates hydraulic fluid degradation and generating sludge and varnish. Cleaning outside the reservoir must include cleaning all other devices that are assembled on the reservoir such as gauges, level indicators, etc.

Clean Inside the Reservoir: Reservoir cleanliness is a vital maintenance task because:
- Any contaminants left in the reservoir will circulate in the system.
- pumps and other components are designed with close clearances to operate in the 2000 to 5000 psi range, and they have become much more sensitive to dirt.

Therefore, as shown in Fig. 9.1, hydraulic reservoirs must be cleaned before assembly. If no specific instructions are found, the information below provides guidelines for cleaning a reservoir:
- **Drainage:** A reservoir must be completely drained in order to make sure no contaminants or debris are left behind. Failure to do so may result in damage to the pump(s) and/or other components on startup.
- **Sandblasting:** is an effective method to remove built in debris *sandblasting* results in some dust left over after the process. This dust must be properly wiped. In some cases, sandblasting may not be permitted, so find out prior doing it.
- **Flushing:** Without flushing, sandblasting may cause more trouble later than if the reservoir was not cleaned at all. After sand blasting, if permitted, reservoirs must be thoroughly flushed with oil or a solvent to remove all traces of sand. For more details about flushing process, refer to Volume 3 of this series of textbooks.
- **Painting:** Unpainted steel surfaces will form rust due to water condensation. The inside and outside surfaces of hydraulic steel reservoirs should be painted. Paint must be compatible with the fluid being used.

Before Cleaning   After Cleaning

Fig. 9.1- Cleaning of Hydraulic Reservoirs

Clean Suction Strainers: Suction strainers (Fig. 9.2) are used to prevent large contaminant particles from entering the pump. If the outer surface of the strainer is coated by brown varnish, this means that the oil has been operated at high temperature and/or exceeds the specified number of working hours. Washable strainers should be cleaned using approved compatible solvent. A strainer can be dried by blowing clean air from inside to outside.

Cleaning Solvents: If using a solvent for cleaning strainers, make sure it's recommended for hydraulic systems and specified by the system manufacture. Even very small amounts of the wrong solvent can affect certain hydraulic fluid additives.

Cleaning Towels: Make sure to use an industry specified and lint-free towels when cleaning out the reservoir.

Fig. 9.2- Clean versus Dirty Suction Strainer

Cleaning/Replace Air Breather Cartridge: The visual indicator shows the blockage level of the air breather element.

Change Gaskets: If the upper cover or the side cover (manhole) is removed to clean inside the reservoir, the gasket must be replaced. Compatible seals and gaskets for fluid in the system will eliminate external leaks. Gaskets below connecting flanges must also be changed.

Clean Around the Reservoir: If there is oil around the reservoir, check for leaks around fittings and make the necessary repairs. Clean up leakage immediately to avoid slippage or fire potential. Do not return any spillage to the reservoir (it causes contamination). Put mopped spillage and associated materials into proper disposal containers.

Space Around:
- Reservoirs must be placed in a well vented space to allow heat dissipation.
- DO NOT locate the reservoir under benches, tables, or in cabinets.
- If needed, use an industrial fan to force air around the reservoir.

Space Below: Mount the reservoir on a horizontally level foundation with the reservoir bottom at least 15 cm (6 inches) above the floor level to facilitate fluid changes and air circulation around the reservoir.

Shield from Heat: It is recommended to shield reservoirs from heat sources such as furnaces, steam pipes, steel mills, etc. Reservoirs also must be shielded from harsh environment such as direct sun light, snow, very cold or very hot weather, rain, flood, dust storm, etc. Connecting pipes should also be shielded.

Refill the Reservoir: Filling a reservoir directly from the barrel, thinking a new fluid is clean, is a common mistake. Never add new fluid or make up exiting fluid without passing it through a filter. If no instructions are found for the filter rating, 10-micron with Beta Ratio 75 is recommended. Figure 9.3 shows typical setup that can be used for such a purpose. The transmission hoses should be clean and have a protective cap between usages and when in storage. If the hose is to be inserted into a drum it must be cleaned externally to avoid contamination of the fluid when it is inserted in the drum. If there is no way but using the traditional method of filling, use clean cloth and clean oil funnel to avoid introduction of contaminants during the servicing process. When adding hydraulic fluid to a reservoir, NEVER FILL TO THE TOP! Oil expanding under working temperatures will overflow the reservoir. So, the assigned air space on top of the fluid surface must be respected.

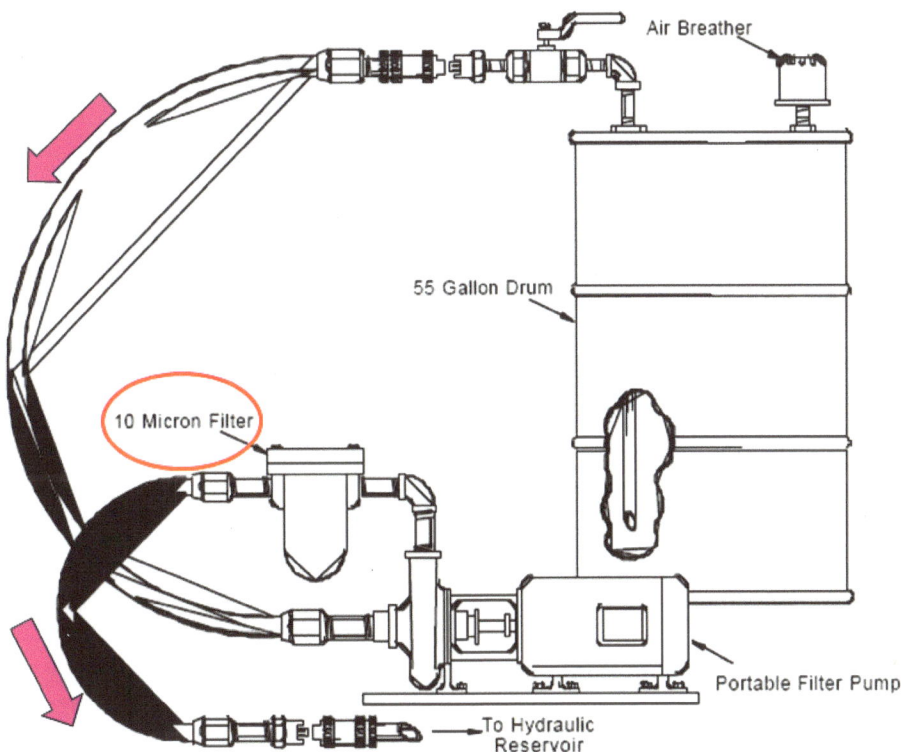

Fig. 9.3- Best Practices for Filling Hydraulic Reservoirs

Chapter 10

Maintenance of Transmission Lines

Objectives

This chapter provides guidelines for **transmission lines** selection, replacement, maintenance scheduling, installation, testing, storage and transportation. This chapter is supported by examples and figures granted by leading fluid power manufacturers.

The following topics are discussed in Chapter 3 in Volume 4 "Hydraulic Fluids Conditioning" of this series of textbooks:
- Basic Types and Contribution of Hydraulic Transmission Lines
- Sizing of Hydraulic Transmission Lines
- Rated Pressures for Hydraulic Lines
- Hydraulic Pipes
- Hydraulic Tubes
- 3.6- Hydraulic Hoses
- Flanges for Transmission Line Connections
- Rubber Expansion Fittings
- Test Points
- Pressure Measurement Hoses
- Manifolds

The following topics are discussed in Chapter 10 in Volume 6 "Troubleshooting and Failure Analysis" of this series of textbooks:
- Hydraulic Transmission Lines Inspection
- Hydraulic Transmission Lines Troubleshooting
- Hydraulic Transmission Lines Failure Analysis

Brief Contents

10.1-BP-Transmission Lines-01-Selection and Replacement
10.2-BP-Transmission Lines-02-Maintenance Scheduling
10.3-BP-Transmission Lines-03-Installation and Maintenance
10.4-BP-Transmission Lines-04-Standard Tests and Calibration
10.5-BP-Transmission Lines-05-Transportation and Storage

Chapter 10: Maintenance of Transmission Lines

The following set of best practices provide general guidelines and may not be applicable for all cases. They are not intended to replace the instructions given by the component manufacturer. It is strongly advisable to adhere to instructions provided by the manufacturer.

10.1-BP-Transmission Lines-01-Selection and Replacement

During system design, *Transmission Lines* were originally specified to consider leak-free connections and laminar flow conditions. During system maintenance, when replacing an existed transmission line, the following specifications shall not be changed:

Size: Line size has a direct effect on laminar flow condition and pressure losses. Therefore, when a line is replaced, the new line must have the same size. It is to be reminded that, if the line is a *flexible hose*, the line size is based on the inner diameter. If the line is a *tube* or *pipe*, then the size is based on the outer diameter.

Temperature: Particularly for hoses, minimum and maximum permissible temperatures must be checked. If the new hose is not within the same temperature rating, there is a good chance it will prematurely fail. If the original hose is a thermoplastic type, new hose must be the same.

Application: If a pressure hose is used to replace a suction hose, there will be a chance for the inner layers to collapse due to vacuum resulting in partial line blockage and pump cavitation. If the original hose has abrasion resistant outer cover, the new hose should be the same.

Media (Fluid Compatibility): Particularly for hoses, the new hose must be compatible with the hydraulic fluids. Otherwise, hose may be deteriorated over the time and become a source of contamination. It also can fail due to losing the attachment force between the hose and the end connections.

Pressure: Two lines that have the same size and look like each other DOES NOT mean they can replace each other. Rated pressure is directly related to the safety of the machine operation. Maximum allowable static and dynamic (pressure spikes and pulses) must be considered when replacing a hose.

Electrically Nonconductivity: Hydraulic hoses that are used nearby high voltage lines, must be electrically nonconductive. When any of these hoses are replaced, the new ones must also be electrically nonconductive. Otherwise, there is a good chance to accumulate static charges resulting in a potential explosion and possible fire.

Length: Generally speaking, pipes, tubes, and hoses are preferred for long, medium, and short line lengths; respectively. However, original line length should be respected. DO NOT reduce the line length because this may result in stretching the line when it is pressurized and may result in detaching the end connections. DO NOT increase the line length because this increases the line losses and may result in pressure hammering.

Hard Line (Pipes and Tubes) Material: Same inner diameter and wall thickness are not enough for tubes to replace each other. The steel grade, from which the tube is made, should also be the same. Otherwise, maximum permissible pressure might be different.

End Joints: Size, type and pressure rating of *End Joints* must be kept the same. Avoid using adaptors because it causes more pressure drop and affects the overall length of the line.

10.2-BP-Transmission Lines-02-Maintenance Scheduling

Unless otherwise is stated by components and systems manufacturer, Table 10.1 provides guidelines for *scheduling* preventive maintenance actions for hydraulic transmission lines.

#	Preventive Maintenance Actions	Daily	Weekly	Monthly	Biannually	Annually
1	Line visual inspection **(Note 1)**	✔	✔	✔	✔	✔
2	Inspect for leakage **(Note 2)**		✔	✔	✔	✔
3	Clean around and the outer surface including the end joints **(Note 3)**		✔	✔	✔	✔

Table 10.1- BP-Transmission Lines-02-Maintenance Scheduling

Note 1: Like a vehicle tire, a transmission line should be replaced based on given service life no matter how it looks like. When it comes to visual inspection, a hose service life is shorter than hard tubing. Therefore, they should be visually inspected more frequently than hard tubing. Any of the following conditions require immediate shut down and replacement of the Hose Assembly no matter what its lifetime is:
- Kinked, crushed, flattened or twisted Hose.
- Cracks from minimum bend radius exceeded and around fittings.
- Brittleness or loss of flexibility.
- Frayed protective layers or broken reinforcement layers.
- Outer cover pulled back from the end of the coupling.
- Fitting slippage on Hose.
- Rusted, broken, cracked, damaged, leaking, or badly corroded Fittings.

Note 2(Refer to Fig. 10.1): All types of transmission lines must be inspected for leakage. As best practices for *leakage* inspection, consider the following actions"

- Never tighten a fitting while the system is under pressure.
- Avoid possible oil *injection*. DO NOT use your hand to check leakage from a line.
- Wear proper industrial gloves that prevent oil injection.
- Use the technique of *fluorescent* dye and *UV light* for leakage detection. For details on leakage detection kit, refer to Volume 6 (Troubleshooting and Failure Analysis, Chapter 2).
- If special leakage detection equipment isn't available, use a piece of wood or cardboard that is 1-2 feet long. Hold one end of the wood or cardboard, place the other end approximately 1-2 inches away from the inspected part of the line, and move it around the line.
- For challenges and best practices if fluid injection occurs, refer to section 1.14 in Chapter 1 (Hydraulic System Safety) of this textbook.

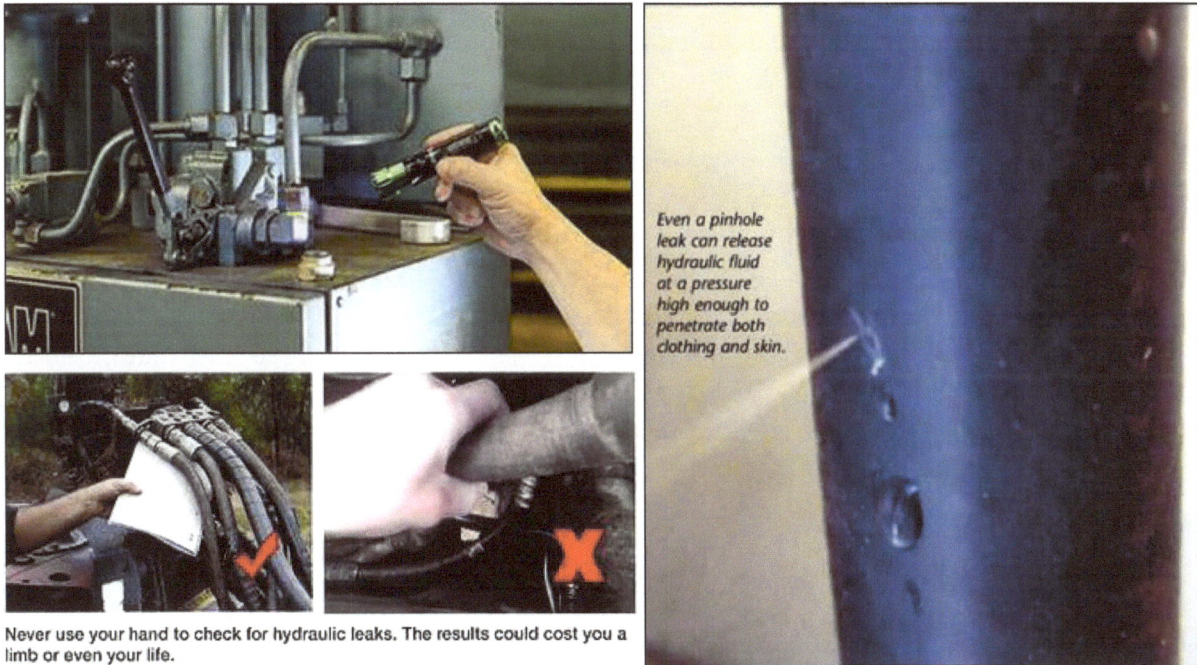

Even a pinhole leak can release hydraulic fluid at a pressure high enough to penetrate both clothing and skin.

Never use your hand to check for hydraulic leaks. The results could cost you a limb or even your life.

Fig. 10.1- Bet Practices for Hydraulic Transmission Line Leakage Inspection (www.bondfluidaire.com)

Note 3 (Refer to Fig. 10.2): Outer surfaces including the end joints of the transmission lines must be frequently cleaned to allow better heat dissipation and to maintain overall acceptable cleanliness level of the hydraulic fluid.

Fig. 10.2- Keep Outer Surfaces including End Joints (mac-hyd.com)

10.3-BP-Transmission Lines-03-Installation and Maintenance

Proper installation and maintenance of hydraulic transmission lines are important processes for trouble-free operation of the system. The following best practices provide guidelines for *installation and maintenance* of transmission lines.

BP-Transmission Lines-03-Installation and Maintenance:
1. Proper Line Cleaning before Assembling.
2. Proper Hose Crimping.
3. Proper Hose Routing.
4. Proper Hose Assembling.
5. Assemble for Leakage Prevention.

The following sections provide detailed interpretations, with examples, for the actions listed in the previous best practices list.

10.3.1- Proper Line Cleaning before Assembly

Contamination settled inside new and used hydraulic lines is very dangerous for hydraulic system operation. Therefore, it is necessary to clean inside the transmission lines before assembly. One or combination the following cleaning methods are used to clean inside the transmission lines, *Pickling, Flushing*, and *Projectile Cleaning.* Detailed information about these processes are contained by Volume 3 of this textbook series (Hydraulic Fluids and Contamination Control, Chapter 10). However, the following sections give just brief definitions.

Pickling (Cleaning by Acids): This process isn't applicable for flexible hoses. Producing hydraulic pipes and tubes involve harsh steps such as, extrusion, trimming, etc. All such manufacturing process result in forming built-in scales, greases, and many other contaminants left inside. The process of pickling means cleaning inside the lines by special chemicals and acids. The following set of bullets presents the best practices for pickling:
- Pickling liquids are dangerous and must be handled in accordance with OSHA requirements.
- Pickling process must be carried out by a special contractor at his location or onsite.
- Pickling process consists of the following sequences:
    1. Degreasing all parts.
    2. De-rusting using commercial de-rusting solution.
    3. Rinsing in cold running water.
    4. Neutralizing in another tank using neutralizing solution.
    5. Rinsing in hot water.
    6. Drying by blowing dry, hot, filtered air.

Flushing (Cleaning by Hydraulic Fluids): This process is applicable for all types of lines that may be used for first time or after major maintenance or predefined working hours. Production, handling, and storage of new hydraulic lines generates contamination that settle inside the lines. Additionally, older system designs collect contaminants in the reservoir. Low fluid velocity allows contaminants to settle in lines as well. Lack of sufficient filtration intensifies the accumulation effect. These layers of contaminants will occasionally rip off and could result in a breakdown and/or failure. Flushing is a process similar to kidney wash where low viscosity fluid is circulated at high velocity and high temperature through the system to scavenge all contaminates.

Projectile Cleaning (Cleaning by Compressed Air): Projectile cleaning is a method of final cleaning of all types of hydraulic transmission lines. However, it mostly used to clean hydraulic hoses. This process is applicable for lines that may be used for first time, during major maintenance and after hose cutting, tube flaring, or pipe threading. The easiness of the process makes it usable in field. The method is simply pushing a projectile inside the line using compressed air. Proper selection of quality of air, projectile size, and projectile type are essential for successful projectile cleaning process.

10.3.2- Proper Hose Crimping

Referring to Fig. 10.3, the following is an example of hose crimping

1. Setup the crimping machine referring to the manufacturer' instructions identifying the crimp specifications shown in the figure.
2. Select the proper die series. The dies are color coded and stamped on the top.
3. Before loading the die, brush the inside surface by a lubricant.
4. Load the selected dies into the crimper.
5. Place the die ring above the die.
6. Determine the hose insertion depth.
7. Insert the hose into the crimpable fitting until the insertion mark aligned with the end of the fitting.
8. Insert the hose into the die and properly align the coupling with the die fingers.
9. Finish the crimping, remove the hose, check the crimping diameter using a caliper.

Compatible Fittings

Hose and Fitting Combinations

Tools Required

Crimp Specifications

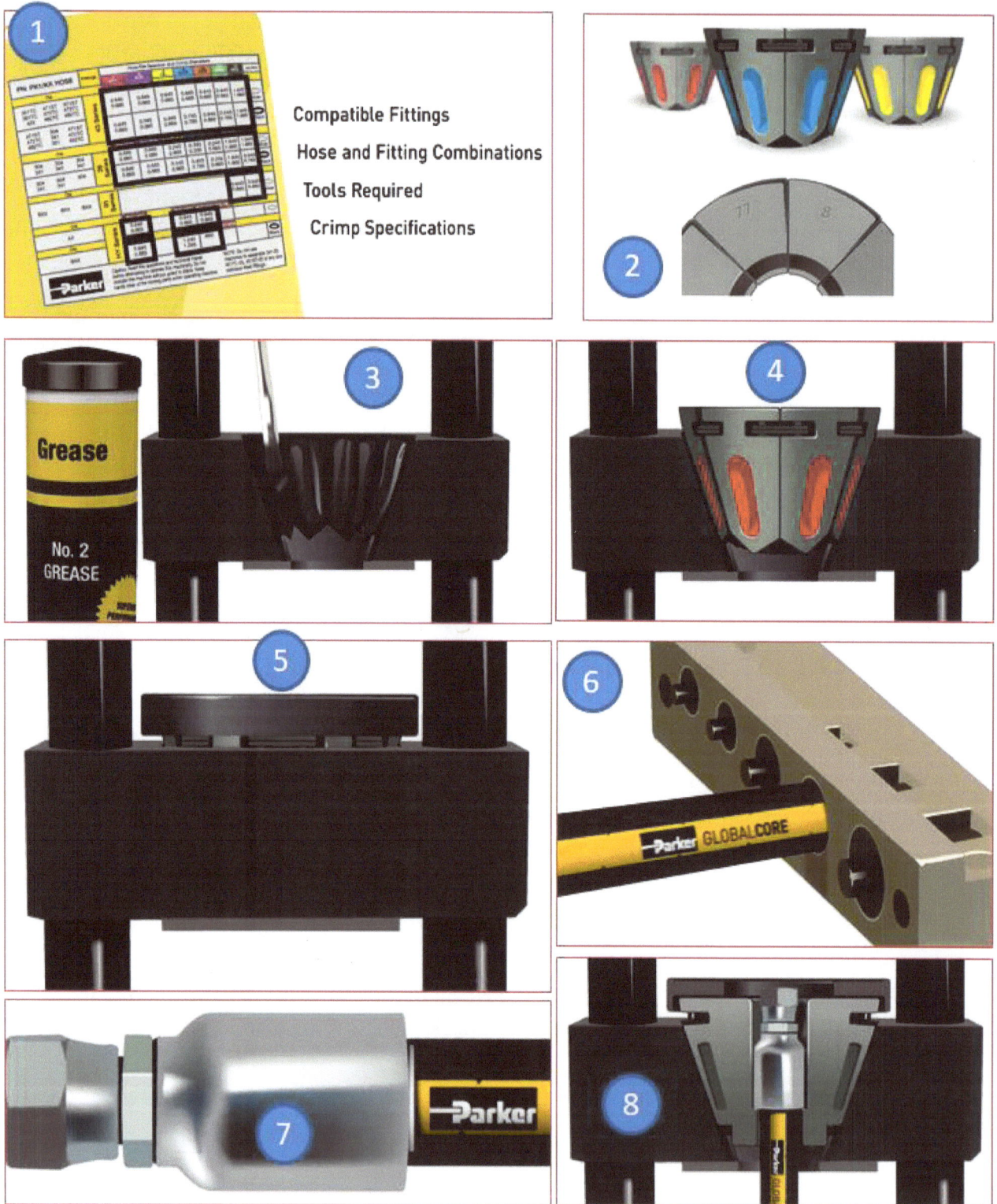

Fig. 10.3- Hose Crimping Process (Courtesy from Parker)

Hose Insertion Depth: Volume 4 of this textbook series provides information about hose *Cut Length* and the hose *Overall Assembly Length*. However, when crimping a hose in a permanent fitting, *Hose Insertion Depth* must be checked and marked. Estimating the depth of the hose by eyes can be extremely dangerous. If the hose isn't inserted into the shell of the fitting per the recommended length, the fitting can blow off. Therefore, as shown in Fig 10.4, it is recommended to check the hose insertion depth using standard gauge blocks.

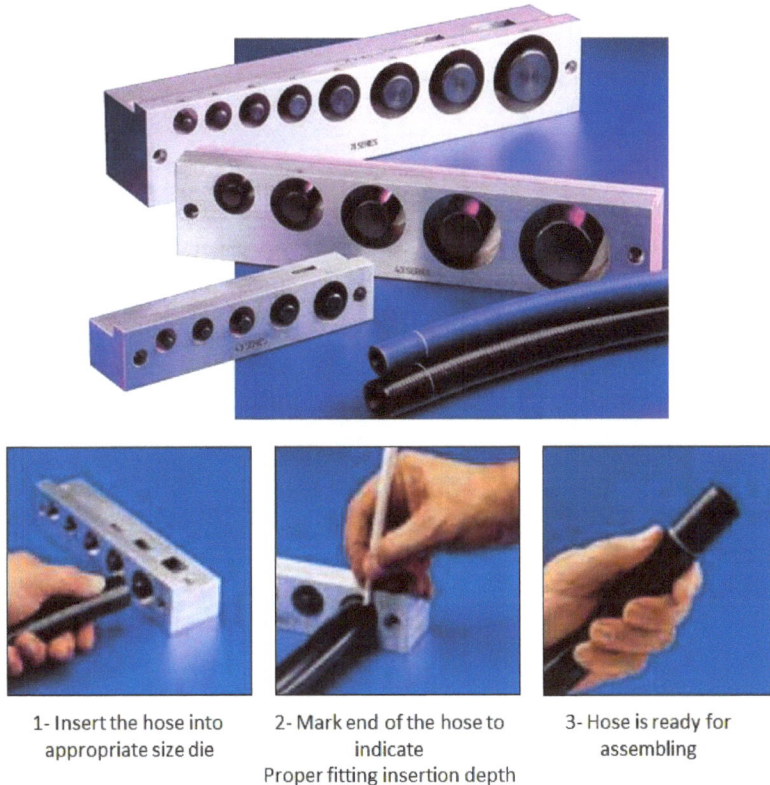

| 1- Insert the hose into appropriate size die | 2- Mark end of the hose to indicate Proper fitting insertion depth | 3- Hose is ready for assembling |

Fig. 10.4- Hose Crimping Machines (Courtesy from Gates)

10.3.3- Proper Hose Routing

Referring to Fig. 10.5, the following guidelines shall be considered when routing hydraulic hoses lines:

Route for Better Appearance and Serviceability (1): Hoses must be routed so that the final assembly looks ordered and not messy. Also, hose ends shouldn't be close to each other for better serviceability. Use 45-degree or 90-degree adapters in order to route hoses directly and to provide clearance for wrenches.

Avoid Abrasion and Sharp Edges (2): Line must be routed to avoid abrasion with moving or sharp elements. Otherwise, the outer surface will continuously be rubbing causing line failure.

Respect Minimum Bend Radius (3): Hose and tubes manufacturers specify minimum bend radius depends on the line size, length, and pressure ratings. Lines must be routed with full respect to the minim bend radius and to avoid tight bend. Otherwise, the outer surface will crack, and the line prematurely fails. Avoid using hoses that are unnecessarily long.

Shield from Heat Sources (4): Lines, particularly hoses, must be protected against heat sources such as exhaust or hot steam pipes. Protection includes isolating using heat resistant shields of baffles. Otherwise, hoses may be burned.

Do Not Twist Hydraulic Hoses (5): Hose ends must be assembled to the last turn without twisting the hose. If a hose is twisted, the hose length is shortened, its pressure carrying capacity is reduced, and hose end connection will loosen with the time.

Allow Clearance for Hose Length Change (6): When a hose is pressurized, it loses (2-4) % of its length. If a hose is stretched during the first assembly, the hose ends likely will detach under pressure pulsation. Therefore, when installing hoses, leave a little extra length to allow for changing in length when pressure is supplied. The other best practices in this regard is not to clamp a low pressure and high-pressure hose together. Otherwise, they will rub against each other.

Allow Clearance for Flexing Component Movement (7): If a hose is connected to a moving component, it must be routed to allow clearance so that the hose doesn't twist and/or exceed its minimum bend radius throughout the range of component movement. Use of swivel joints could be a solution for such a case.

Use Clamps (8): Use clamps or other devices t secure hoses so they do not rub against machine parts or they do not vibrate during machine operation.

Reduce Connections (9): Reduce number of pipe threads joints by using hydraulic adapters instead of pipe fittings.

Separate Hydraulic Lines from Electric Lines: Hoses must be routed to avoid intersection or contact with electrical lines. That is to avoid possibility of explosion or ignition in case of developing spark and oil spray.

Shield Hoses from Heat Sources: Used guards and shields to keep Hoses away from extreme temperatures or heat sources.

Protect the Hoses from Hazardous Material and Conditions: Use armures, guards, or sleeves to protect the hoses as needed. Also, keep hoses away from chemicals where possible

Quick-Disconnect Couplings: When assembling quick connect couplings, make sure they are matched. Ensure they are connected/disconnected properly. If in doubt, disconnect and re-connect the couplings again.

1- Appearance and Serviceability

2- Abrasion

3A- Min. Bend Radius

3B- Tight Bend

4- Heat Shield

5- Twist

6- Length Change

7- Movement and Flexibility

8- Clamps

9- Reduce Connections

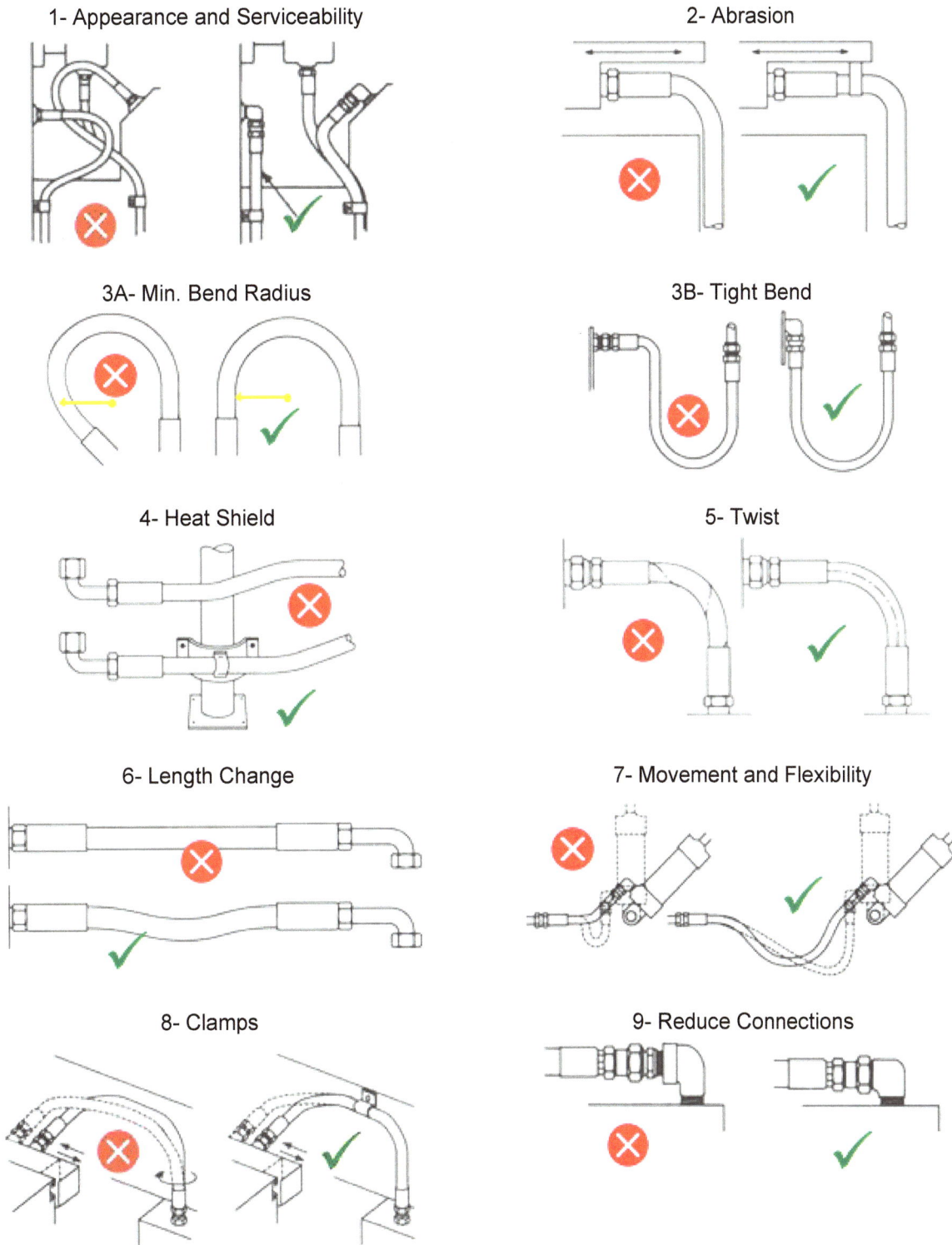

Fig. 10.5- Best Practices for Hydraulic Hose Routing

10.3.4- Proper Hose Assembling

Best Practices for Hose Assembling: Referring to Fig. 10.6, the following steps provides best practices for installing hose assembly:

1. Clean the surrounding area where connections are to be made. Make sure no dirt or contamination gets into hydraulic openings.
2. Install adapters into ports (if used). Torque to manufacturers specifications.
3. Lay the hose assembly into routing position to verify length and correct routing.
4. Thread one end of hose assembly onto port (or adapter). If the hose assembly uses an angled fitting, always install it first to ensure proper positioning.
5. Thread other end of the assembly without twisting the hose. Use a wrench on the backup hex on the fitting while tightening.
6. Properly torque both ends with manufacturer's torque considered.
7. Run the hydraulic system to circulate oil under low pressure and safely reinspect for leaks and potentially damaging contact.

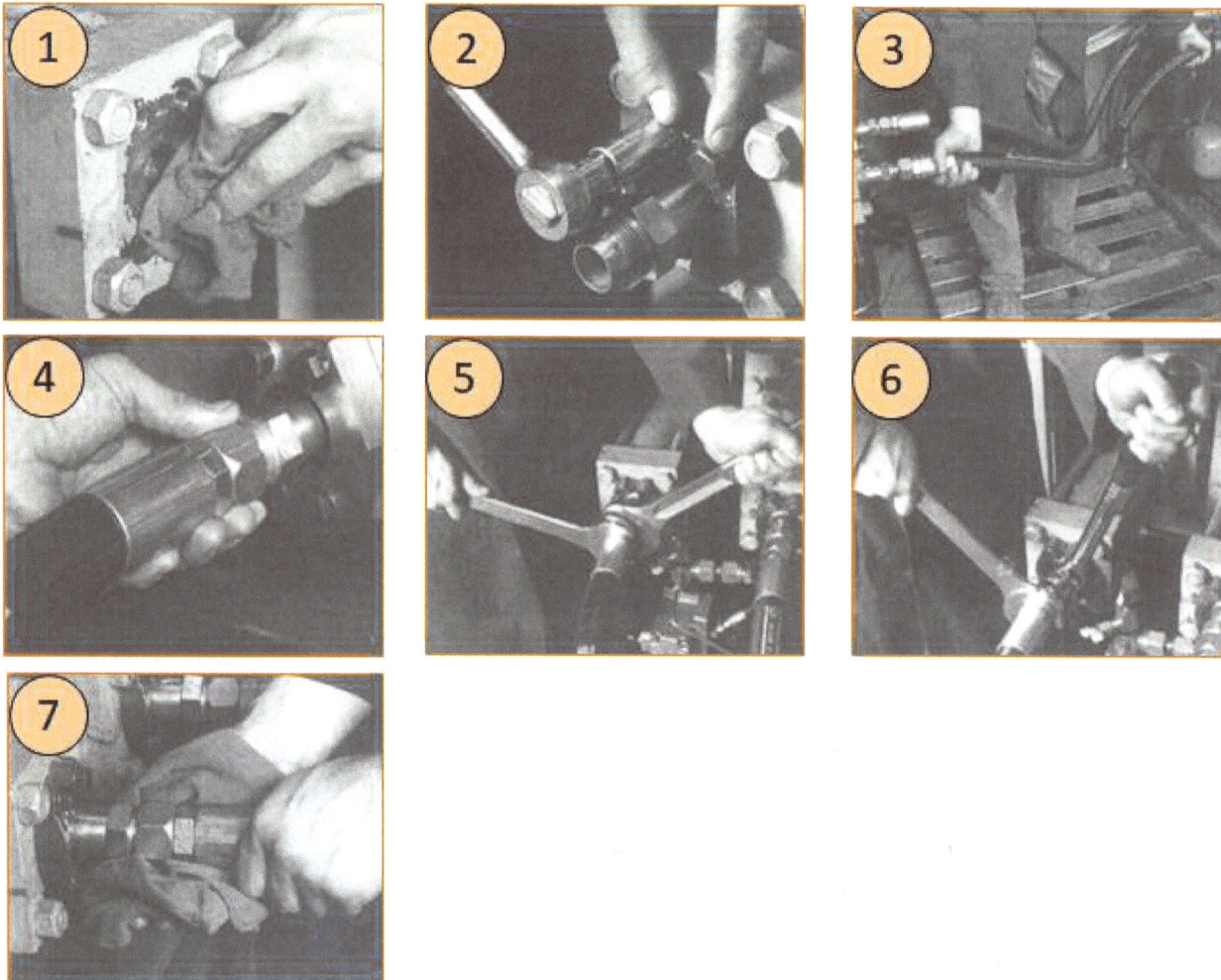

Fig. 10.6- Steps to Install Hose Assembly (Courtesy of Gates)

10.3.5- Assemble for Leakage Prevention

As shown in Fig. 10.7, fluid stains or puddles under hydraulic equipment or transmission lines indicate the presence of a leak in the line.

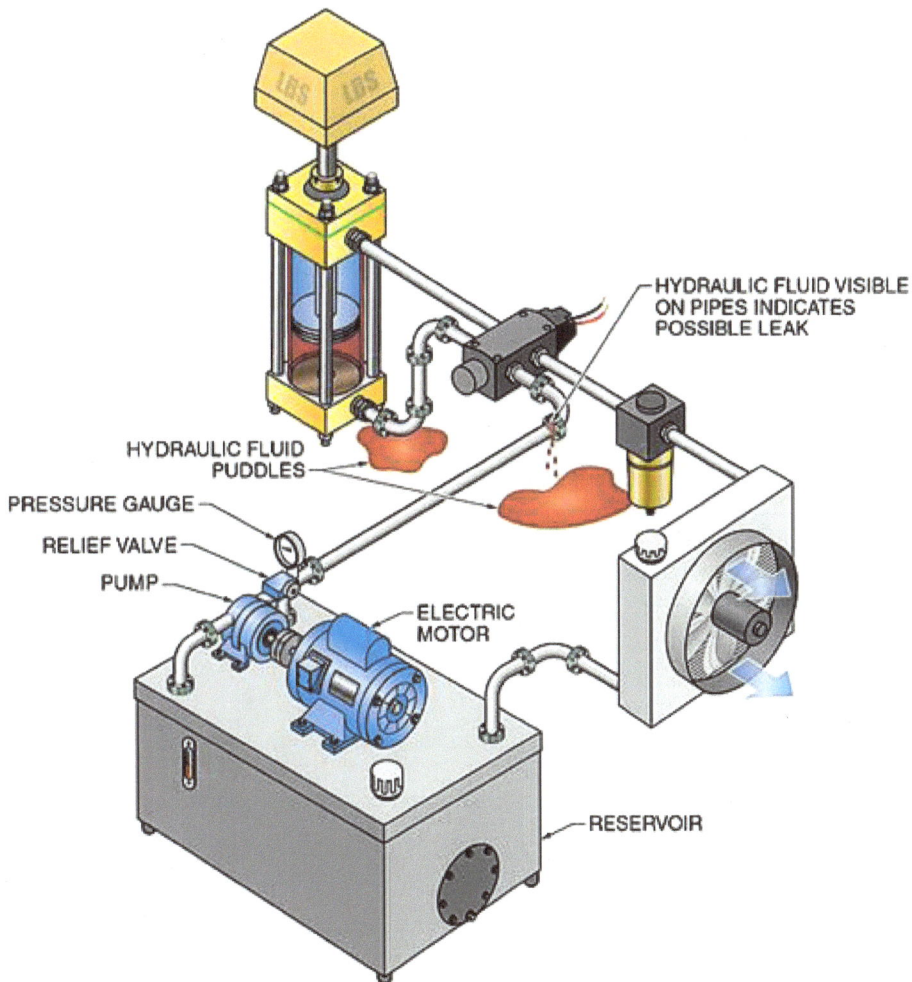

Fig. 10.7- Transmission Line Signs of Leak (Courtesy of American Technical Publisher)

The following best practices help minimize the possibility of line leakage when considered at the time of line selection and assembly:

❖ **Quality of Product:**
- Never use galvanized steel or commercial "from-off-the-shelf" fittings.
- Since fittings, and crimpers are designed to work together as a system, avoid mixing of fittings and hoses form different suppliers even if they follow same standard (SAE, ISO, etc.). This can result in hose assembly leakage, hose separation or other failures which can cause serious bodily injury or property damage from spraying fluids, flying projectiles, or other incidences.

❖ **Determining the Thread Type (Fig. 10.8):** As shown in the figure, threads of various fittings look similar. Mixing fittings of different thread types causes leakage. Therefore, it is advisable to assure correct thread identification before using them. Threads are measured as follows:

▪ **Thread Gauge:** Using a *Thread Gauge*, the number of threads per inch can be determined. Holding the gauge and coupling threads in front of a lighted background helps to obtain an accurate measurement.

▪ **Caliper Measure:** A *Vernier Caliper* should be used to measure the thread OD and ID diameters.

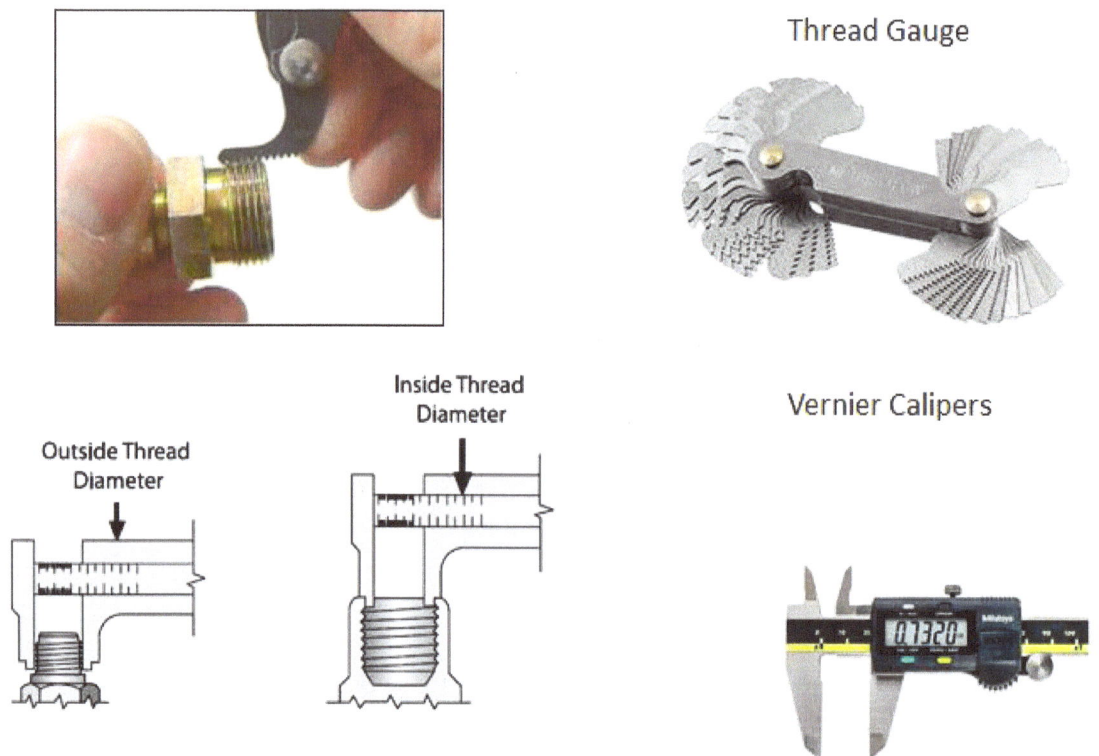

Thread Gauge

Inside Thread Diameter

Outside Thread Diameter

Vernier Calipers

Fig. 10.8- Determining the Thread Type of Hydraulic Fittings

❖ **Fittings Assembly Torque:**
▪ Over torqueing a fitting DOES NOT mean it seals better. Overtightening means overstressing or cracking. Manufacturers specify assembly torque which depends on the type, size and pressure rating. Instructions from the manufacturers should be fully respected.

▪ Never tighten a fitting while the pump is running. For your safety, release the system pressure firs in accordance with manufacturer's recommendations.

■ Tables 10.2, 10.3, and 10.4 show examples of typical data provided by a manufacturer. The torque values in the following tables give minimum and maximum torque recommendations. The minimum value will create a leakproof seal under most conditions. Applying torque values greater than the maximum recommendation will distort and/or crack the fitting.

Size		Steel			
		Ft. Lbs.		Newton-Meters	
Dash	Inches	Min	Max	Min	Max
-4	1/4	10	11	13	15
-5	5/16	13	15	18	20
-6	3/8	17	19	23	26
-8	1/2	34	38	47	52
-10	5/8	50	56	69	76
-12	3/4	70	78	96	106
-16	1	94	104	127	141
-20	1-1/4	124	138	169	188
-24	1-1/2	156	173	212	235
-32	2	219	243	296	329

Table 10.2- Recommended Tightening Torque for 37° & 45° (Machined or Flared Fittings) (www.new-line.com)

Size		Newton-Meters	
Dash	Inches	Min	Max
-4	1/4	14	16
-6	3/8	24	27
-8	1/2	43	54
-10	5/8	60	75
-12	3/4	90	110
-14	7/8	90	110
-16	1	125	240
-20	1-1/4	170	190
-24	1-1/2	200	245

Table 10.3- Recommended Tightening Torque for Flat-Face O-Ring Seal (Steel) (www.new-line.com)

Size		Ft.Lbs. Working Pressures 4,000 Psi (27.5 Mpa) And Below		Newton-Meters Working Pressures 4,000 Psi (27.5 Mpa) And Below		Ft.Lbs. Working Pressures Above 4,000 Psi (27.5 Mpa)		Newton-Meters Working Pressures Above 4,000 Psi (27.5 Mpa)	
Dash	Inches	Min	Max	Min	Max	Min	Max	Min	Max
-3	3/16	–	–	–	–	8	10	11	13
-4	1/4	14	16	20	22	14	16	20	22
-5	5/16	–	–	–	–	18	20	24	27
-6	3/8	24	26	33	35	24	26	33	35
-8	1/2	37	44	50	60	50	60	68	78
-10	5/8	50	60	68	81	72	80	98	110
-12	3/4	75	83	101-1/2	113	125	135	170	183
-14	7/8	–	–	–	–	160	180	215	245
-16	1	111	125	150	170	200	220	270	300
-20	1-1/4	133	152	180	206	210	280	285	380
-24	1-1/2	156	184	212	250	270	360	370	490

Table 10.4- Recommended Tightening Torque for SAE Steel O-Ring Boss (www.new-line.com)

❖ **Proper Assembling of Flareless Fitting:** Figure 10.9 shows an example of best practices of assembling of *flareless* fitting as follows:

1- Properly burr the tube end, clean the burring products, and visually inspect the tube end to make sure it is burred properly.
2- Check the 90^0 tube end trimming using proper tool.
3- Apply lubricant to the fitting.
4- Assemble the nut and the cutting ring with the tube.
5- Assemble the tube assembly with the fitting.
6- Tight by hand.
7- Tight by a wrench to torque specified by manufacturer.
8- Disassemble the tube assembly and check the proper attachment between the tube and cutting ring. The cutting ring may rotate in place around the tube but must not move axially along the tube. A special light source tool is used to make sure light isn't seen from between the tube and the cutting ring. If so, fluid leak will not occur.

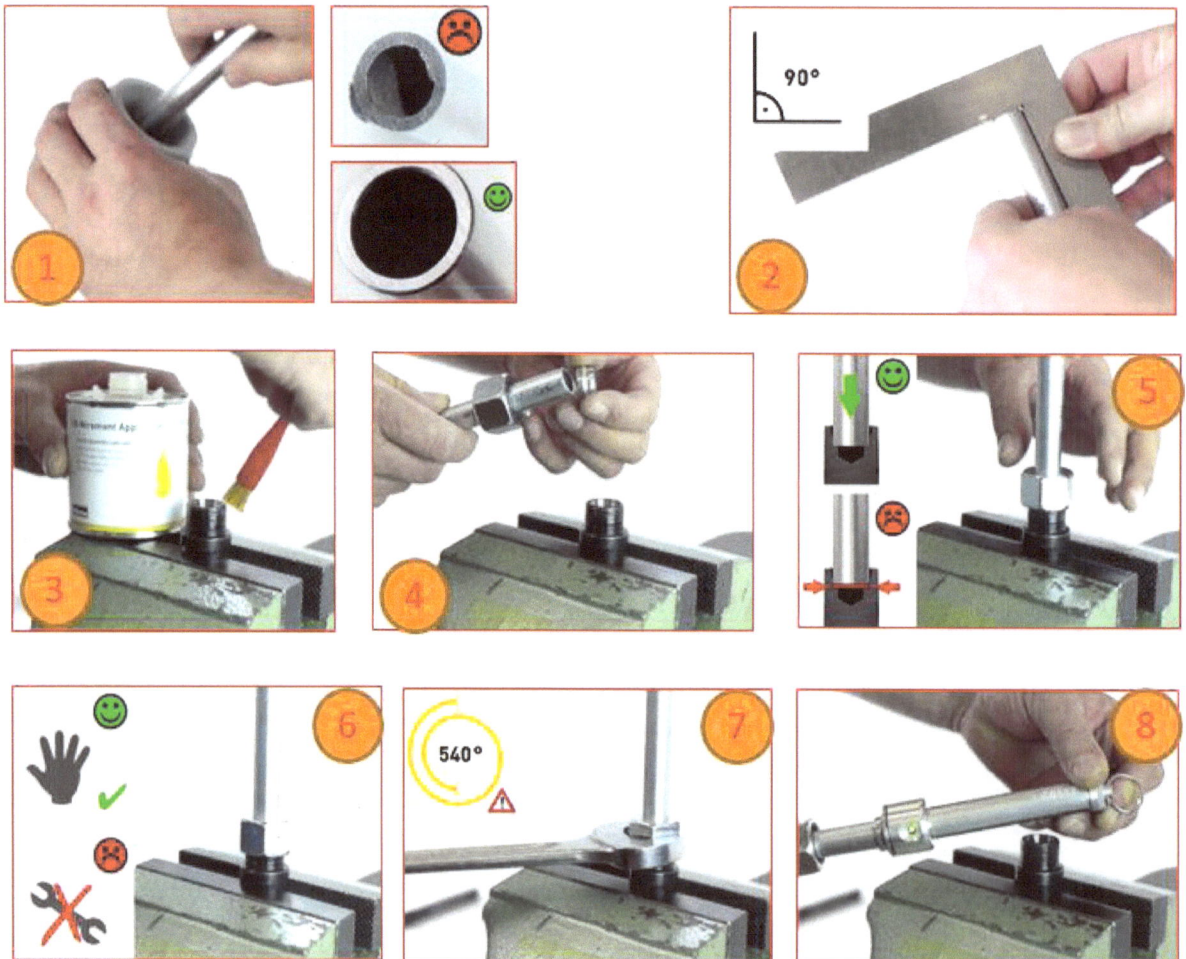

Fig. 10.9- Best Practices for Assembling Flareless Fitting (Courtesy from Parker)

❖ **Proper Tube Flaring (Fig. 10.10):** As shown in the figure, improper tube flaring results in cracked tube nose and fluid leakage. Follow the standard tube flaring procedure and dimensions for every tube size. The following steps provide best practices for tube flaring:

1. Select proper tubing based on (fluid compatibility, operating pressure, etc.).
2. Review the standard flaring dimensions based on single or double flare. For detailed information about the dimensions of tube flaring, refer to Volume 4 (Hydraulic Fluids Conditioning – Chapter 3).
3. Cut the tube to required length.
4. End should be trimmed square within +/- 2° tolerance.
5. Properly burr O.D. and l.D. of tube.
6. Clean tube to remove all dirt from both O.D. and I.D. of tube.
7. Assemble tube nut and sleeve on tube. The threaded end of nut and flared end of sleeve must point toward the end of the tube to be flared.
8. Flare the tube end using the correct flaring tool for the tube size and desired flare angle.
9. Inspect flare to the dimensions indicated in the standard tables.
10. In addition, flare should be checked for concentricity, thin out, cracks, nicks, loose slivers, burrs, pits or other defects which may prevent sealing.

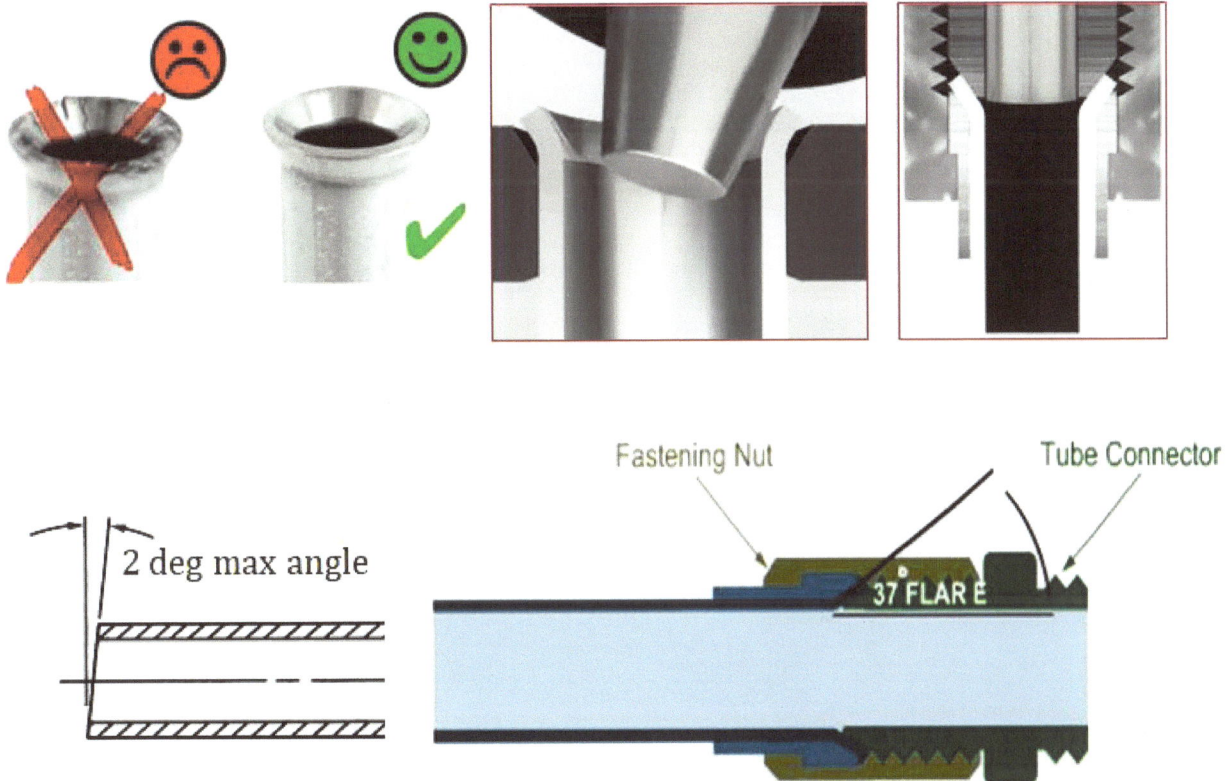

Fig. 10.10- Best Practices for Tube Flaring (Courtesy from Parker)

❖ **Proper Assembling of Flared Fitting:** Figure 10.11 shows an example of simple three steps of assembling a 37º *flared* tub and tightening to torque specified by manufacturer. The catalogue indicates either the tightening torque or the how many faces the nut should rotate after tightening by hand.

1- Clap the body on a Vice

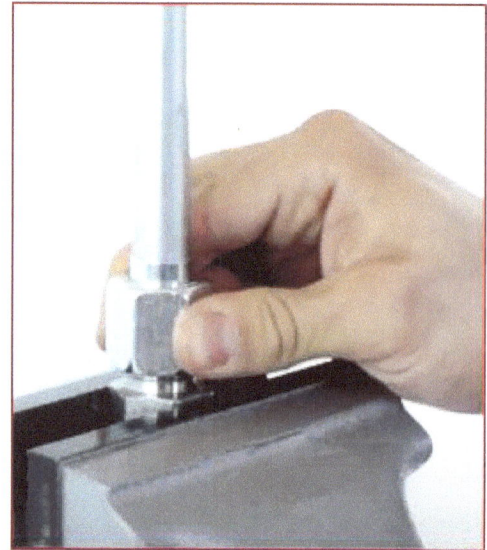

2- Tightened by Hand

3- Tightened by Wrench to Specified Torque

Fig. 10.11- Best Practices for Assembling 37º Flared Fitting (Courtesy from Parker)

❖ **Flared Tube Sealing:** Sealing between a fitting and a transmission line is an additional challenge. Imperfection of Tube flaring may cause leakage. Figure 10.12 shows a patentable sealing element named *"Flaretite"* and pronounced "Flare Tight". Such sealing element conceptually compressed to match the misalignment between the flared surfaces.

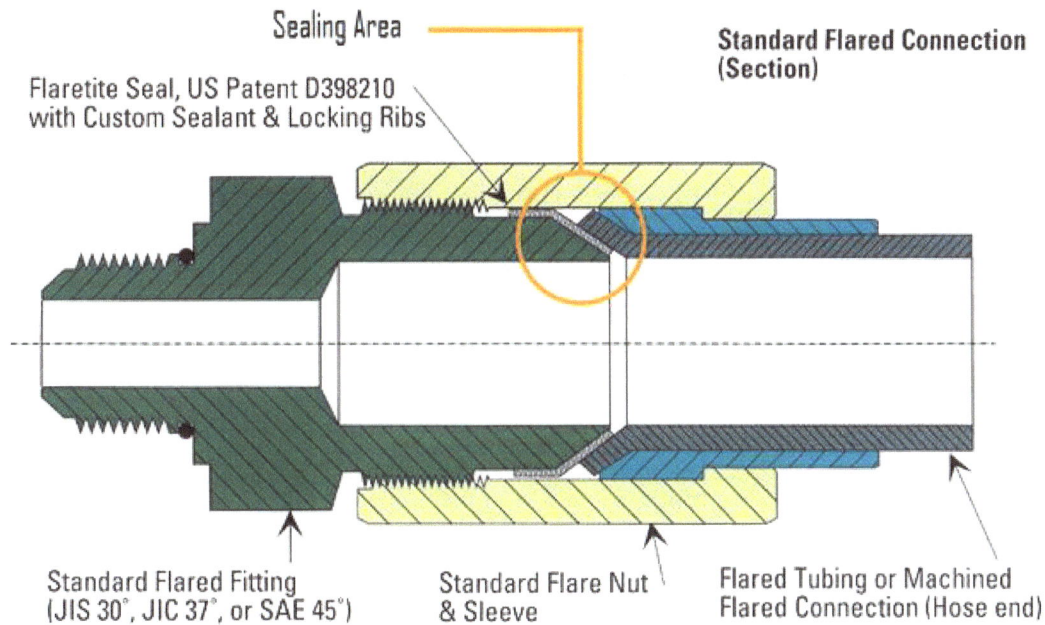

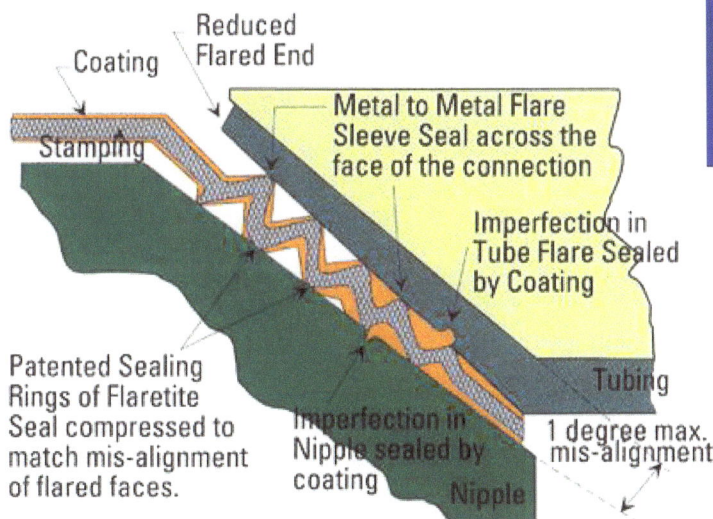

Fig. 10.12- Flare Tight Seal (www.Flaretite.com)

❖ **Proper Tube Bending:** Tubes should be bent in accordance with minimum bend radius instructed by the tube manufacturer based on the tube size, pressure ratings, etc. Tubes can be bent by hand, or by press depending on the tube size. Figure 10.13 shows perfect bend versus other incorrect bend cases.

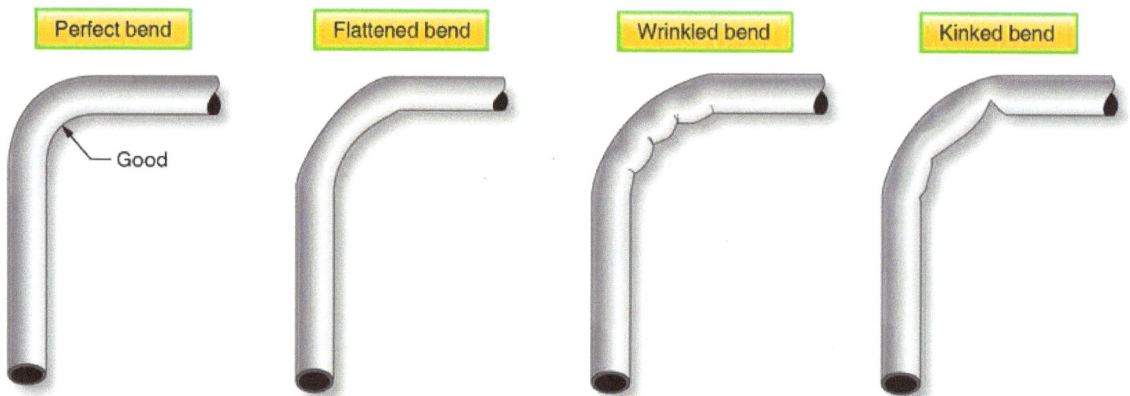

Fig. 10.13- Perfect and Incorrect Tube Bends (www.aircraftsystemstech.com)

❖ **Proper Pipe Welding:** The following shall be meticulously considered when welding pipes, particularly high-pressure ones. Otherwise, pipes may leak.

- Pipe assemblies must be properly prepared for welding to avoid excessive contamination in the line.
- Components which can be affected by the extreme heat must be properly isolated.
- Welding personnel must be certified and well trained.
- Inspect the pipe after welding and before installation, the sealing surfaces and internal surfaces of piping shall be free of any visible detrimental foreign matter such as scale, burrs, swarf, etc.

❖ **Proper Pipe Assembly:** As shown in Fig 10.14, when assembling a pipe, leave the end pipe threads free from pipe thread sealant or Teflon® tape to ensure that it does not contaminate the system.

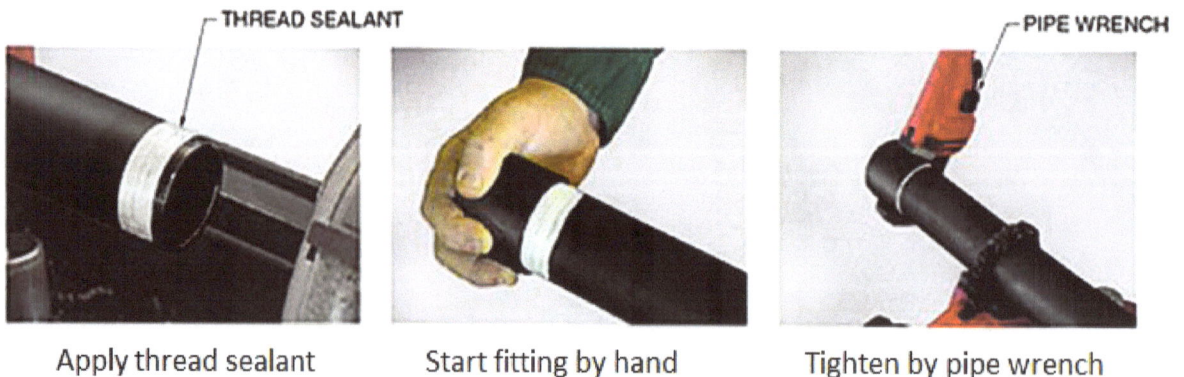

Fig. 10.14- Pipe Assembly (Courtesy from American Technical Publishers)

❖ **Transmission Line Proper Clamping:**
▪ **Why should a hydraulic line be clamped?**
  o Unclamped transmission lines transmit noise and vibration.
  o If a pressurized transmission line assembly blows apart, the fittings can be thrown off at high speed causing risks of injuries.
  o If an unclamped pressurized hose is accidentally detached from one of its ends, it whips like a snake with great force.

▪ **Best practices for hose clamping:**
  o Do Not temporarily drape a return line hose into the reservoir.
  o Do Not allow hoses to drag in the dirt or lay on the ground.
  o Clamp hoses to the structure; not to components to minimize vibration.
  o Clamp far from moving parts.
  o Clamp to avoid surface abrasion as shown in Fig. 10.15.
  o Use hose restraints, shown in Fig. 10.16, to protect against injury when a hose is blown.

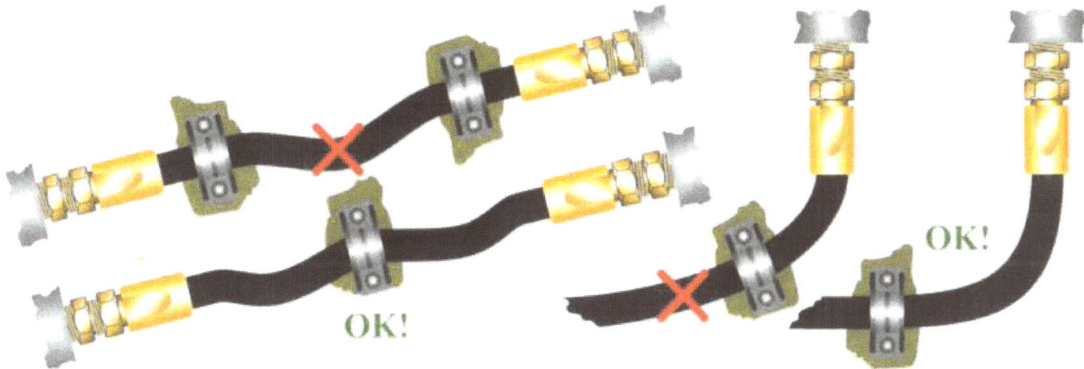

Fig. 10.15- Proper Hose Clamping

Fig. 10.16- Restrains for Hoses Clamping

▪ **Best practices for tubes/pipes clamping:** As shown in Fig. 10.17, different standard tube clamping collars are available in the market. Collars can be made of polypropylene, self-extinguish polyamide, aluminum or zinc-coated steel with plastic supports and they are fastened via a weld plate, a thread, or a guide. Figure 10.18 provides guidelines for clamping distance. Random clamping distance could lead to generating of noise and vibration.

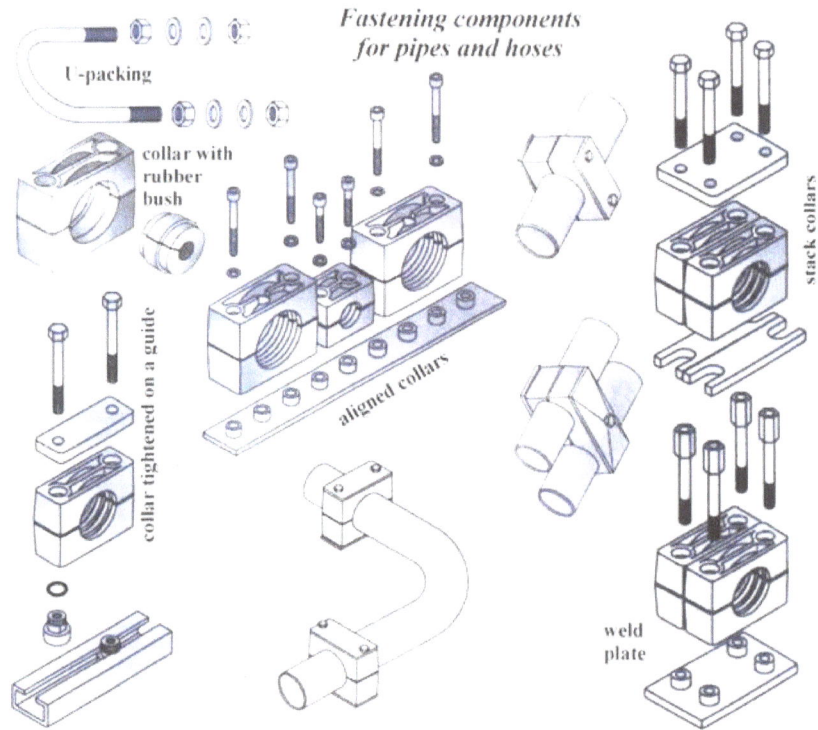

Fig. 10.17- Clamping Collars for Tubes and Pipes (Courtesy of Assofluid)

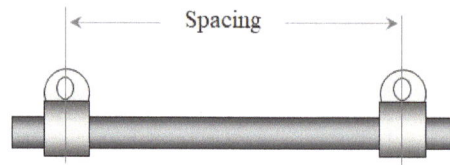

Table-11-1			
TUBE O.D."	EQUIVALENT TUBE (mm)	FOOT SPACING BETWEEN SUPPORTS	SPACING IN METERS (Approx.)
1/4" - 1/2"	6 - 13 mm	3 ft.	.9 m
3/8" - 7/8"	14 - 22 mm	4 ft.	1.2 m
1"	23 - 30 mm	5 ft.	1.5 m
1-1/4" & up	31 & up mm	7 ft.	2.1

Fig. 10.18- Guidelines for Tube Clamping Distances

Avoid Mechanical Stresses on Hard Transmission Lines: Referring to Fig. 10.19, the following set of examples provide guidelines to avoid developing mechanical stresses and consequently reduce the chance of leakage and permanent failure.

Example 1: Dead weight of the components stressing the transmission lines. Avoid excessive stress on joints by selecting appropriate clamping spots for plumbing and components.

Example 2: Hard transmission lines shall not be used for equipment support. If a hard line connects to a component, an additional support is required to reduce the stresses on the line.

Example 3: Avoid poor tube bending that may result in stressing the fittings. DO NOT heat to ease tube bending. The tube may lose its ductility and thereby be subject to failure under high-pressure conditions.

Example 4: Do NOT lay or stand on transmission lines during system servicing.

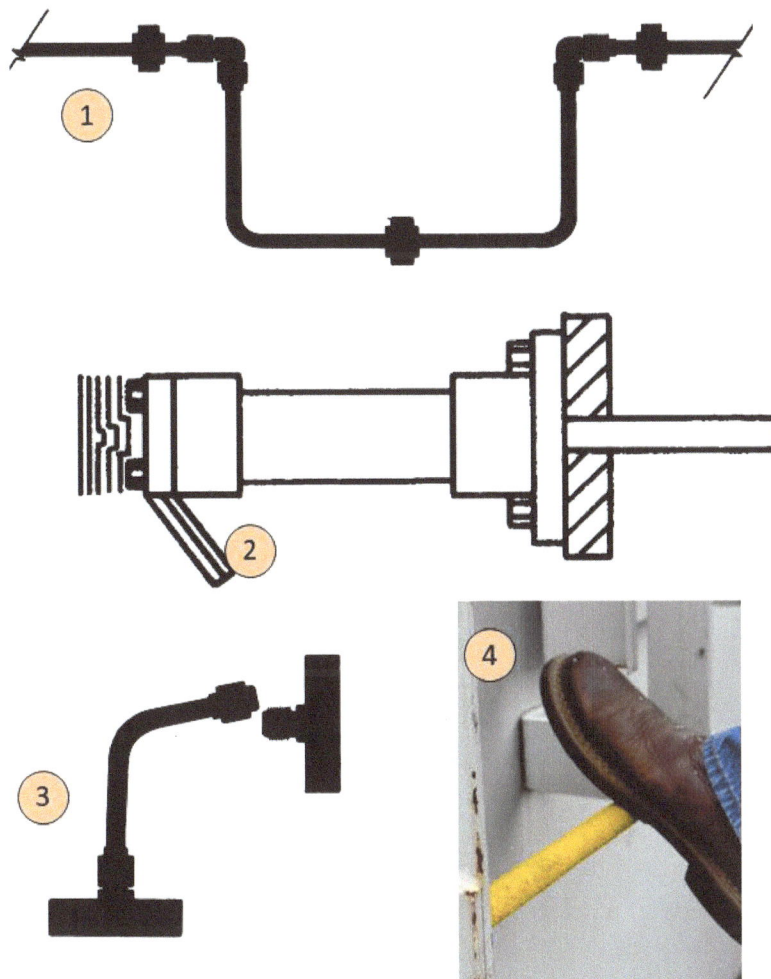

Fig. 10.19- Guidelines to Avoid Mechanical Stresses (Courtesy from Parker Hannifin)

10.4-BP-Transmission Lines-04-Standard Tests and Calibration

Hoses: ISO 6605 2002 Hydraulic Hoses and Hose Assemblies-Test Procedures. This international *standard* specifies test methods for evaluating the performance of hoses and hose assemblies used in hydraulic fluid power systems. The following are common tests:

Pressure Proof Test (ISO 1402): *Pressure Proof Test* is a nondestructive test. The hose assembly is hydrostatically tested to the specified proof pressure in accordance with the relevant product specification using the method specified in ISO 1402 for a period 30-60 seconds. The assembly is considered as passing the test with no signs of leakage or failure.

Burst Pressure Test (blog.parker.com): This is a destructive test. A *Burst Test* is a hydrostatic pressure test of a hose assembly that determines the actual burst strength of the assembly. Any signs of leakage, bulging, coupling ejection or hose burst below the specified minimum rated burst pressure of the assembly are considered a failure. Minimum burst values are used as one factor in the establishment of a reasonable and safe maximum work pressure and hose lifetime. As shown in Fig. 10.20, maximum rated working pressure is based on safety factor of 4.

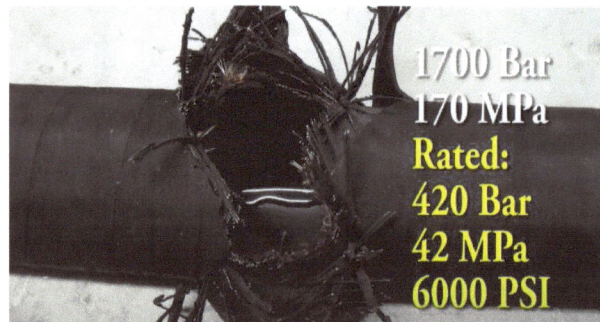

Fig. 10.20- Hose Burst Test

Cold Flexibility Test (ISO 10619): A *Cold Flexibility Test* is a destructive test specified in ISO 10619. It is used to determine the ability of a hose to operate under cold weather. Condition hose assembly at a temperature equal to the minimum application temperature of the relevant product specification in a straight position for 24 hours. While still at the minimum application temperature, bend the sample on a mandrel over a time of 8-12 seconds. For hose (ID) sizes <= 22 mm bend them through 180^0 mandrel. For hose (ID) sizes > 22 mm bend them through 90^0 mandrel. After bending, allow the sample to warm to room temperature, visually examine it. When tested, the sample's hose or cover should not crack; and when warmed to ambient temperature, the test piece should not leak or crack when subjected to proof pressure.

Impulse Test: An *Impulse Test* for 100Rxx hydraulic hose is one of the key predictions of hose life. Impulse testing involves pressurizing the hose cyclically at 133% of working pressure, at rates up to 1 cycle per second while the hose is held in either a 90° or 180° configuration. To pass the test, the hose must meet or exceed the minimum number of impulse cycles based on the applicable industry standard.

Salt Spray Test (ASTM B117): *Salt Spray Test* (also called *Salt Fog Test*) is a standardized and popular corrosion test method for hydraulic fittings used to check corrosion resistance of materials and surface coatings. Salt spray testing is a destructive test. It is an accelerated corrosion test that produces a corrosive attack to coated samples in order to evaluate and compare the suitability of the coating for use as a protective finish.

Most commonly, the time taken to oxides the samples under test is considered to determine whether the test is passed or failed. Figure 10.21 shows SAE J516 standard hydraulic fittings after the 144-hour salt spray test in accordance with ASTM B117. The appearance of red rust within this period indicates failure.

Fig. 10.21- Hose Burst Test (Courtesy from Gates)

Gravimetric Cleanliness Measurement (ISO 4405): Gravimetric measurement is a reporting method that references the total mass of contaminant found inside a hydraulic component as normalized by the total internal component surface area of a hydraulic component. A fluid is used to dislodge contamination in a hose assembly and is then poured through a membrane catch filter. The filtered particles are dried and weighed, and results are typically stated in milligrams of contaminant per square meter of internal surface area. Figure 10.22 shows an example determining the cleanliness level when 52 mg of contaminant is found in a hose assembly:

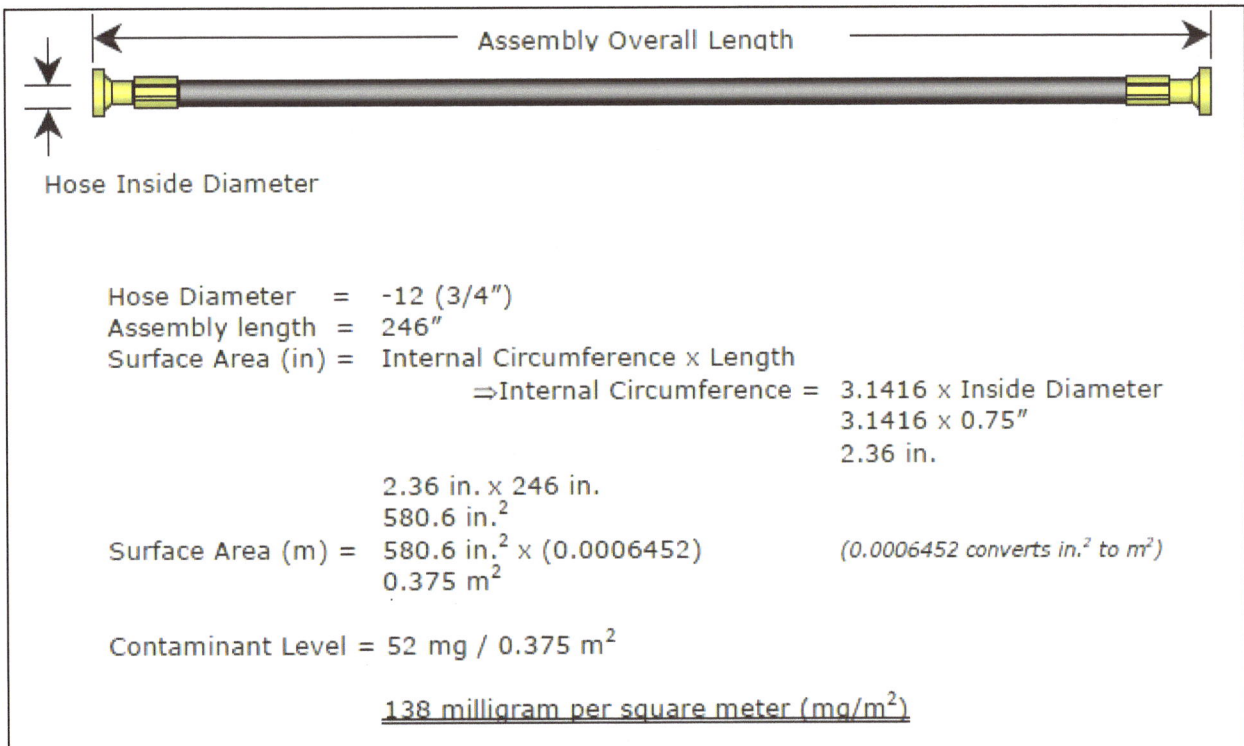

Hose Inside Diameter

Hose Diameter   =   -12 (3/4")
Assembly length =   246"
Surface Area (in) =   Internal Circumference x Length
                       $\Rightarrow$Internal Circumference = 3.1416 x Inside Diameter
                                                             3.1416 x 0.75"
                                                             2.36 in.
                       2.36 in. x 246 in.
                       580.6 in.^{2}
Surface Area (m) =   580.6 in.^{2} x (0.0006452)   *(0.0006452 converts in.^{2} to m^2)*
                       0.375 m^2

Contaminant Level = 52 mg / 0.375 m^2

                       138 milligram per square meter (mg/m^2)

Fig. 10.22- Hose Cleanliness Measurement (Courtesy from Gates)

Reliability Assessment Test (Fig. 10.23):   The SAE and ISO standards for hydraulic hose impulse tests are ran for a finite number of cycles under standardized conditions. If the hose passes this test, it meets the minimum requirements of the applicable standard. Comparable hoses from various manufacturers will usually always pass this test due to the limited number of cycles. Therefore, there is no way to determine from this test which manufacturer will last longer in an application. To do so, a test protocol must be developed using application specific parameters and then test to failure or a specific number of cycles required by the application.

The *Hose Reliability Assessment Test* is an application-based laboratory test program developed by Milwaukee School of Engineering (MSOE). One specific application example was to assess the reliability of hydraulic hoses used on refuse vehicles under various operating conditions. The test program considered investigating the effect of the following operating conditions:
- Four different hose manufacturers and two different hose construction types (2W and 4W spiral) multiple specimens from each manufacturer and hose type.
- Level of continuous pressure: based on vehicle measured data.
- Magnitude and frequency of pulsating pressure: based on vehicle measured data.
- Working temperature: based on vehicle measured data.
- Bend radius: one-half of the minimum specified by the applicable hose construction.
- Twisting angle: 45 Degrees between hose couplings when assembled on the manifold.
- Minimum test length: 2,000,000 cycles.

The test results concluded the following:
- 4W spiral hose lasts up to 5 times longer than the 2W hose in this specific application
- Increased vehicle productivity.
- Reduced environmental costs of cleanups.
- Estimated company annual savings was $10M.

Initial Setup 16-June-03

Fig. 10.23- Hose Reliability Assessment Test

10.5-BP-Transmission Lines-05-Transportation and Storage

Referring to Fig. 10.24, the following best practices are guidelines for transportation and storage of transmission lines:

Hose Storage (1): Should be coiled on reels that may rotate manually or using a motor.

Tube/Pipe Storage (2): They should be placed on supports with distances based on the size and length to avoid bending. Pipe manufacturers must be reviewed for such information.

Dust Caps (3): All lines ends should be closed by proper dust caps.

Storage Space: It should be clear of high temperature, corrosive liquids, humidity, ultraviolet light, ozone, oils, fumes, solvents, high humidity, insects, electromagnetic fields or radioactive materials.

Storage Time: Maintain a system of age control that shows a hose must is used before the shelf life is expired and facilitate first-in first-out usage based the date of manufacturing. Follow the manufacturer's shelf life. However, **SAE J517** specifies the shelf life of hydraulic hoses.

Transportation: transmission lines of all types must be sealed with protective caps.

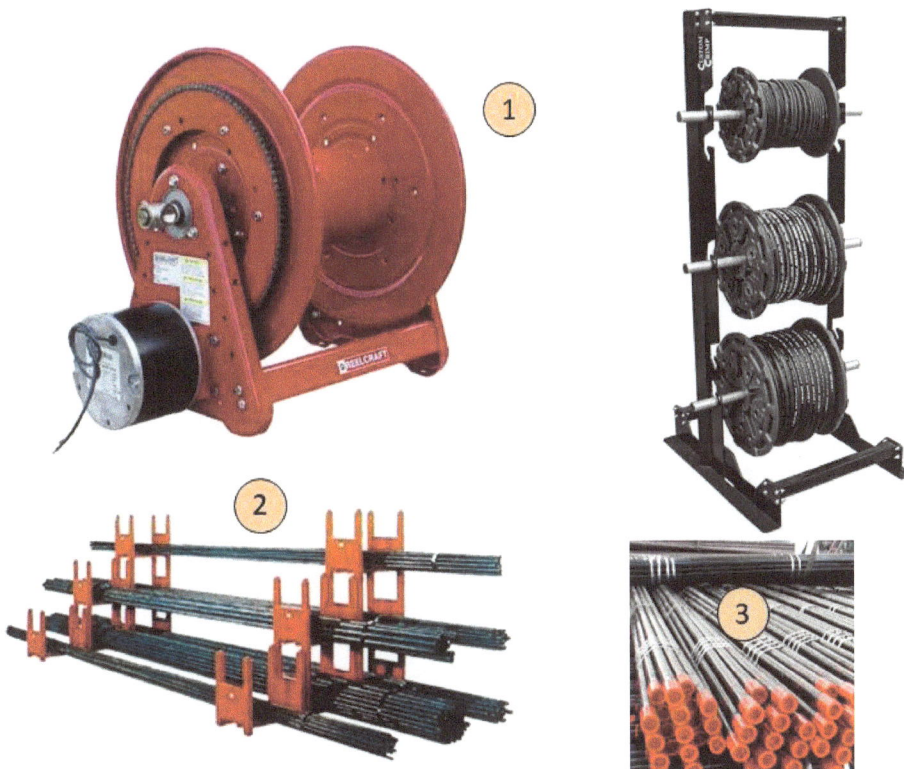

Fig. 10.24- Best Practices for Storage of Transmission Lines

Chapter 11

Maintenance of Heat Exchangers

Objectives

This chapter provides guidelines for **heat exchangers** selection, replacement, maintenance scheduling, installation, testing, storage and transportation. This chapter is supported by examples and figures granted by leading fluid power manufacturers.

The following topics are discussed in Chapter 5 in Volume 4 "Hydraulic Fluids Conditioning" of this series of textbooks:
- Contribution of Heat Exchangers:
- Air-Type versus Water-Type Oil Coolers
- Determination of Cooling Capacity for an Oil Cooler
- Air-Type Oil Coolers
- Shell-and-Tube Water-Type Oil Coolers
- Plat-Type Oil Coolers
- Cooling-Filtration Units
- Oil Cooling Circuit Diagram
- Oil Temperature Automatic Control Solutions
- Electrical Oil Heaters

The following topics are discussed in Chapter 11 in Volume 6 "Troubleshooting and Failure Analysis" of this series of textbooks:
- Heat Exchangers Inspection
- Heat Exchangers Troubleshooting
- Heat Exchangers Failure Analysis

Brief Contents

11.1-BP-Heat Exchangers-01-Selection and Replacement
11.2-BP-Heat Exchangers-02-Maintenance Scheduling
11.3-BP-Heat Exchangers-03-Installation and Maintenance
11.4-BP-Heat Exchangers-04-Standard Tests and Calibration

Chapter 11: Maintenance of Heat Exchangers

The following set of best practices provide general guidelines and may not be applicable for all cases. They are not intended to replace the instructions given by the component manufacturer. It is strongly advisable to adhere to instructions provided by the manufacturer.

11.1-BP-Heat Exchangers-01-Selection and Replacement

Heat Exchangers shall be selected to consider removing the heat and ensure reliable system operation. When a heat exchanger is replaced, the most important factor to be checked is the size. Size of a heat exchangers means certain cooling or heating capacity. If the size is changed, one or combination of the following problems may occur:
- Hydraulic system temperature instability.
- All consequences of hydraulic fluid viscosity increasing/decreasing.
- Possible pump cavitation.
- Reduced system reliability.

11.2-BP-Heat Exchangers -02-Maintenance Scheduling

Unless otherwise is stated by components and systems manufacturer, Table 11.1 provides guidelines for *scheduling* preventive maintenance actions for heat exchangers.

#	Preventive Maintenance Actions	Daily	Weekly	Monthly	Biannually	Annually
1	Check for external leaks	✔	✔	✔	✔	✔
2	Clean the dust on outer surfaces		✔	✔	✔	✔
3	Check hydraulic connections			✔	✔	✔
4	Check coolant connection			✔	✔	✔
5	Check proper connection with electrical instrumentation			✔	✔	✔
6	Check Zinc Anodes for corrosion **(Note 1)**			✔	✔	✔
7	Inside deep cleaning **(Note 2)**				✔	✔
8	Check proper setting of temp. control system **(Note 3)**				✔	✔
9	Check for internal leaks					✔

Table 11.1- BP-Heat Exchangers-02-Maintenance Scheduling

Note 1: Cooling water may contain mineral salts that cause corrosion of water connections between the heat exchanger and water supply lines. That is why they need frequent inspection. The figure shows a new and used zinc anodes that were disassembled from a hydraulic system heat exchanger after seen seven months of service.

As shown in Fig. 11.1, each heat exchanger has its own *Zinc Anodes*. These need to be replaced periodically, some more often than others. How periodically this is needs to be done, it depends on the size of the cooler, the material of the tube and shell, the type of the cooling water.

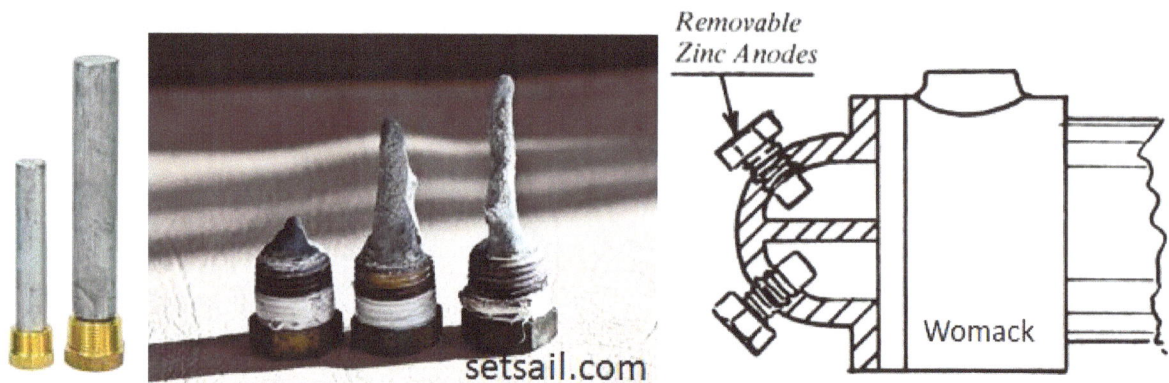

Fig. 11.1- New and Used Zinc Anodes

Note 2: Inside deep cleaning of a heat exchangers includes cleaning tube and shell of a water-cooled heat exchanger and fins of an air-cooled heat exchanger. Fins on air cooled heat exchangers may need to be cleaned more frequently due to dusty operating conditions such as in mobile equipment.

Note 3: Check proper setting of temperature for either On/Off or analog control system. This act should be applicable for heaters and coolers. If an air cooler is used, verify that the cooler fan is turned on at the specified temperature.

11.3-BP-Heat Exchangers-03-Installation and Maintenance

Heat exchanger failure or malfunction cause a complete shutdown of operations. Therefore, they should be kept well maintained to avoid unplanned shutdowns. The following set of bullets provides guidelines about the best practices for hydraulic heat exchangers maintenance and installation.

Proper Placement of a Water-Type Heat Exchangers: The following bullets define the proper location of a heat exchanger:

- Should be placed in a well-ventilated area.
- Should be placed apart or shielded from external heat sources.
- Should be shielded from harsh weather such as direct sunlight or freezing temperature.
- Should be protected from moving machinery such as swinging booms, lift trucks, etc.
- As shown in Fig. 11.2, it is more convenient to install heat exchangers horizontally so that the gravity of the oil and the coolant won't affect the speed of circulation. However, requirements for draining the circuits should be taken into consideration.

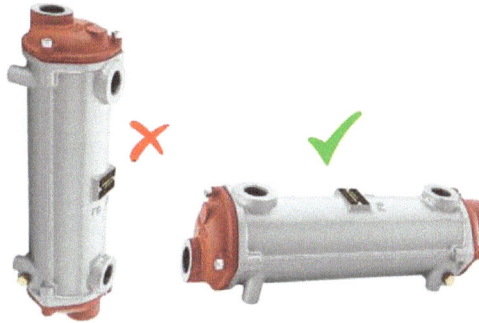

Fig. 11.2- Correct Orientation when Installing Water-Cooled Heat Exchangers

Proper Filter-Cooler Assembly: As shown in Fig. 11.3, the cooler is installed downstream of a filter for the following reasons:

- Capturing oil contaminants is easier when the oil is hot.
- Pressure drop across the filter is less when the oil is hot.
- If cold oil is forced into a filter, back pressure may damage the tubes in the cooler.
- Forcing oil into a cooler before filtering it reduces the cooler efficiency as dirt can accumulate inside the heat exchanger.

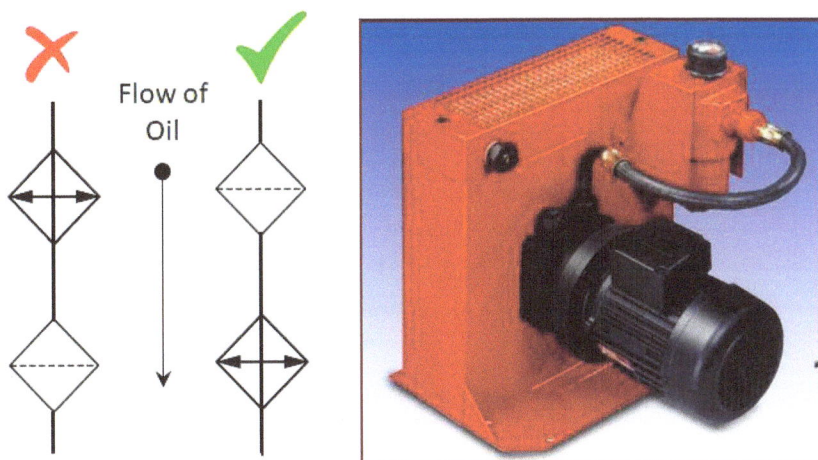

Fig. 11.3- Proper Filter-Cooler Assembly

Fouling of Water-Type Oil Cooler: The deposition of any undesired material on heat transfer surfaces is called *fouling*. Fouling may significantly impact the thermal and mechanical performance of heat exchangers. Fouling also impedes fluid flow, accelerates corrosion and increases pressure drop across heat exchangers. Fouling is a dynamic phenomenon which depends on the cleanliness level of hydraulic fluids, the source of the supplied cooling water, and the duration of heat exchanger operation.

As shown in Fig. 11.4, different types of fouling mechanisms have been identified. They can occur individually but often occur simultaneously. Descriptions of the most common fouling mechanisms are provided below:

Corrosion Fouling (1): Results from a chemical reaction which involves the heat exchanger surface material. Many metals such as copper and aluminum form adherent oxide coatings which serve to passivate the surface and prevent further corrosion. *Metal oxides* which are corrosion products exhibit quite a low thermal conductivity and even relatively thin coatings of oxides may significantly affect heat exchanger performance.

Particulate/Sedimentation Fouling (2): Sedimentation occurs when particles (e.g. dirt, sand or rust) in the solution settle and deposit on the heat transfer surface. Like scale, these deposits may be difficult to remove mechanically depending on their nature.

Scaling/Crystallization Fouling (3): Scaling is the most common type of fouling and is commonly associated with inverse solubility salts such as *calcium* carbonate ($CaCO3$) found in water. Reverse solubility salts become less solute as the temperature increases and thus deposit on the heat exchanger surface. Scale is difficult to remove mechanically. Chemical cleaning may be required.

Biological Fouling (4): Occurs when biological *organisms* grow on heat transfer surfaces. It is a common fouling mechanism where untreated water is used as the coolant. Problems range from algae to other microbes such as barnacles and zebra mussels. During seasons when these microbes are said to bloom, colonies several millimeters deep may grow across the surface within hours, impeding circulation near the surface wall and impacting heat transfer.

Varnish Fouling (5): *Varnish* formation occur as a result of oil degradation due to increased working temperature.

Salt Fouling (6): Occurs as a result of *salt* depositing if sea water is directly used without desalination.

Fig. 11.4- Fouling of Heat Exchangers (hcheattransfer.com/fouling1.html)

Cleaning of Water-Cooled Heat Exchangers: Routine cleaning of tube and shells of water-cooled heat exchangers reduces water usage, unexpected shutdowns, and hence improves heat exchanger efficiency, system reliability and operation cost effectiveness.

Assuming clean oil is flowing through the shell, then only inside the water tubes are required to be cleaned. However, if varnish has built up in oil, cleaning the outside surfaces of water tubes and inside surfaces of the shell is required. The following sections provide different technologies in cleaning inside tubes:

<u>Small Heat Exchangers:</u> For small or lightly contaminated heat exchangers, traditional *Flushing* inside a water-cooled heat exchanger can be done by circulating hot wash oil or light distillate through the tube side or shell side will usually effectively remove sludge or similar soft deposits. Soft salt deposits can be washed out by circulating fresh, hot water. A mild alkaline solution, such as 1.5% solution of sodium hydroxide or nitric acid can also be used. The tubes should be flushed in the opposite direction of the normal oil flow. Drying after flushing by hot and dry air is useful to remove any liquid residue from flushing process. As shown in Fig. 11.5, flushing can be done tube-by-tube by a flushing gun.

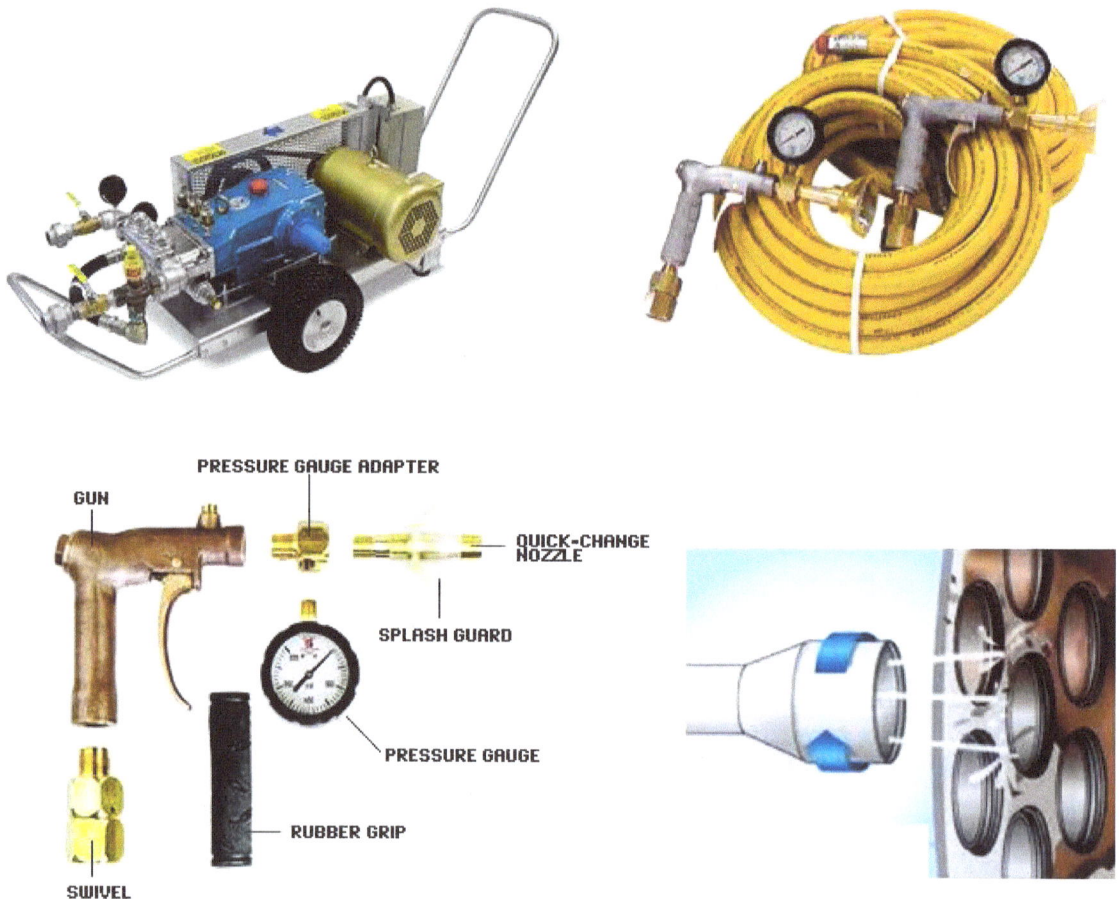

Fig. 11.5- Traditional Flushing Equipment (Courtesy from Conco Systems)

Large Heat Exchangers: For large or heavily contaminated heat exchangers, power flushing systems are used to push cleaning plugs through the tube at 10 to 20 feet per second (at 200-300 PSI), removing deposits including microbiological fouling, organic scales, obstructions and corrosion. DO NOT use AIR to push the cleaning plugs because the water provides lubrication. Depending on the type of heat exchanger fouling, one or combination of the following techniques shall be followed. Figure 11.6 shows cleaning tubes by plastic cleaners.

Cleaning Straight Tubes by Plastic Tube Cleaner (1): This is designed for removal of only the softest types of deposits such as mud, silt and microbiological fouling in condensers and heat exchangers. They are available in sizes 5/8" to 1-1/4" diameter. The cleaner's fins are slit horizontally and vertically which, along with a small diameter hole in the core, allows water to bypass the cleaner to lubricate and flush out deposits.

Cleaning U-Tubes by Plastic U-Tube Cleaner (2): This is designed to navigate very tight u-tube radiuses previously unreachable by conventional cleaners. The plastic u-tube cleaner is available for tube sizes 5/8" to 1-1/4" I.D.

**Type P Plastic
Tube Cleaner (1)**

**Type U Plastic
U-Tube Cleaner (2)**

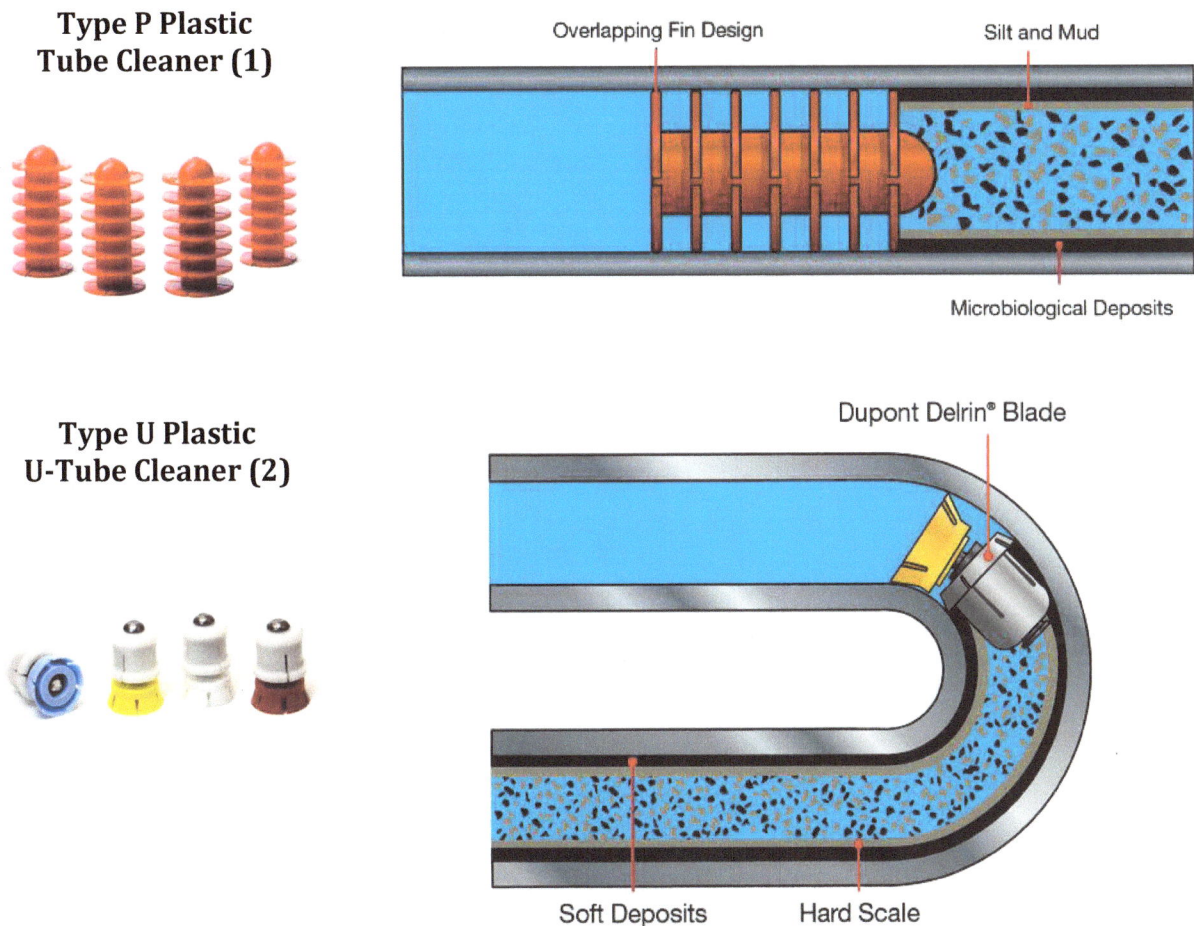

Fig. 11.6- Cleaning Tubes using Plastic Tube Cleaners (Courtesy from Conco Systems)

Figure 11.7 shows cleaning tubes by Brush cleaners.

<u>Cleaning Tubes by Plastic H-Brush Tube Cleaner (1):</u> This is typically utilized for removing light deposits in condensers and heat exchangers with tube sizes 5/8" to 1-1/2" I.D. They remove micro/macro fouling, soft organic deposits, some corrosion by-products, mud and silt, and most types of obstructions. It can be used in applications with enhanced tube surfaces and is safe on all inserts and epoxy coatings.

<u>Cleaning Tubes by Plastic XL-Brush Tube Cleaner (2):</u> The XL-Brush tube cleaner with similar characteristics to the H-Brush tube cleaner features an extra-long design for tubes that require additional cleaning.

<u>Cleaning Tubes by Stainless Steel Tube Cleaning Brush (3):</u> This is designed for use on all types of tube materials in sizes 5/8" to 1-1/4" I.D. , and can also be used on tubes with inserts and epoxy coatings. It is particularly effective on manganese, iron and silica deposits. They are also effective on all types of obstructions, macrofouling and debris. The SSTB will restore the tube surface to its original heat transfer characteristics, while providing absolutely the best protection from under-deposit corrosion.

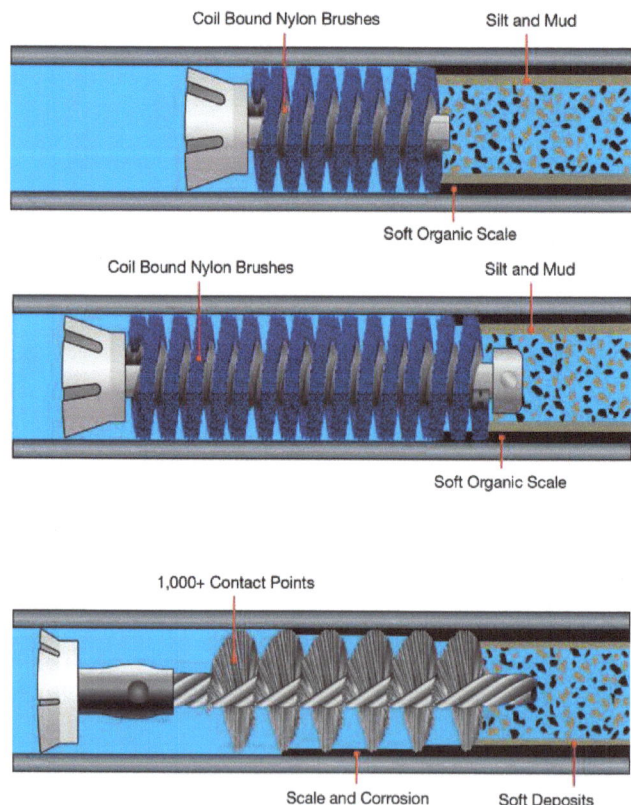

Fig. 11.7- Cleaning Tubes using Brush Cleaners (Courtesy from Conco Systems)

Figure 11.8 shows cleaning tubes by Metallic cleaners.

Cleaning Straight Tubes by C2x Metallic Tube Cleaner (1): This technology features a two-stage design with six points of cleaning contact per blade and is available for condensers and heat exchanger tube sizes ¾", 7/8", 1" and 1-1/8" I.D. . This tube cleaner is designed to remove thin tenacious deposits such as manganese, silica, iron and calcium, and is also effective in removing corrosion product and other types of debris and obstructions. For tubes that are extremely fouled, consider the C3X tube cleaner featuring a three-stage design.

Cleaning U-Tubes by Metallic Tube Cleaner (2): This technology features robust, carbon steel construction, and is designed to navigate U-tube configurations in heat exchangers and feed water heaters, removing tough deposits that softer cleaners can't touch. The metal U-Tube cleaner is available for tube sizes 5/8" to 1-1/4" I.D.

C2X
Tube Cleaner (1)

Metal
U-Tube Cleaner (2)

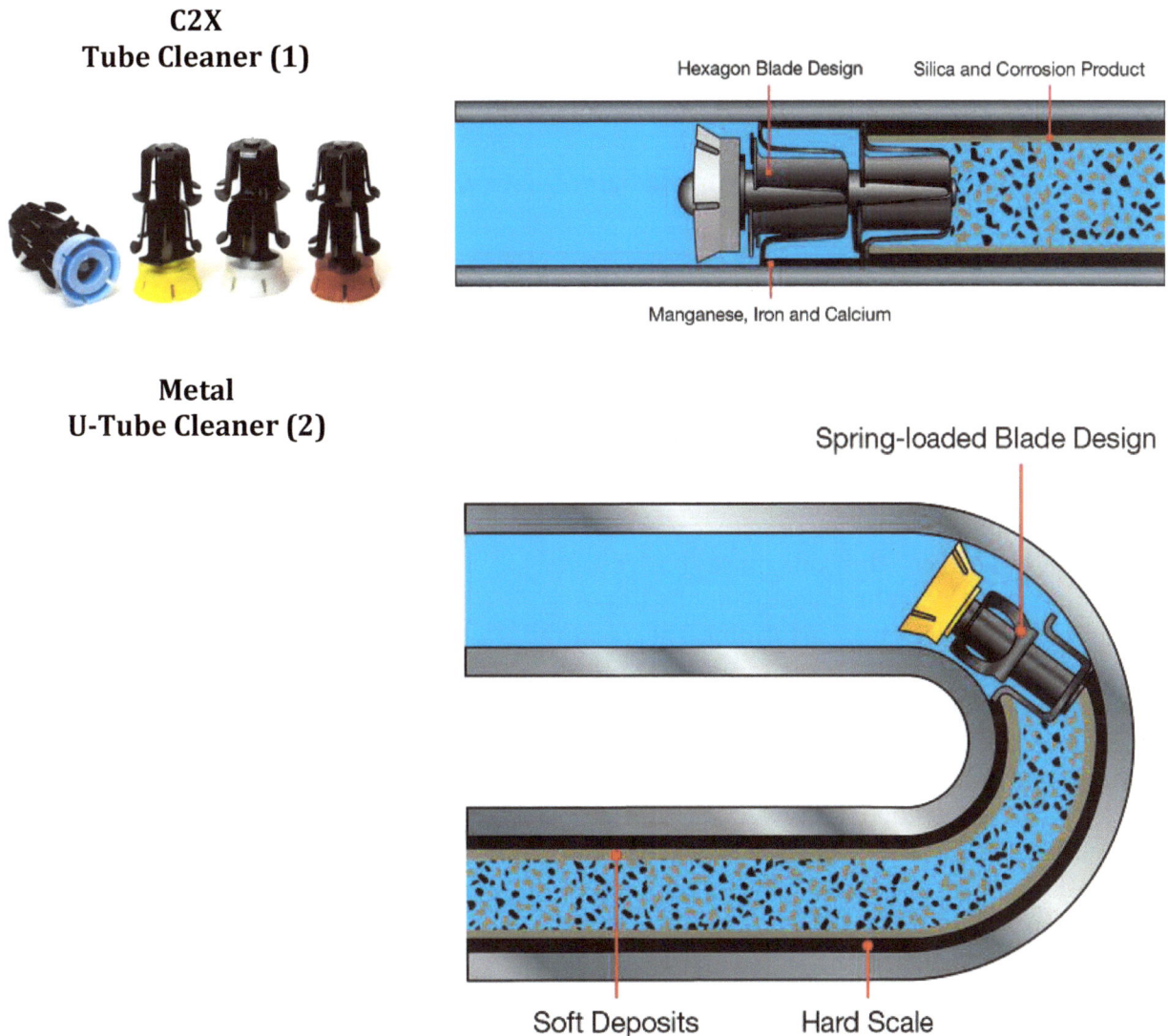

Fig. 11.8- Cleaning Tubes using Metallic Tube Cleaners (Courtesy from Conco Systems)

Cleaning of Air-Cooled Heat Exchangers: As shown in Fig. 11.9, air-cooled (fin-fan) heat exchangers are fouled with dust, dirt, debris, pollen, leaves and other deposits which can significantly reduce cooler efficiency. Keeping radiators clean is imperative to maintaining optimal performance and reliability. Fins of the radiator and the outside surfaces of the tubes are cleaned to remove dirt which may stick between the fins/tubes by a brush, compressed air, or by steam.

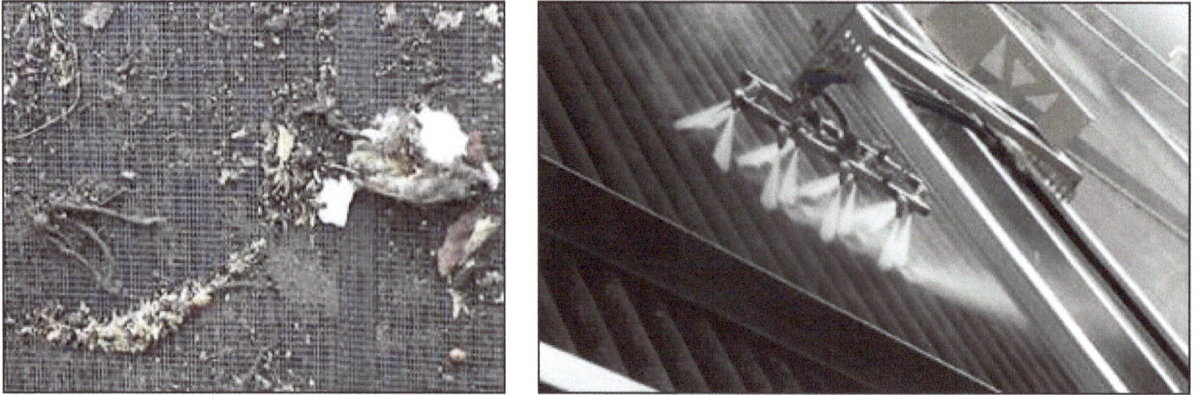

Fig. 11.9- Cleaning Radiators of Air-Cooled Heat Exchangers

11.4-BP-Heat Exchangers-04-Standard Tests and Calibration

Test for Internal Leaks: Internal leaks can occur between the oil and water chambers in a heat exchanger if the tubes are corroded or the pressure of either chambers exceeds the specification by the manufacturer. Therefore, water content in the oil must be routinely monitored. This can be investigated by drawing a sample of the oil and performing the required laboratory test. In this regard, refer to Volume 3 (Hydraulic Fluids and Contamination Control). Contents of oil in water should also be frequently checked, particularly if the cooling water circulates in a cooling tower. Withdraw the sample from the top of the cooling water because the oil has less density than oil. As shown in Fig. 11.10 shows a setup of a pressure test using compressed air to identify the place of leakage.

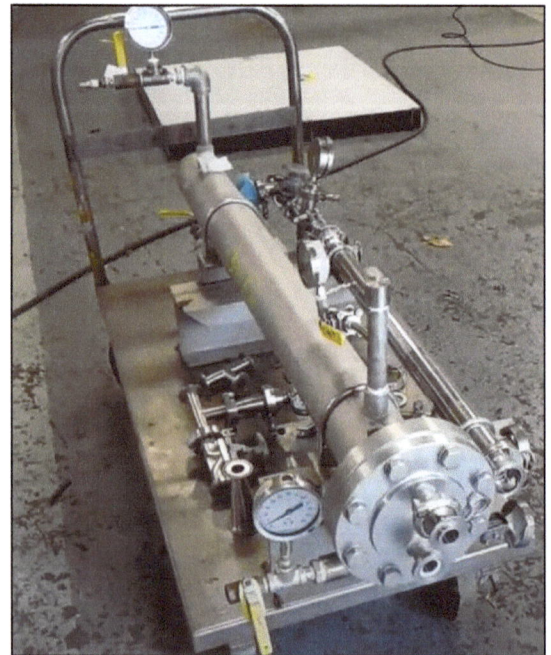

Fig. 11.10- Pressure Test of Heat Exchangers

Chapter 12

Maintenance of Filters

Objectives

This chapter provides guidelines for **Filters** selection, replacement, maintenance scheduling, installation, testing, storage and transportation. This chapter is supported by examples and figures granted by leading fluid power manufacturers.

The following topics are discussed in Volume 3 "Hydraulic Fluids and Contamination Control" of this series of textbooks:
- Chapter 2: Hydraulic Fluids
- Chapter 3: Energetic Contamination
- Chapter 4: Gaseous Contamination
- Chapter 5: Fluidic Contamination
- Chapter 6: Chemical Contamination
- Chapter 7: Particulate Contamination
- Chapter 8: Hydraulic Fluid Analysis
- Chapter 9: Hydraulic Filters Performance Ratings
- Chapter 10: Contamination Control in Hydraulic Transmission Lines

The following topics are discussed in Volume 4 "Hydraulic Fluids Conditioning" of this series of textbooks:
- Chapter 01: Introduction to Hydraulic Filters
- Chapter 02: Filter Media and Filtration Mechanisms
- Chapter 08: Filter Selection Criteria

The following topics are discussed in Chapter 12 in Volume 6 "Troubleshooting and Failure Analysis" of this series of textbooks:
- Filters Inspection
- Filters Troubleshooting
- Filters Failure Analysis

Brief Contents

12.1-BP-Filters-01-Selection and Replacement
12.2-BP-Filters-02-Maintenance Scheduling
12.3-BP-Filters-03-Installation and Maintenance
12.4-BP-Filters-04-Standard Tests and Calibration
12.5-BP-Filters-05-Transportation and Storage

Chapter 12: Maintenance of Filters

The following set of best practices provide general guidelines and may not be applicable for all cases. They are not intended to replace the instructions given by the component manufacturer. It is strongly advisable to adhere to instructions provided by the manufacturer.

12.1-BP-Filters-01-Selection and Replacement

Filters are originally specified based on cleanliness level specified by the system designer, placement in the circuit, size, dirt holding capacity, static and dynamic working conditions (pressure, temperature, and flow), and mechanical mounting method. A filter acts like a kidney in a human body. So, replacing an existing filter with a new one of different specifications impacts the hydraulic system reliability on short term operation. So, when replacing an existed filter, none of the originally specified design and operating specifications shall be changed.

The shown below example explores the importance of maintaining the specification of the filter. Table 12.1 shows the amount of the dirt that passes through a pump as function the oil cleanliness level based on conditions of (200 lit/min flow, 18 hours a day, and 340 working days per year). The figure also shows that, even new oil is typically contaminated with particles to ISO 19/17/14.

ISO Code	NAS 1638	Description	Suitable for	Dirt/year
ISO 14/12/10	NAS 3	Very clean oil	All oil systems	7.5 kg *
ISO 16/14/11	NAS 5	Clean oil	Servo & high pressure hydraulics	17 kg *
ISO 17/15/12	NAS 6	Light contaminated oil	Standard hydraulic & lube oil systems	36 kg *
ISO 19/17/14	NAS 8	New oil	Medium to low pressure systems	144 kg *
ISO 22/20/17	NAS 11	Very contaminated oil	Not suitable for oil systems	> 589 kg *

**Table 12.1- Amount of Dirt Pass through a Filter based on Oil Cleanliness Level
(Courtesy of C.C. Jensen Inc.)**

12.2-BP-Filters-02-Maintenance Scheduling

Unless otherwise is stated by components and systems manufacturer, Table 12.1 provides guidelines for *scheduling* preventive maintenance actions for hydraulic filters.

#	Preventive Maintenance Actions	Daily	Weekly	Monthly	Biannually	Annually
1	Clean the dust on outer surface		✔	✔	✔	✔
2	Check hydraulic connections			✔	✔	✔
3	Check status of the filter through clogging indicator **(Note 1)**		✔	✔	✔	✔
4	Check electrical connections (if found)			✔	✔	✔
5	Disassemble and inspect/clean/wash/replace filter element and filter housing **(Note 2)**			✔	✔	✔
6	Check valve performance through standard tests				✔	✔

Table 12.2- BP-Filters-02-Maintenance Scheduling

Note (1): Continuous monitoring of filter conditions in a hydraulic system provide good insight about how healthy the system is. All filter assemblies shall be equipped with a device that indicates when the filter requires servicing. The indication shall be readily visible to the operator or maintenance personnel. When this requirement is not available, scheduled filter element replacement shall be addressed routinely in the operator manual.

Note (2): A disassembled filter can be visually inspected by taking a good look at the filter and check if the filter has any of the following signs:
- Damage such as pleats have signs of cuts or bunched together.
- Filter elements aren't properly sealed on both end caps.
- Center tube is collapsed or buckled.
- Oil degradation products accumulated on the filter element or the end caps.
- High concentration of debris as a sign of metal wear.
- Nonmetallic particles such as paint chips, fibers, seal wear products, etc.

12.3-BP-Filters-03-Installation and Maintenance

One of the keys to consistent filtration performance is good maintenance practices. Remember, any contamination induced by a filter change goes directly into the system. Electro-hydraulic systems utilizing servo valves are the most sensitive and very susceptible to any contamination or air. Referring to Fig. 12.1, the following set of bullets provide common best practices for installation and maintenance of hydraulic filters considering a spin-on filter as an example:

Review available instruction (1): leading manufacturers provide step-by-step servicing guidelines. Where possible, follow the filter service instructions supplied by the original equipment manufacturer. Also, check if there are any servicing instructions that given by filter pictogram. Filter *Pictogram* is a method of printing symbols on the side of filters to indicate specific type of servicing instructions. Figure 12.2 shows an example of filter pictograph.

Check the service indicator (2): Verify that the OEM specified service interval has been reached by any of the service indicators shown in the figure.

Turn off system pressure (3): Be sure the system is turned off and that there is no pressure present in the system. At least isolate the filter under service if isolation setup is found.

Remove the used filter and gasket (4): Remove the spin-on filter, properly dispose of the filter in accordance with local regulations or recycle it. Recycling used hydraulic filters is environmentally friendly. Check your local disposal regulations for proper disposal and recycling.

Clean the filter mounting head and bowel (5): Clean the surfaces of the filter head or cover. Flush sediments from filter bowls only with a pre-filtered solvent. Use only lint free wipes or filtered air to dry the bowl.

Lubricate the filter gasket (6): Lubricate threads and spin on seal with clean system oil.

Inspect new filter (7): Check the new filter you will be installing for any shipping or handling damage. DO NOT install any filter or filter element that shows any signs of damage. The exterior of the filter housing should be cleaned, preferably with a compatible solvent wash. Avoid touching or handling a new element if possible. If the element is supplied in a plastic bag, remove the bag after the filter is in place. After replacing the new filter element, secure the bowl immediately.

Install new filter for instructions (8): Install the spin-on filter until the top of the gasket first contacts the sealing surface. Then for final tightening in accordance with the given instructions. Do not overtighten. Once the filter housing is secured, attempt to bleed air from the housing on initial system start-up, to prevent the air from entering the system, particularly cylinders.

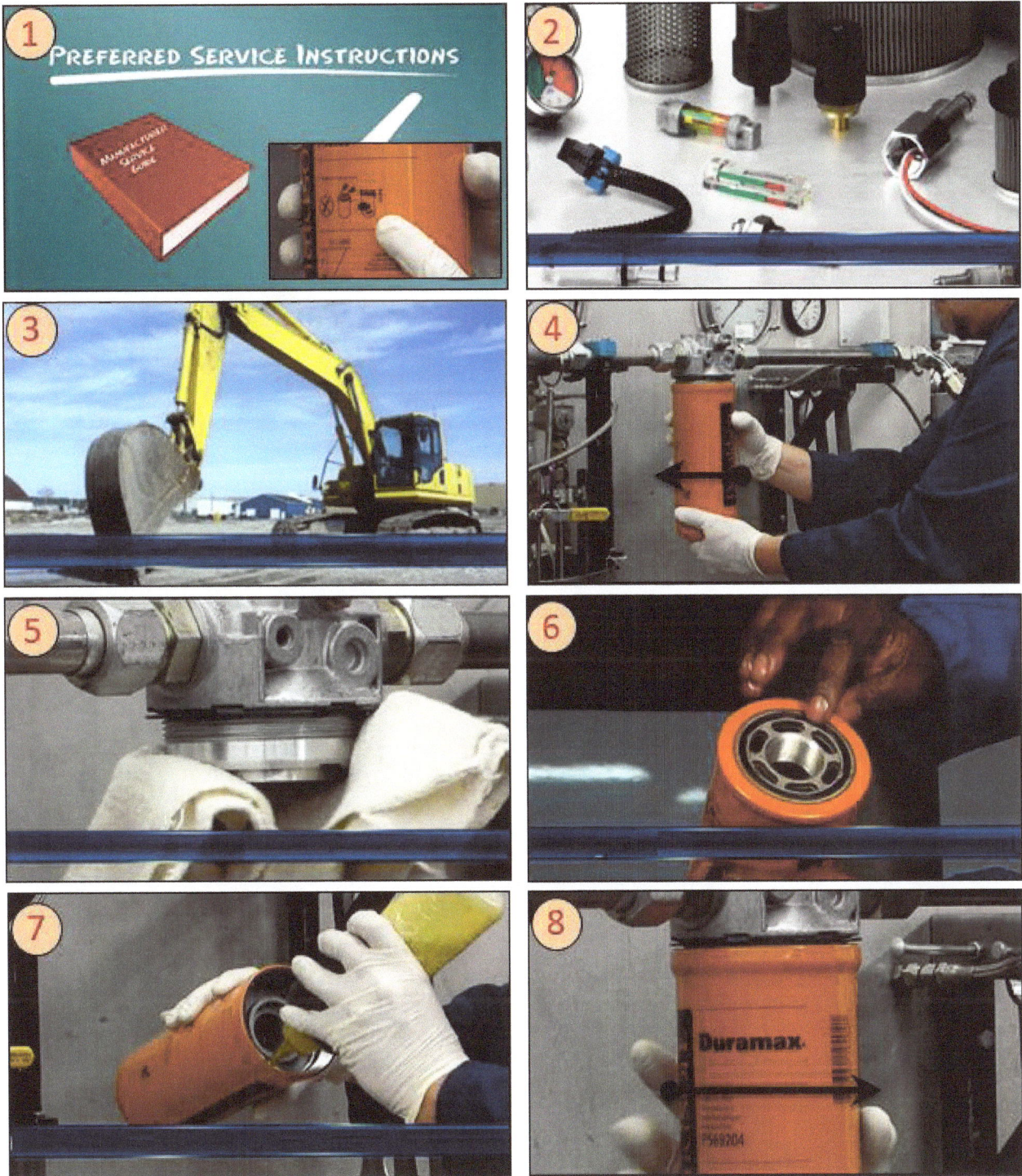

Fig. 12.1- Hydraulic Spin-On Filter Replacement Steps (Courtesy of Donaldson)

Fig. 12.2- Spin-On Hydraulic Filters Service Pictograms (Courtesy from Donaldson)

12.4-BP-Filters-04-Standard Tests and Calibration

Performance characteristics of hydraulic filters are evaluated by several test. The following is a list of standard test methods:

- **ISO 2942:** Filter Element Structural Integrity (Bubble Point) Test.
- **ISO 2943:** Hydraulic Fluid Compatibility Test
- **ISO 16889:** Efficiency and Capacity (Multipaas) Test
- **ISO 3723-2015:** End Load Test
- **ISO 3968:** Differential Pressure Test
- **NFPA (T-2.6.1):** Rated Burst Pressure (RBP) of a Filter Housing
- **ISO 1077-1:** Rated Fatigue Pressure (RFP) of a Filter Housing
- **Cyclic Test Pressure** (CTP) of a Filter Housing
- **ISO 2941:** Collapse Pressure of a Filter Element
- **ISO 3724 OR ISO 23181:** Flow Fatigue Test for Filter Element

Tests shall be run in accordance with the sequence given in Fig 12.3. The table shows the logic sequence of conducting these tests. For example, there is no meaning of conducting burst pressure test if the filter isn't compatible with the fluid.

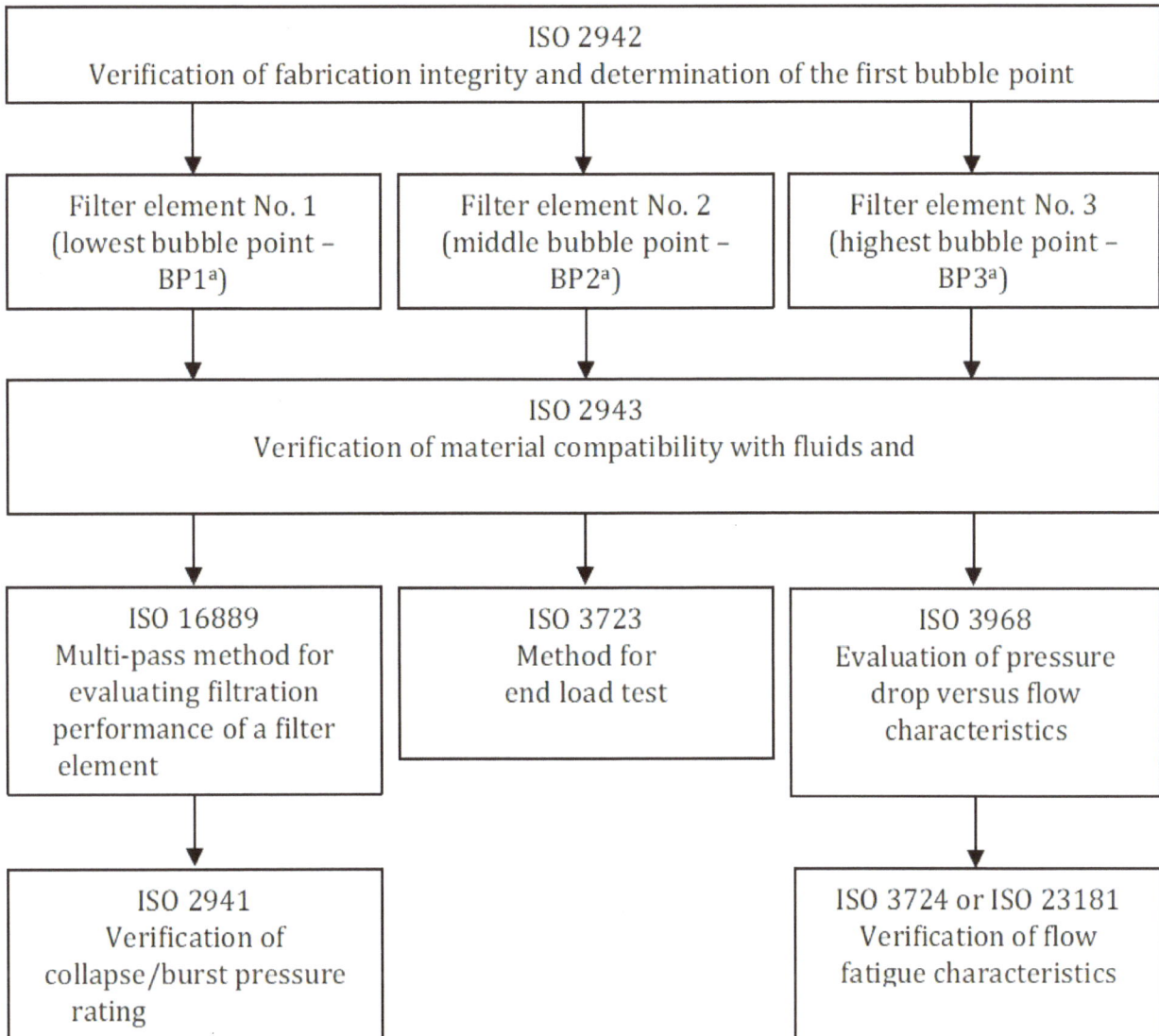

```
┌──────────────────────────────────────────────────────────────────────────┐
│                                ISO 2942                                    │
│    Verification of fabrication integrity and determination of the          │
│                        first bubble point                                  │
└──────────────────────────────────────────────────────────────────────────┘
        │                          │                          │
        ▼                          ▼                          ▼
┌──────────────────┐   ┌──────────────────┐   ┌──────────────────┐
│ Filter element   │   │ Filter element   │   │ Filter element   │
│ No. 1 (lowest    │   │ No. 2 (middle    │   │ No. 3 (highest   │
│ bubble point –   │   │ bubble point –   │   │ bubble point –   │
│ BP1[a])          │   │ BP2[a])          │   │ BP3[a])          │
└──────────────────┘   └──────────────────┘   └──────────────────┘
        │                          │                          │
        ▼                          ▼                          ▼
┌──────────────────────────────────────────────────────────────────────────┐
│                                ISO 2943                                    │
│            Verification of material compatibility with fluids and          │
└──────────────────────────────────────────────────────────────────────────┘
        │                          │                          │
        ▼                          ▼                          ▼
┌──────────────────┐   ┌──────────────────┐   ┌──────────────────┐
│   ISO 16889      │   │   ISO 3723       │   │   ISO 3968       │
│ Multi-pass       │   │ Method for       │   │ Evaluation of    │
│ method for       │   │ end load test    │   │ pressure drop    │
│ evaluating       │   │                  │   │ versus flow      │
│ filtration       │   │                  │   │ characteristics  │
│ performance of a │   │                  │   │                  │
│ filter element   │   │                  │   │                  │
└──────────────────┘   └──────────────────┘   └──────────────────┘
        │                                                 │
        ▼                                                 ▼
┌──────────────────┐                       ┌──────────────────┐
│   ISO 2941       │                       │ ISO 3724 or      │
│ Verification of  │                       │ ISO 23181        │
│ collapse/burst   │                       │ Verification of  │
│ pressure rating  │                       │ flow fatigue     │
│                  │                       │ characteristics  │
└──────────────────┘                       └──────────────────┘
```

Fig. 12.3- Sequence of Conducting Standard Tests for Hydraulic Filters
(Courtesy from Donaldson)

The following sections present the tests based on the sequence they shall be run.

12.4.1- ISO 2942: Filter Element Structural Integrity (Bubble Point) Test.

Verification of fabrication integrity is used to define the acceptability of filter elements for further use or testing. That is why this test is conducted first.

Bubble Point Resistance Test is used to verify the fabrication integrity of a filter element (by checking the absence of bubbles). The fabrication integrity test determines whether a filter element meets the manufacturer's prescribed maximum allowable pore size. Damage created during shipping or manufacturing is identified using the bubble point test.

As shown in Fig. 12.4, the bubble point test is performed by submerging the test element in isopropyl alcohol, or other suitable fluid, while applying air pressure to the inside of the element through a special adapter fitted into the open end of the element. No evidence of a steady stream of bubbles should be detected at the minimum bubble point level, designated by the manufacturer.

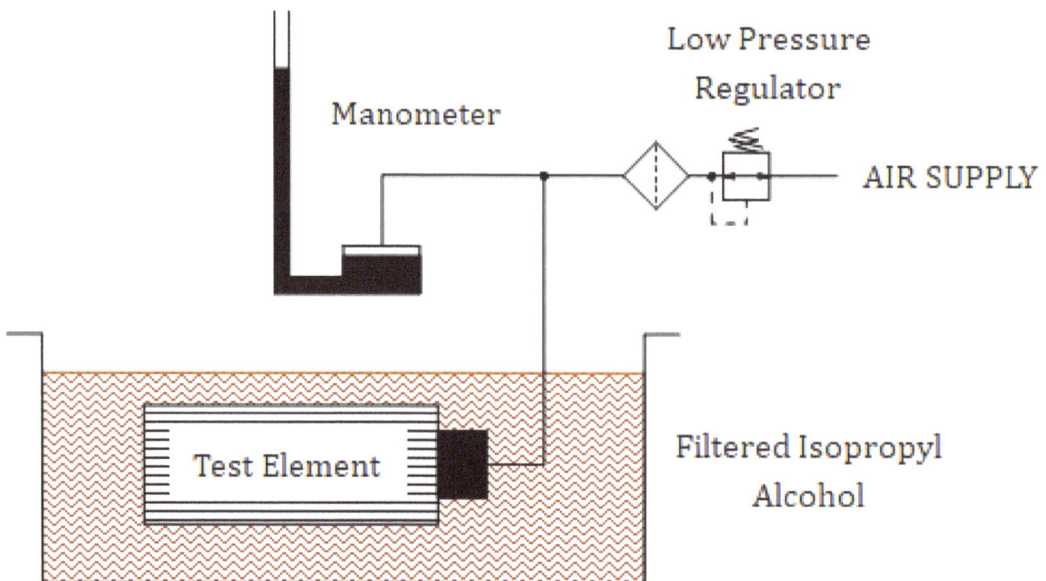

Fig. 12.4- Typical Bubble Point Test Setup

12.4.2- ISO 2943: Hydraulic Fluid Compatibility Test

After passing the structural integrity test, the element compatibility with the hydraulic fluid must be verified. *Fluid Compatibility Test* is conducted according to ISO 2943 standard to verify the compatibility of a specific hydraulic fluid with the component materials of a filter element at maximum expected fluid temperature. Filter elements are submerged in the fluid of interest and subjected to a temperature 15° C (59° F) above the recommended maximum operating temperature of the fluid for a 72-hour period. The element passes the test when no visual evidence of structural failure or material degradation should be present.

12.4.3- ISO 16889: Efficiency and Capacity (Multipaas) Test

12.4.3.1- Multipass Test Purpose and Procedure

The *Multipass Test* is the worldwide recognized method of characterizing hydraulic filter element filtration performance including efficiencies and dirt capacity. To obtain the beta ratio, particulate contaminants of specific sizes of interest must be counted at the upstream (Nu) and downstream (Nd) sizes of the filter. In this test, as shown in Fig. 12.5, hydraulic fluid (Mil-H-5606) is injected with a uniform amount of contaminant (such as ISO 12103-A3 MTD, ISO Medium Test Dust). As shown in the figure, the contaminated fluid is pumped through the filter unit being tested. An automatic particle counter is used to count the particles of certain sizes in both upstream and downstream sides of the filter to determine the *Beta Ratio*.

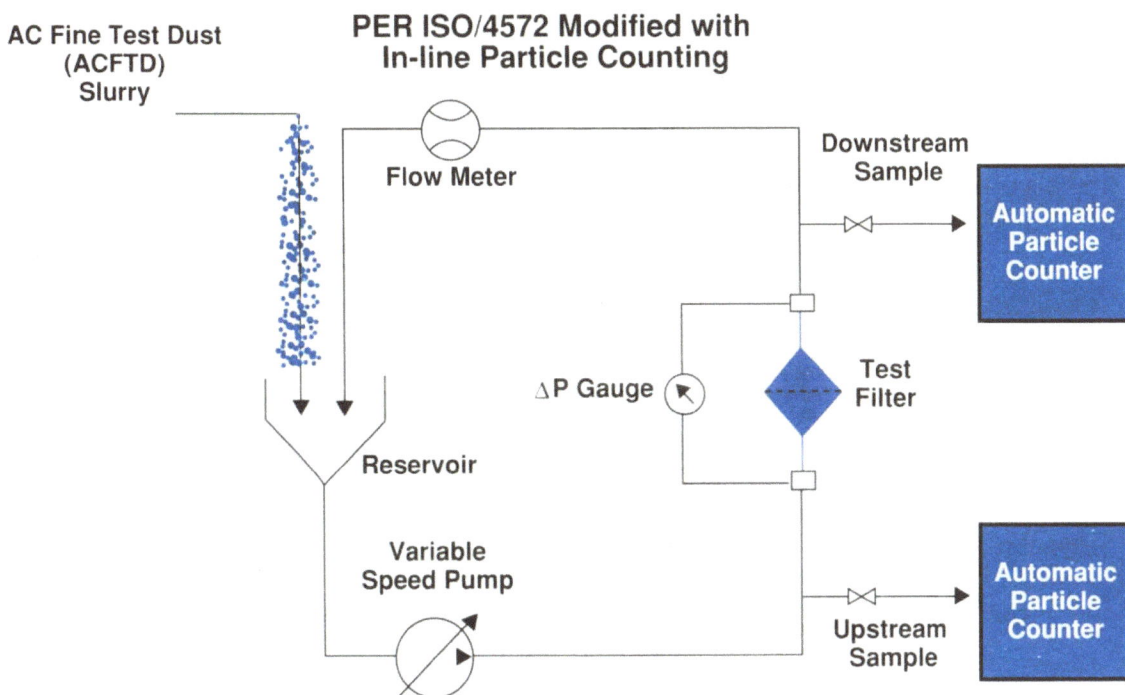

Fig. 12.5- Typical Multipass Performance Test Setup (Courtesy from Pall)

12.4.3.2- Calculation of Beta Ratio

As shown in Fig. 12.6 and Eq. 12.1, *Beta Ratio* is calculated by dividing the number of particles greater than a given size (x) that enter the filter (Nu) by the number of the particles of that same size that leave the filter (Nd). Figure 12.7 shows an example of calculating the beta ratio.

$$\beta_x = \frac{N_U}{N_D}$$

12.1

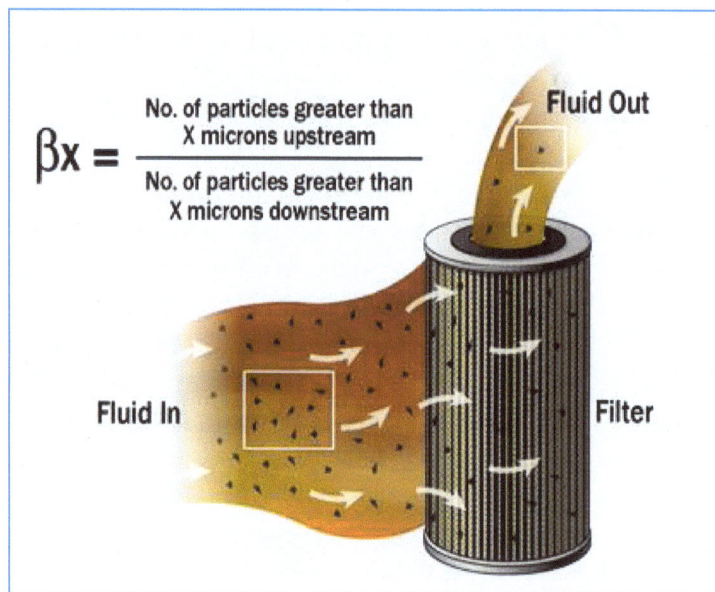

Fig. 12.6- Calculation of Beta Ratio (www.magneticfiltration.com)

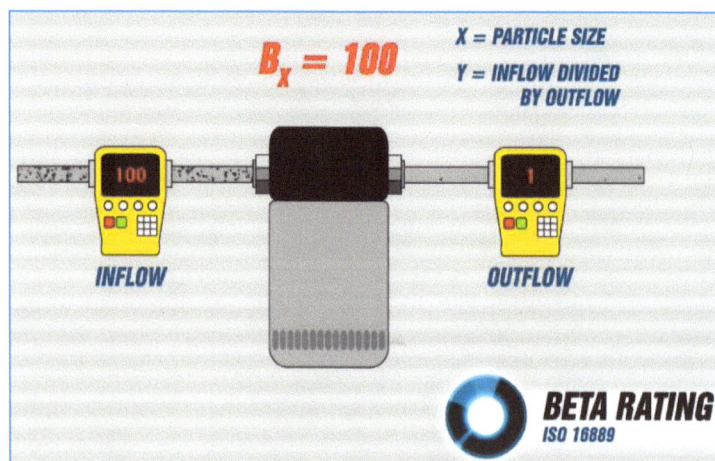

Fig. 12.7- Example of Beta Ratio Calculation (Courtesy of Noria Corporation)

Table 12.3 shows some typical data from a Multipass test. The data is interpreted as follows, the filter has a beta ratio equal to 12 for particle size > 2 μm, a beta ratio equal 100 for particle size > 5 μm, and a beta ratio equal 3000 for particle size > 10 μm.

Particle Size (μm)	Particle Counts (#/ml)		Beta Ratio
2	upstream downstream	15,200 1,267	$\beta_2 = 12$
5	upstream downstream	8,000 80	$\beta_5 = 100$
10	upstream downstream	3,000 1	$\beta_{10} = 3000$

Table 12.3- Typical Multipass Test Data (Courtesy of Pall)

Obviously, as shown in Fig. 12.8, improving the filter rating may double the bearing service life.

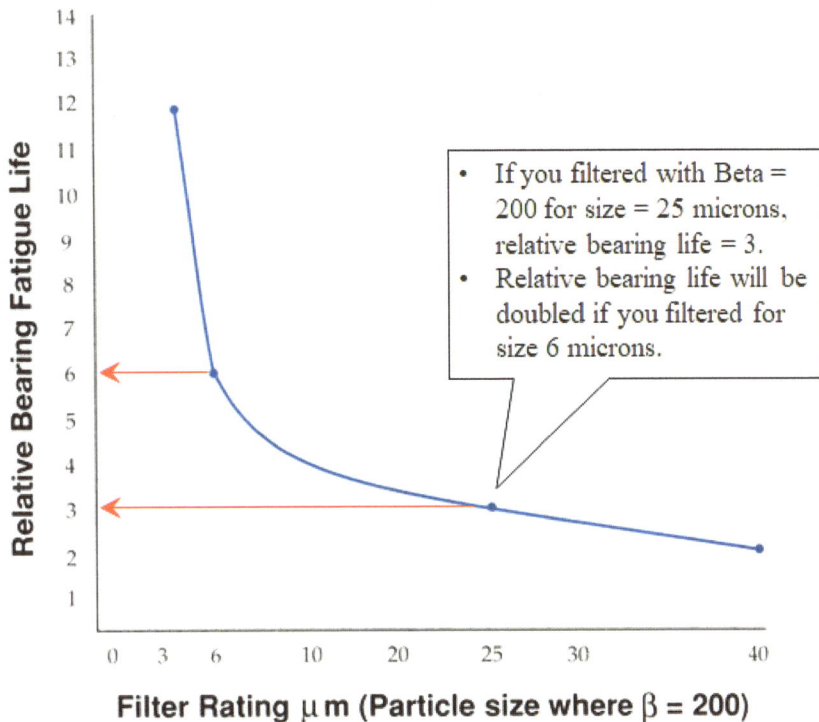

Ref: Macpherson, P.B., Bhachu, R., Sayles, R., "The Influence of Filtration on Rolling Element Bearing Life"

Fig. 12.8- Effect of Beta Ratio on Bearing Life

12.4.3.3- Beta Ratio Stability

The Multipass test is performed under controlled laboratory conditions and does not take into account some of the challenges an inline pressure filter will experience in most hydraulic systems, such as air bubbles, vibrations, pressure and flow surges. Surge pressure and flow can occur during normal operation, e.g. during start-stop, and when pressure compensated pumps are used.

However, Beta Ratio stability is important because it relates to how well a filter element will perform in service over time. Therefore, beta ratio of a filter should be defined within range of working temperature (such as in cold start) and differential pressure across the filter element.

As shown in Fig. 12.9, cyclic or *Surge Flow* affects the Beta ratio and degrades filter performance dramatically unless the filter is properly designed to resist this action. Such design involves filter medium support and resin bonding, as well as smaller pores.

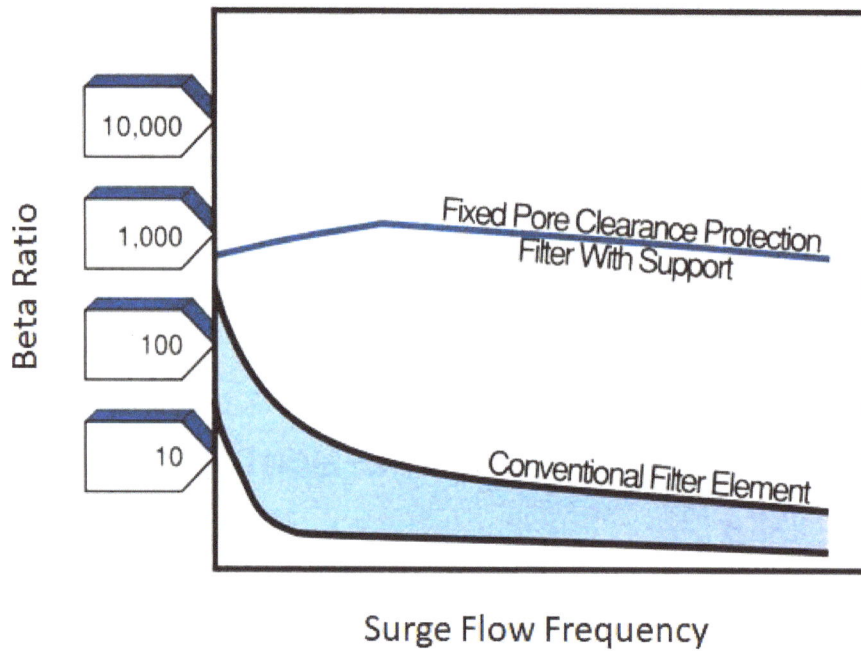

Fig. 12.9- Effect of Surge Flow on Beta Ratio (Courtesy of Pall)

12.4.3.4- Filter Efficiency

Considering of *Filter Efficiency* as being straight forward and easier than beta ratio, it can be calculated using Eq. 12.2. For a beta ratio equal 2, this means that the filter holds 50% of the number of particles introduced to the filter. If Beta Ratio equals 3, then the Filter Efficiency is 67%.

$$E_x = \left[1 - \frac{N_D}{N_U}\right] \times 100 = \left[1 - \frac{1}{\beta_x}\right] \times 100 = \left[\frac{\beta_x - 1}{\beta_x}\right] \times 100 \qquad 12.2$$

Figure 12.10 shows the results of applying the previous equation for a filter element. The figure shows that, in spite of the large change in the beta value from 200 to 1000, the corresponding change in the efficiency is very small (0.4%). Therefore, differentiating between two filters based on beta ratio above 200 is somewhat deceiving.

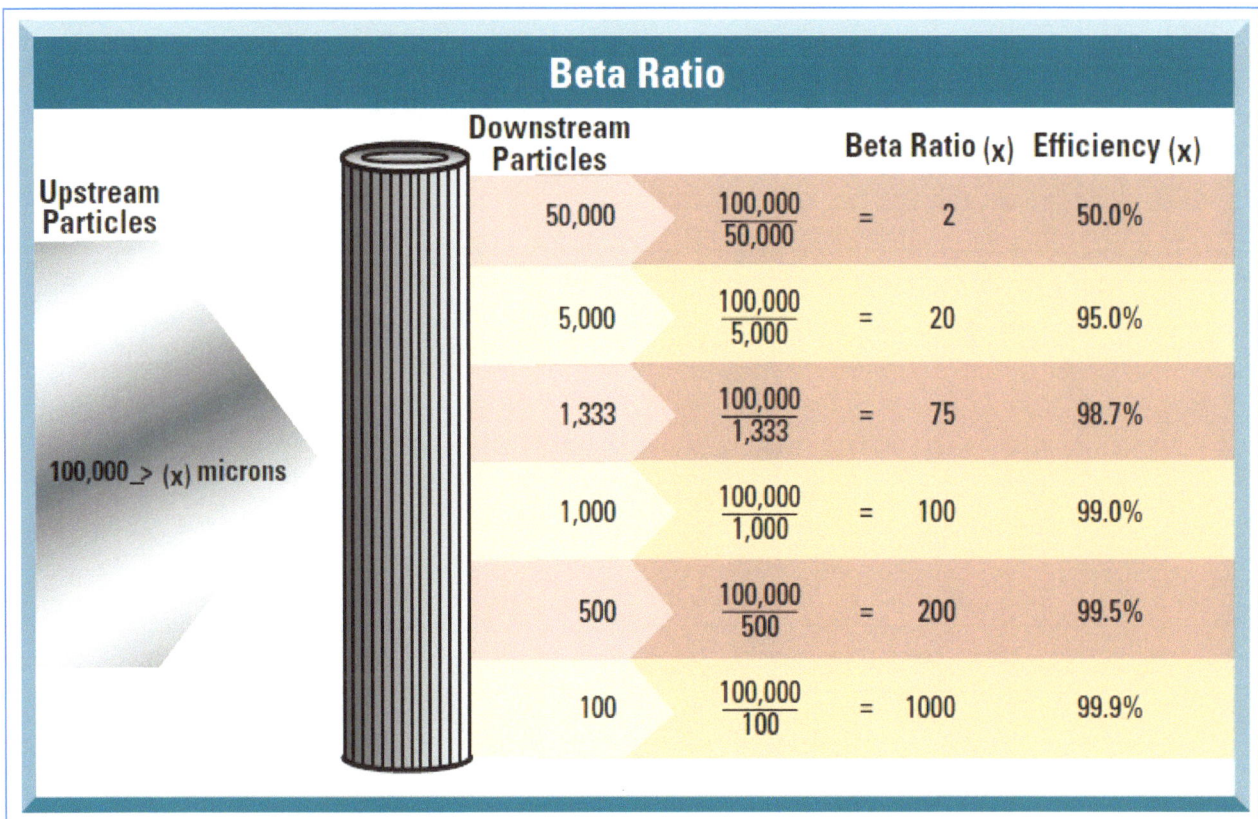

Beta Ratio

Upstream Particles	Downstream Particles	Beta Ratio (x)		Efficiency (x)
100,000 > (x) microns	50,000	$\frac{100,000}{50,000}$ =	2	50.0%
	5,000	$\frac{100,000}{5,000}$ =	20	95.0%
	1,333	$\frac{100,000}{1,333}$ =	75	98.7%
	1,000	$\frac{100,000}{1,000}$ =	100	99.0%
	500	$\frac{100,000}{500}$ =	200	99.5%
	100	$\frac{100,000}{100}$ =	1000	99.9%

Fig. 12.10- Filter Efficiency vs. Beta Ratio (Courtesy of Parker)

Filter efficiency versus fitter beta ratio can be graphically represented as shown in Fig. 12.11.

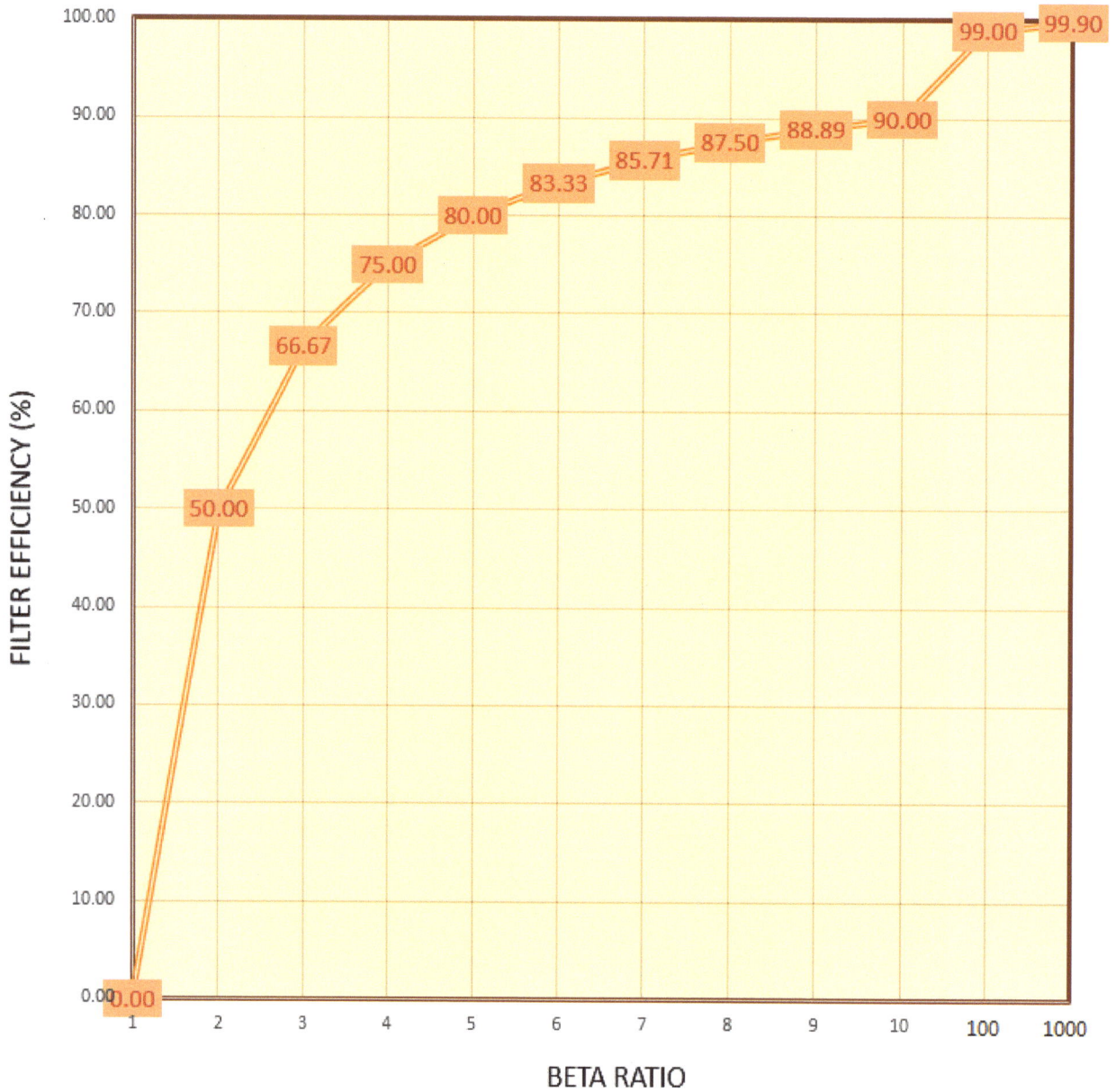

Fig. 12.11- Filter Efficiency versus Beta Ratio

12.4.3.5- Nominal and Absolute Ratings

As shown in Table. 12.4, *Nominal Rating* is the particle size (x) where the filter has 50% efficiency, i.e. $\beta x = 2$ (Ex = 50 %).

Absolute Rating, based upon a historical military standard, is particle size (x) where the filter has 98.7% efficiency, i.e. $\beta x = 75$ (Ex = 98.7 %).

Filters can be rated for various particle sizes as B3/6/15 = 2, 10, 75. This means:

- Filter is nominal at 3 microns.
- Filter is 90% efficient at 6 microns.
- Filter is absolute for 15 microns.

As it has been stated previously for Beta Ratios above 75%, the corresponding increase in the filter efficiency is very slight.

Filtration Ratio (at a given particle size)	Capture Efficiency (at the same particle size)
2	← Nominal → 50 %
5	80%
10	90%
20	95%
75	← Absolute → 98.7 %
100	99%
200	99.5%
1000	99.9%

Table 12.4- Nominal and Absolute Ratings

12.4.3.6- Filter Dirt Holding Capacity

Another important characteristic on which a filter is evaluated is the *Dirt Holding Capacity (DHC)*. It is defined as the weight of dirt that a filter element can hold before the pressure drop (*Terminal Pressure*) across the filter element reaches a predetermined (saturation) limit. As shown in Fig. 12.12, to measure the DHC of a filter element, ISO MTD Test Dust is added to the system to bring the test filter element to a specified maximum differential pressure drop. The total grams of dirt that a filter held is measured. This is part of ISO 16889.

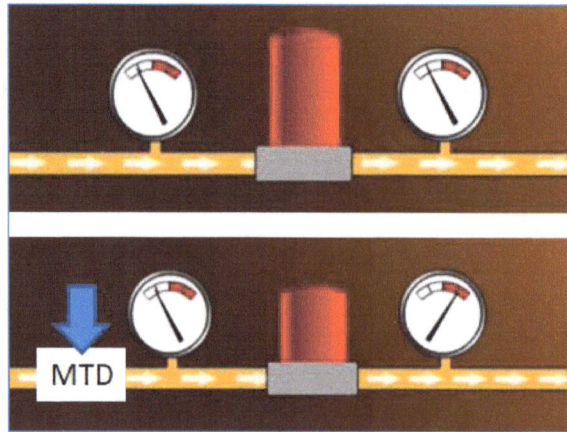

Fig. 12.12- Dirt Holding Capacity Test (Courtesy of Parker)

Since elements with higher DHC need to be changed less frequently, DHC has a direct impact on the overall cost of operation. Equation 12.3 shows the calculation of the cost of removing 1 kg of dirt.

$$\textbf{Cost of Removing 1 kg or lb of Dirt} = \frac{\textbf{Cost of Filter Element (Installation \& Disposal)}}{\textbf{Dirt Holding Capacity in kg or lb}} \qquad \textbf{12.3}$$

Table 12.5 shows the results of applying the previous equation on two different filters as follows:

- While most conventional pressure line filter elements can retain less than hundred grams of dirt (<0.2 lbs), they may be fairly inexpensive to replace. However, if the cost of removing 1 kg or pound of oil contamination is calculated, these conventional pressure filter elements will suddenly appear quite expensive.

- A good quality cellulose based, microfiber, or synthetic offline filter elements can retain up to several kgs/lbs of dirt, so even though the purchase price is higher, the calculated cost for removing one kg or pound of contamination will be considerably lower than that of a pleated pressure filter element, giving lower lifetime costs.

	Example 1	Example 2
Filter type	Glass fiber based pressure filter insert	Cellulose based offline filter insert
Cost of element/insert	€ 35 / $ 50	€ 200 / $ 300
Dirt holding capacity	0.085 kg / 0.18 lbs	4 kg / 8 lbs
Cost per kg/lb removed dirt	€ 412 / $ 278	€ 50 / $ 40

Table 12.5- Cost of Removing Dirt (Courtesy of C.C. Jensen Inc.)

Figure 12.13 shows a stacked disc filter element that is 3 µm nominal and 8 µm absolute. This means that 50% of all particles larger than 3 µm and 98.7% of all solid particles larger than 8 µm are retained. The filter can hold anywhere from 1.5-8 kg of dirt depends on the filter size. Such types of filter elements have high efficiency and DHC but their flow is very low. That is why they commonly used for offline filtration in parallel with the main filter in the system.

Before   After

**Fig. 12.13- Example of Stack Disc Elements
(Courtesy of C.C. Jensen Inc.)**

12.4.3.7- Filter Flow Rate

The fluid flow in the system is a major factor in determining the appropriate filter to use. It is important to have the right size filter to meet the system's requirements. Because fluid can only travel through the filter media so fast, a system with a higher flow rate will need physically larger filters compared to a system with a lower flow rate. If the filter is too small, it will not be able to handle the system flow rate and will create excessive pressure drop, possibly even opening the bypass valve allowing unfiltered fluid through.

Therefore, the filter shall be selected such that the initial differential pressure recommended by the filter manufacturer is not exceeded at the intended flow rate and maximum fluid viscosity.

It is to be noted that, in some hydraulic systems, the maximum flow rate in a return line filter can be greater than the maximum pump flow rate. Examples of these systems are when using differential area cylinders, large single acting cylinders that retract faster than extending, and rapid accumulator discharges.

12.4.3.8- Filter Capacity versus Efficiency

Generally speaking, normal size filters in the system are not designed to adequately deal with large quantities of dirt that occur in connection with component machining, system assembly, system filling, system commissioning, or repair work. Such a large amount of dirt is handled by special large filters during system flushing or offline filtration process.

As shown in Fig. 12.14, a highly restrictive media has better efficiency, but it will be plugged by a small amount of dirt. So, it has low DHC. On the other side, a less restrictive media has lower efficiency, but it can retain more dirt before it gets blocked. Therefore, when selecting a filter, a balance between DHC and efficiency of a filter must be considered.

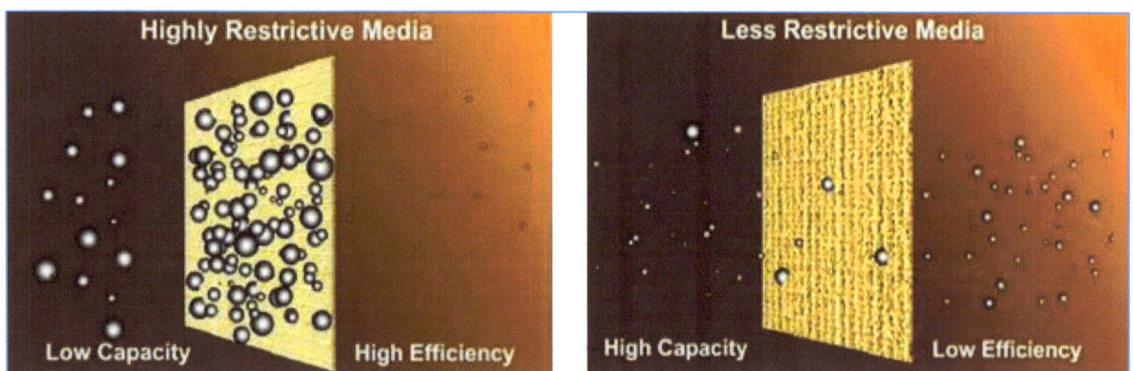

Fig. 12.14- Filter Efficiency vs. DHC (Courtesy of Parker)

12.4.4- ISO 3968: Differential Pressure Test

Differential Pressure, as shown in Fig. 12.15, is the difference between the pressure at the upstream and the downstream sides of the filter. Filter differential pressure depends on:
- Construction of the filter housing.
- Construction and type of filter element.
- Filter size and flow rate through the filter.
- Viscosity and specific gravity (SG) of the fluid flowing through the filter.

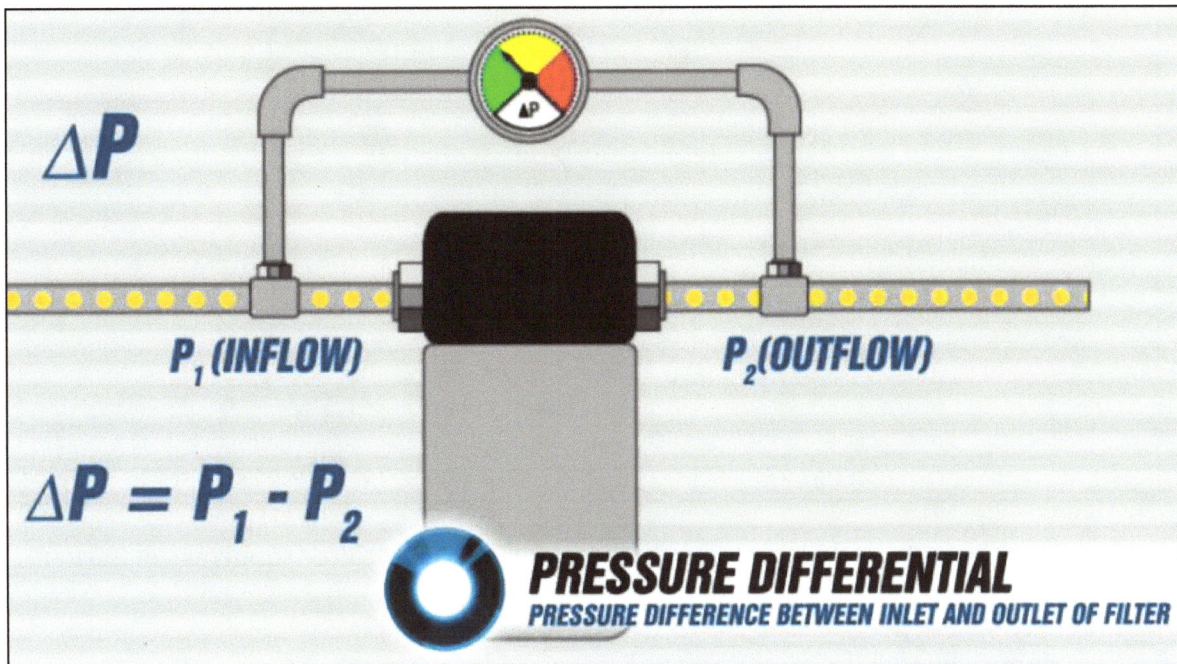

Fig. 12.15- Typical Filter Differential Pressure Test Setup (Courtesy of Noria Corporation)

As shown in Eq. 12.4, pressure drop across a filter is due to both the filter housing and the element.

$$\Delta p_{total} = (\Delta p_H + \Delta p_E) \qquad\qquad 12.4$$

Where, for a specific filter size, fluid flow, viscosity, and specific gravity:
- Δp_{total} is the total differential pressure across the filter assembly.
- Δp_H is the differential pressure across the filter housing (corrected based on SG).
- Δp_E is the differential pressure across the filter element (corrected based on SG & viscosity).

Figure 12.16 shows a typical flow-pressure drop curve for a specific filter size, a specific clean filter media, and a specific fluid.

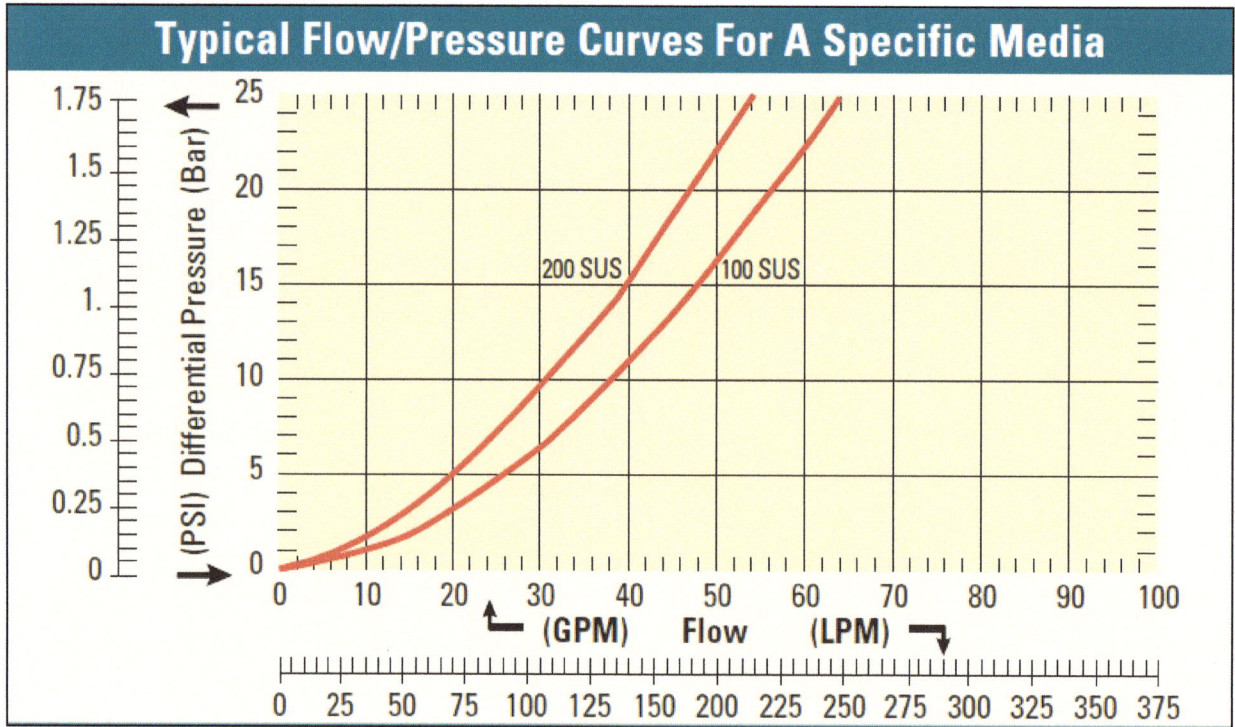

Fig. 12.16- Typical Flow-Pressure Curve for a Specific Filter (Courtesy of Parker)

Catalog data is generally provided for "clean" filter elements for specified media and at a given viscosity. Pressure drop is highly dependent on viscosity, so corrections should be made to the actual fluid being used. In addition, the worst-case viscosity condition is at the coldest anticipated operating temperature, which will need to be considered.

There is no one equation that is applicable for all brands of filters. However, filter manufacturers provide instructions on how to calculate the pressure drop and correct it based on the actual operating conditions. The following examples show different ways to calculate the differential pressure for various filter brands.

Example 1 (Ref. Donaldson):

Given Data:
- Filter Data Sheet for a spin on filter (5 μm) shown in Fig. 12.17.
- Test fluid viscosity = 32cSt [150 SSU] at 100°F (37.7°C).
- Test fluid specific gravity = 0.9 at 100°F (37.7°C).

Exercise:
Find the filter head pressure drop for an actual hydraulic oil of 64 cSt viscosity and 1.1 specific gravity. Estimated flow rate is 150 gpm.

Solution:

$$\Delta p_{\text{Fiter Head}} = 3 \ \times \frac{64}{32} \ \times \frac{1.1}{0.9} \ = \ 7.33 \text{ psid}$$

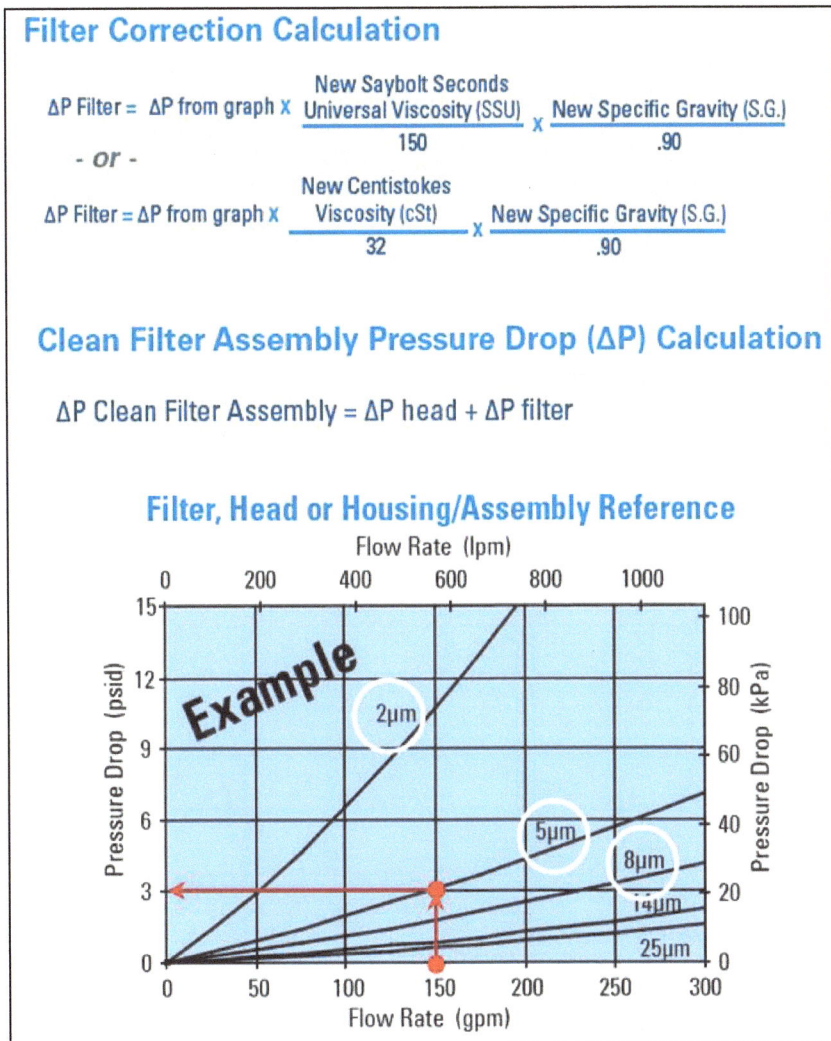

Filter Correction Calculation

$$\Delta P \text{ Filter} = \Delta P \text{ from graph} \times \frac{\text{New Saybolt Seconds Universal Viscosity (SSU)}}{150} \times \frac{\text{New Specific Gravity (S.G.)}}{.90}$$

- or -

$$\Delta P \text{ Filter} = \Delta P \text{ from graph} \times \frac{\text{New Centistokes Viscosity (cSt)}}{32} \times \frac{\text{New Specific Gravity (S.G.)}}{.90}$$

Clean Filter Assembly Pressure Drop (ΔP) Calculation

ΔP Clean Filter Assembly = ΔP head + ΔP filter

Filter, Head or Housing/Assembly Reference

Fig. 12.17- Example of Pressure Drop Calculation (Courtesy of Donaldson)

Example 2 (Ref. Schroeder):

Given Data:
- Filter Data Sheet shown in Fig. 12.18.
- Test fluid viscosity = 32cSt [150 SSU] at 100°F (37.7°C).

Exercise: For a filter NZ25-1N series, find the filter assembly total pressure drop for an actual hydraulic oil of 44 cSt (200 SUS) and 0.86 specific gravity. Estimated flow rate is 15 gpm.

Solution: See the figure below.

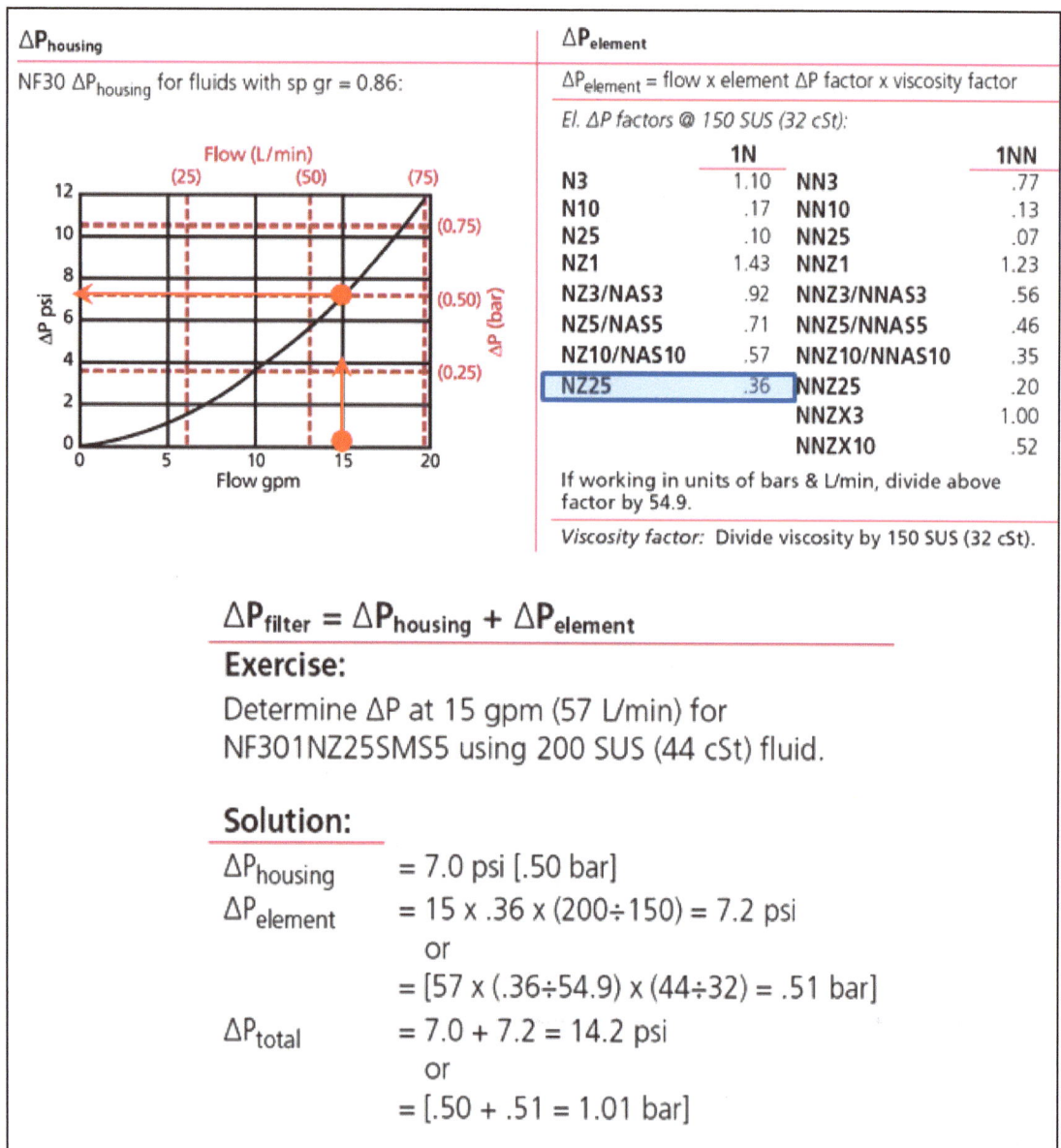

$$\Delta P_{filter} = \Delta P_{housing} + \Delta P_{element}$$

Exercise:

Determine ΔP at 15 gpm (57 L/min) for NF301NZ25SMS5 using 200 SUS (44 cSt) fluid.

Solution:

$\Delta P_{housing}$	= 7.0 psi [.50 bar]
$\Delta P_{element}$	= 15 x .36 x (200÷150) = 7.2 psi
	or
	= [57 x (.36÷54.9) x (44÷32) = .51 bar]
ΔP_{total}	= 7.0 + 7.2 = 14.2 psi
	or
	= [.50 + .51 = 1.01 bar]

Fig. 12.18- Example of Pressure Drop Calculation (Courtesy of Schroeder)

Example 3 (Ref. Hydac):

Figure 12.19 shows another example for calculating the total pressure drop of a filter including the given catalog data, the application data, and the solution.

EXAMPLE - an application with the following criteria would be sized as shown.

Conditions: **Fluid** – Hydraulic Oil (ISO-32)

Specific Gravity – 0.86

Viscosity – 141 SSU

Flow Rate – 30 GPM

Fluid Temperature - 104°F normal

Filter Type Selected - Pressure Filter

HYDAC Model No. **DF ON 240 TE 10 D 1.0 / 12 V -B6**

HOUSING

$$\Delta P \text{ Housing} = \Delta P \text{ Calculation } \textit{(From Curve in catalog)} \times \frac{\text{Actual Specific Gravity}}{0.86}$$

$$\Delta P \text{ Housing} = 1.5 \text{ psid} \times \frac{0.86}{0.86} = 1.5 \text{ psid}$$

ELEMENT

$$\Delta P \text{ Clean Element} = \Delta P \text{ Calculation } \times \frac{\text{Actual Specific Gravity}}{0.86} \times \frac{\text{Actual Viscosity}}{141 \text{ SSU}}$$

$$\Delta P \text{ Clean Element} = 30 \text{ GPM} \times 0.175 \times \frac{0.86}{0.86} \times \frac{141 \text{ SSU}}{141 \text{ SSU}}$$

$$\Delta P \text{ Clean Element} = 5.25 \times 1 \times 1 = 5.25 \text{ psid}$$

FILTER ASSEMBLY

$$\Delta P \text{ Filter Assembly} = \Delta P \text{ Housing} + \Delta P \text{ Clean Element}$$
$$1.5 \text{ psid} + 5.25 \text{ psid} = 6.75 \text{ psid}$$

Fig. 12.19- Example of Pressure Drop Calculation (Courtesy of Hydac)

Example 4 (Ref. Pall):

Given Data:
  - Filter Data Sheet shown in Fig. 12.20.
  - Test fluid viscosity = 32cSt [150 SSU] at 100°F (37.7°C),
  - Test fluid specific gravity = 0.9 at 100°F (37.7°C).
  - Fluid flow = 100 l/min.

Exercise: Find the filter assembly pressure drop for a Series UH210 housing with -20 port sizes housing and an AN grade element of 13" length. Actual hydraulic fluid used has 50 cSt and specific gravity of 1.2. Estimated flow rate is 100 l/min.

Solution: see the figure below.

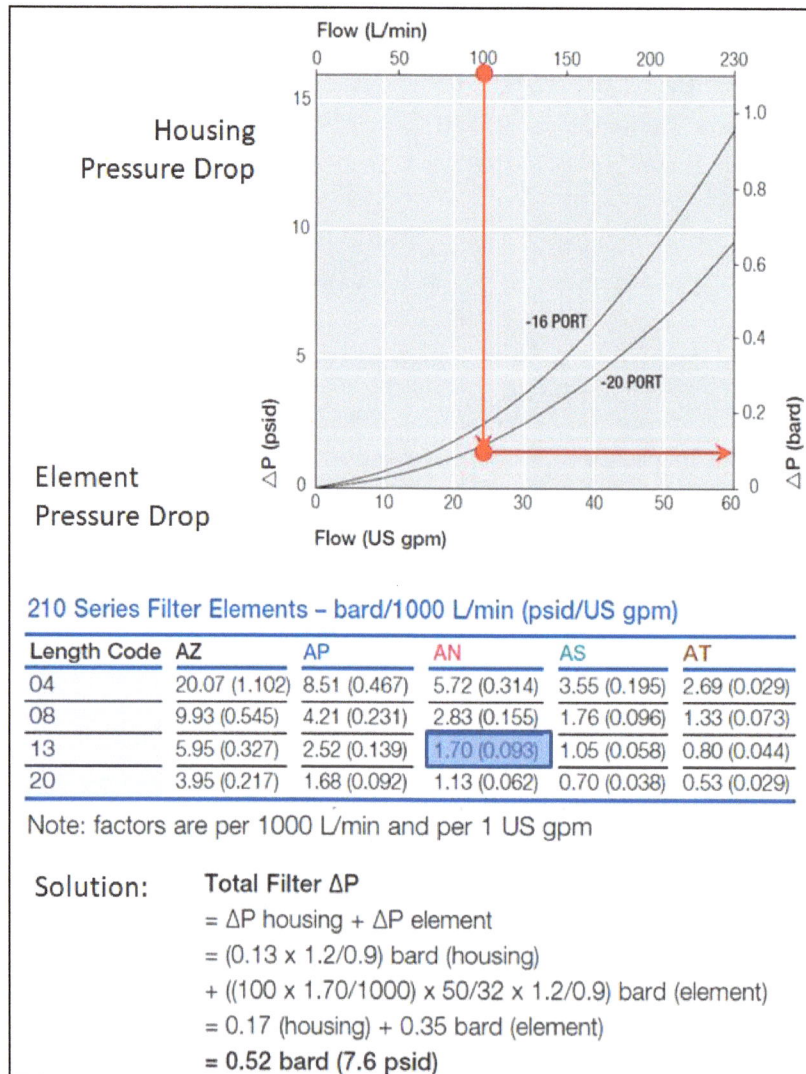

Flow (L/min)

Housing Pressure Drop

Element Pressure Drop

-16 PORT
-20 PORT

ΔP (psid)
ΔP (bard)
Flow (US gpm)

210 Series Filter Elements – bard/1000 L/min (psid/US gpm)

Length Code	AZ	AP	AN	AS	AT
04	20.07 (1.102)	8.51 (0.467)	5.72 (0.314)	3.55 (0.195)	2.69 (0.029)
08	9.93 (0.545)	4.21 (0.231)	2.83 (0.155)	1.76 (0.096)	1.33 (0.073)
13	5.95 (0.327)	2.52 (0.139)	1.70 (0.093)	1.05 (0.058)	0.80 (0.044)
20	3.95 (0.217)	1.68 (0.092)	1.13 (0.062)	0.70 (0.038)	0.53 (0.029)

Note: factors are per 1000 L/min and per 1 US gpm

Solution:

Total Filter ΔP
= ΔP housing + ΔP element
= (0.13 x 1.2/0.9) bard (housing)
+ ((100 x 1.70/1000) x 50/32 x 1.2/0.9) bard (element)
= 0.17 (housing) + 0.35 bard (element)
= 0.52 bard (7.6 psid)

Fig. 12.20- Example of Pressure Drop Calculation (Courtesy of Pall)

12.4.5- NFPA (T-2.6.1): Rated Burst Pressure (RBP) of a Filter Housing

Rated Burst Pressure (RBP) of a filter housing is the static pressure at which <u>filter housing</u> structural failure occurs. Burst Pressure is determined by a test according to NFPA (T-2.6.1) standards.

12.4.6- ISO 1077-1: Rated Fatigue Pressure (RFP) of a Filter Housing

Rated Fatigue Pressure (RFP), as shown in Eq. 12.5, is the maximum allowable pressure for a <u>filter housing </u>according to <u>(</u>ISO 10771-1). Safety factor is typically 4 – 6.

$$\mathbf{RFP} = \frac{\mathbf{RBP}}{\mathbf{Safety\ Fator}}$$
                                        12.5

12.4.7- Cyclic Test Pressure (CTP) of a Filter Housing

There are many hydraulic systems that use highly repetitive functions such as plastic injection molding machines, die-casting machines, and hydraulic presses. In such systems, *Cyclic Test Pressure (CTP)* should be considered when selecting a filter. CTP is the maximum pressure applied for certain number of cycles (typically 1 million cycles) before housing failure occurs. CRP is experimentally tested. However, as shown in Eq. 12.6, CTP can be mathematically calculated. CTP equals RFP multiplied by a factor K that is obtained from tables associated with the above-mentioned standard based on confidence, assurance levels, materials of construction, and number of units tested.

$$\mathbf{CTP} = \mathbf{RFP} \times \mathbf{K}$$
                                        12.6

Example:
- RBP = 20,000 psi.
- Safety Factor = 4 →
- RFP = 20,000/4= 5,000 psi.
- K = 1.1 – 1.4 →
- CTP = 5,000 X 1.5 = 7,500 PSI

12.4.8- ISO 2941: Collapse Pressure of a Filter Element

As shown in Fig. 12.21, differential pressure is used as indicator for the state of the filter. It indicates whether the filter is ok to continue to operate or if it should be replaced.

When a filter reaches a level of plugging or a cold start occurs or a combination of both, an increase in pressure is seen between the inlet (dirty side) and the outlet (clean side). If this differential pressure is high enough, the filter element and/or center tube can rupture or collapse. This is serious because unfiltered fluid and damaged filter components can then be routed back into the system.

A filter assembly whose element cannot withstand, without damage, the maximum differential pressure in its part of the system shall be equipped with a filter bypass valve. Ideally, a filter element should be sized so that the initial differential pressure across the clean element (plus the filter housing drop) is less than half the bypass valve setting in the filter housing.

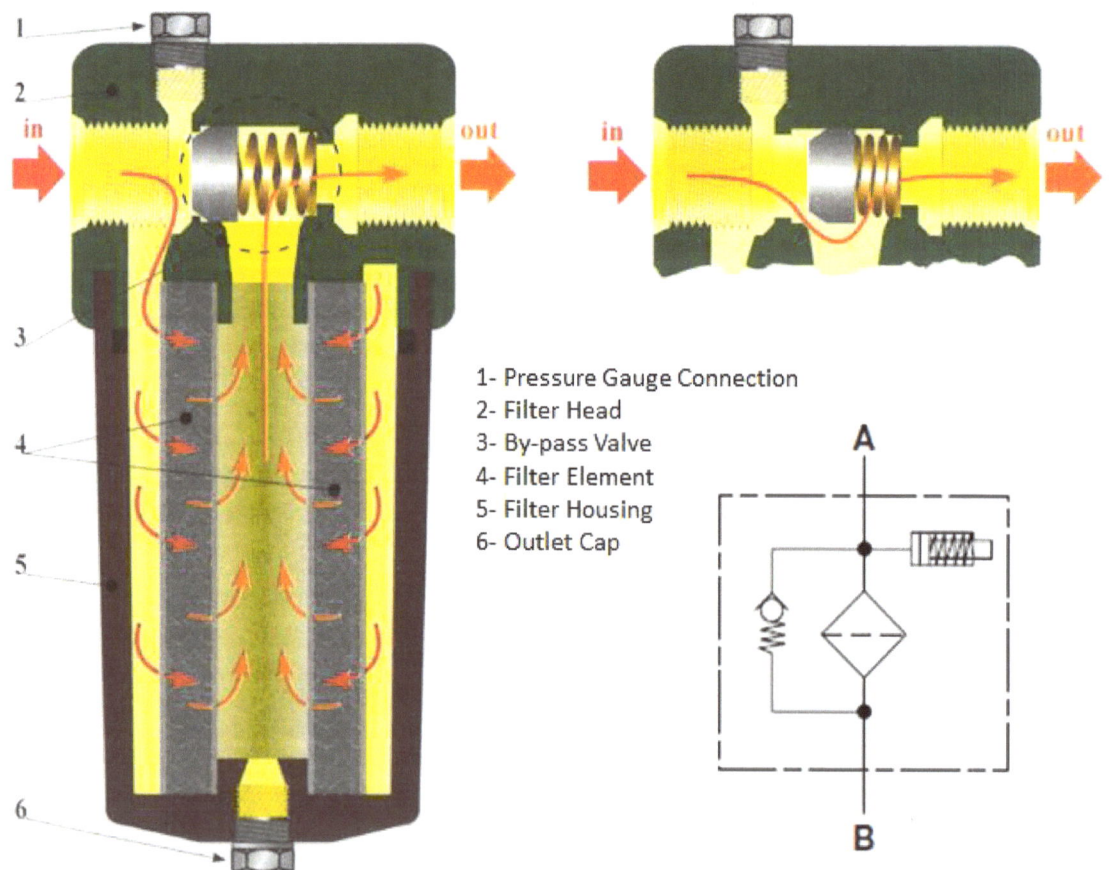

1- Pressure Gauge Connection
2- Filter Head
3- By-pass Valve
4- Filter Element
5- Filter Housing
6- Outlet Cap

**Fig. 12.21- Filter Housing Equipment with Bypass Valve and Clogging Indicator
(Courtesy of Assofluid)**

Collapse Pressure of a <u>filter element</u> is the differential pressure at which a structural failure of the filter element and/or center tube occurs. Collapse pressure is determined by ISO 2941 standards.

As shown in Fig. 11.22, the collapse pressure rating of a filter element installed in a filter housing, with a bypass valve, should be at least two times greater than the full flow bypass valve pressure drop.

Pressure filters with no bypass are recommended with the use of servo valves. The collapse pressure rating for filter elements used in filter housings with no bypass valves must be at least the same as the setting of the system relief valve upstream of the filter high-crush element. When a high-pressure collapse element becomes clogged with contamination all functions downstream of the filter will become inoperative.

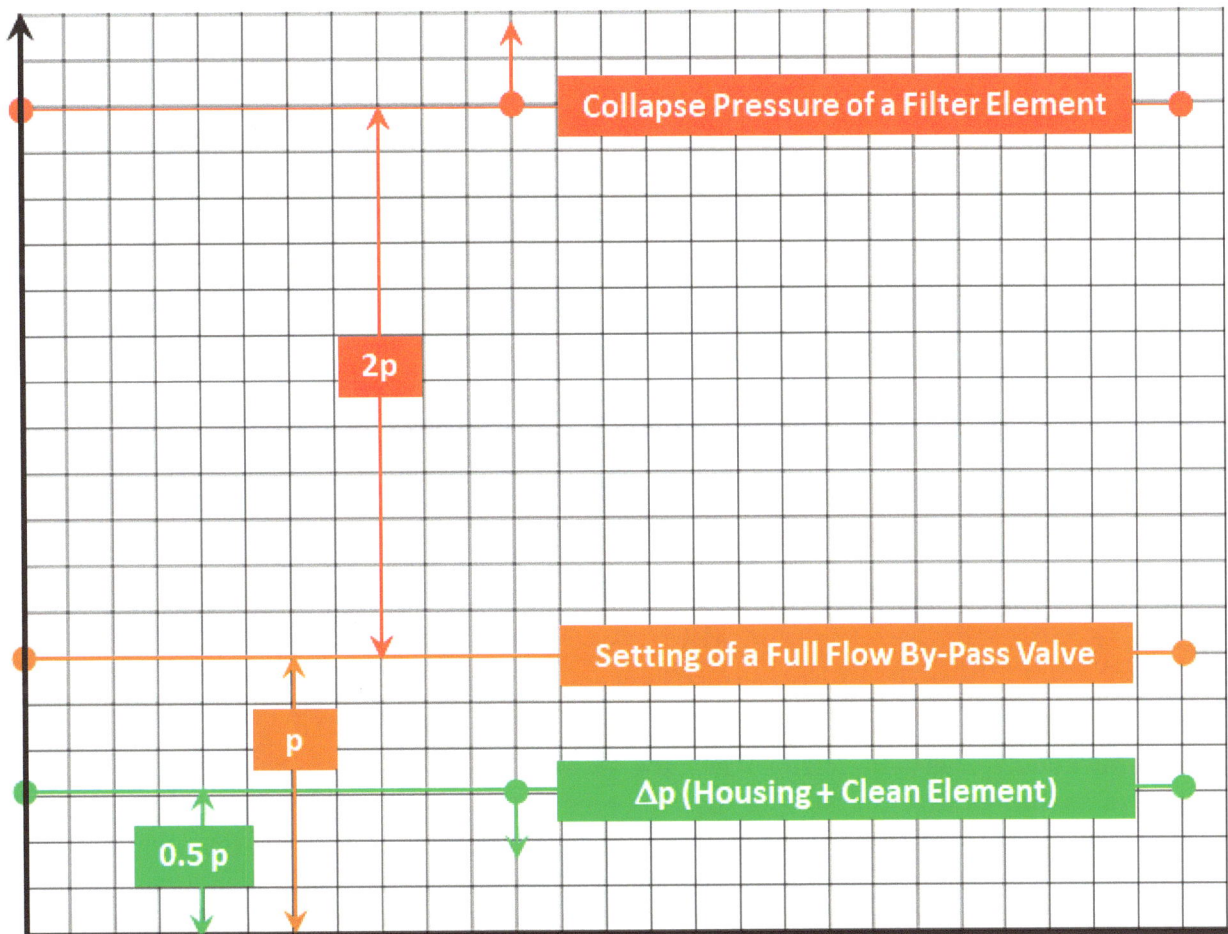

Fig. 12.22- Collapse Pressure of a Filter Element versus By-Pass Setting

Contamination Loading Curve: As shown in Fig. 12.23, as dirt is trapped by the filter, differential pressure (ΔP) increases. As shown in the figure, an alarming pressure must be activated before the bypass valve open, the bypass valve must open before the collapse pressure of the filter media is achieved, and the collapse pressure (1) of the filter media must be lower than the core collapse pressure (2). The *core collapse pressure* is the pressure at which the supporting center tube is crashed or ruptured.

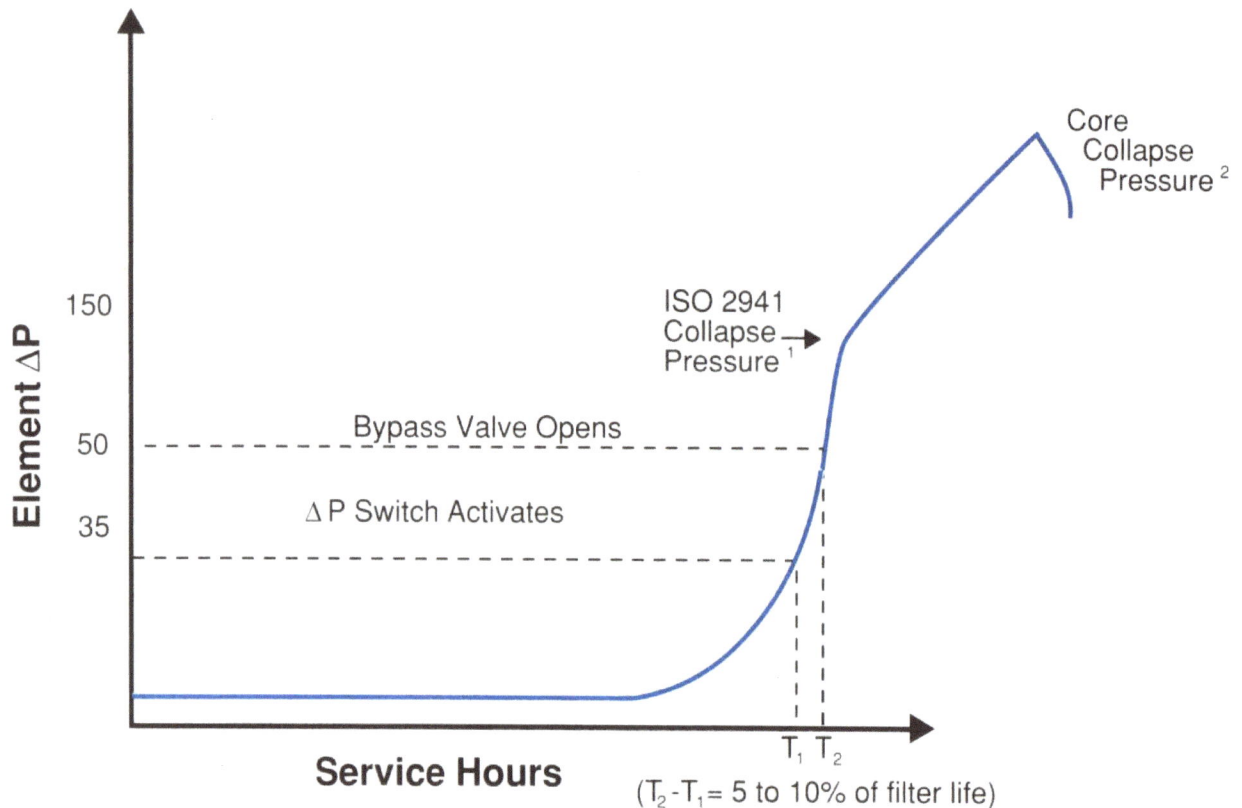

Fig. 12.23- Filter Service Life versus Pressure Drop

12.4.9- ISO 3723-2015: End Load Test

The *End Load Test* is conducted to heck the ability of a filter element to resist axial deformation caused by differential pressure across the element. Weights or other load generating means are used to simulate an axial force on the same surfaces receiving the load when installed in the proper housing. The element is considered pass the test when it is subjected to the maximum axial load specified by the manufacturer without permanent deformation, structural damage, or seal failure.

12.4.10- ISO 3724 OR ISO 23181: Flow Fatigue Test for Filter Element

Due to pulsations of flow, filter media may fail prior to replacement time. High fatigue stability is achieved by better filter element design including supporting both sides of the element and high inherent stability of the filter materials.

Flow Fatigue Test is used to predict the ability of a filter element to withstand structural failure due to flexing of the pleats caused by cyclic flow. Flow Fatigue Test ran in accordance with (ISO 3724 OR ISO 23181).

As show in Fig. 12.24, This test requires that an element be contaminated to its terminal differential pressure. The element is then subjected to a cyclic change in flow from zero to its maximum rated flow and then back to zero for the number of cycles prescribed by the element manufacturer, usually based on (10-200) thousand cycles. An element is considered to have passed the test if there is no visual evidence of structural, seal or filter medium failure.

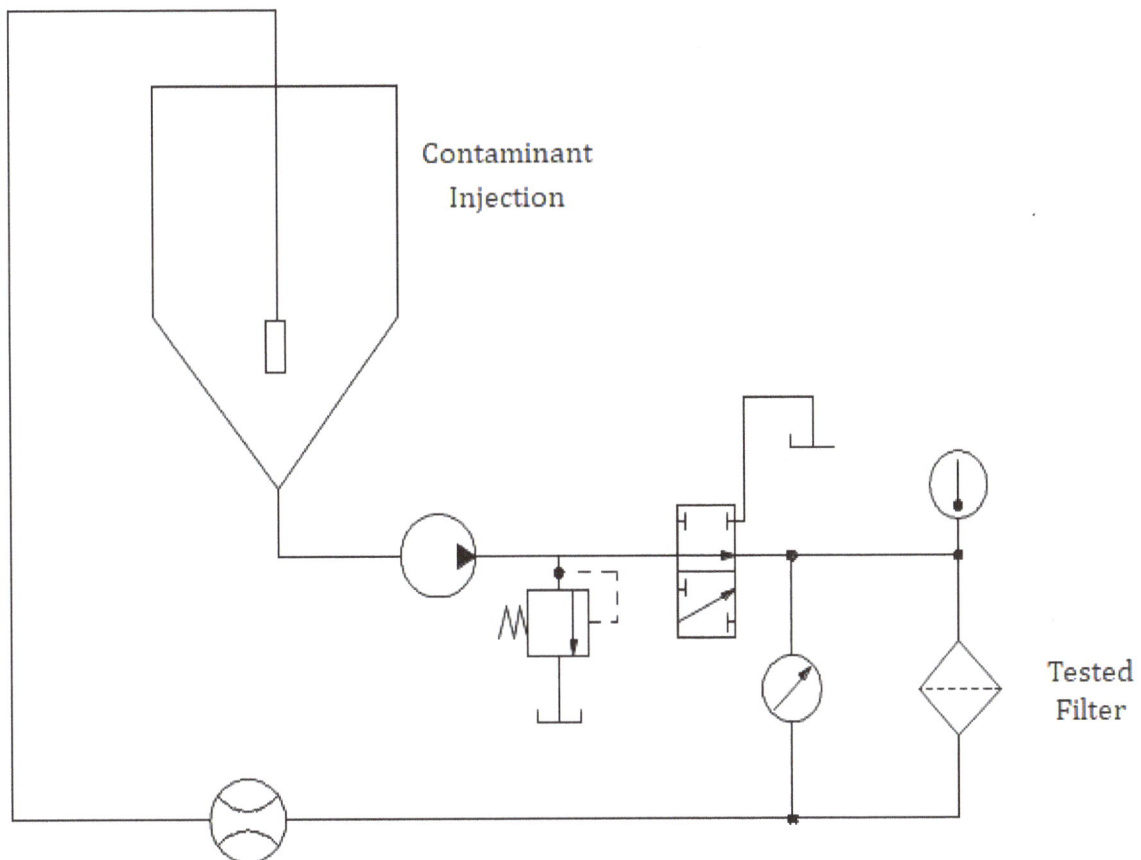

Contaminant
Injection

Tested
Filter

Fig. 12.24- Typical Flow Fatigue Test Setup

12.4.11- Filter Tests Pictogram

As shown in Fig. 12.25, filter tests may be presented in form of pictograms in the filter datasheet or service manual. The shown pictogram shows the standards and the measure of the dirt holding capacity and efficiency via multipass test (1), filter element collapse pressure (2), filter element differential pressure (3), and filter element structural integrity test via bubble point test (4).

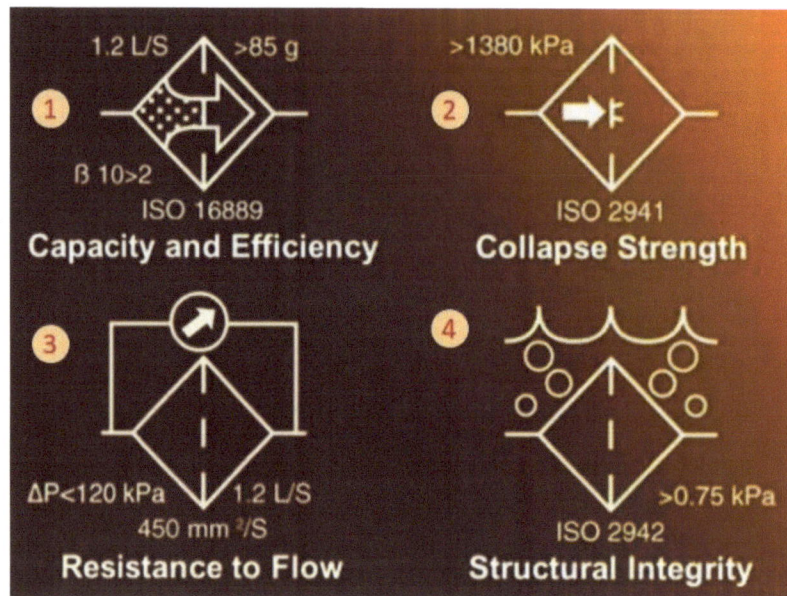

Fig. 12.25- Hydraulic Filters Tests Pictograms (Courtesy from Parker)

12.5-BP-Filters-05-Transportation and Storage

It's important to practice good storage and handling techniques when it comes to filters. As reported by and excerpted from Donaldson service manuals, the following tips are considered best practices for storage and transportation of filters for the sake of contamination control:

Storage:
- Never store a filter on a shelf without it being in a box or totally sealed from outside contaminant.
- When you see an open box of filters on the shelf, tape it shut–unless the filters inside the box are individually sealed.
- Make sure labels with product information and manufacturing dates are visible to personnel selecting from the shelves.
- Metal storage shelves may cause condensation to form on filters if sitting directly on metal. So, it is recommended to use wooden or plastic shelves.
- Over time the filter may get rusty. This is another good reason to store filters in boxes.

Shipping:
- If transporting filters from one job site to another, don't let them roll around on the floorboard or in the back of a truck as it may damage the filter.

Handling:
- Practice "first-in, first-out" with your inventory. When possible, always use the oldest inventory first.
- If a product box has layers of contaminant, take care that the contaminant doesn't get on the new filter as you remove it from the box.
- Handle filters with care to prevent filter damage. For example, don't throw filters into the back of a truck.

APPENDIXES

APPENDIX A: LIST OF FIGURES

Chapter 1 - Hydraulic System Safety

Chapter 02: Basic Concepts of Hydraulic System Maintenance

Chapter 03: Hydraulic Measuring Instruments

Chapter 04: Maintenance of Pumps

Fig. 6.36 – Best Practices for Storing Hydraulic Cylinders

Chapter 07: Maintenance of Valves

Fig. 7.1- Symbols for Counterbalance Valve (Left) and Sequence Valve (Right)
Fig. 7.2- Different Transitional Conditions for a Directional Valve
Fig. 7.3- Proper Housekeeping when disassembling a Hydraulic Valve
Fig. 7.4- Improper Maintenance Practices Kept the Valve Dirty
Fig. 7.5- Last Chance Filters in Servo Valves
Fig. 7.6- Best Practices for Proper Valve Mounting
Fig. 7.7- Example of Manufacturer's Instructions for Valve Installation (Courtesy of Hydraforce)
Fig. 7.8- Main Ports on a 4-way Directional Valve
Fig. 7.9- Supply and Drain of Pilot Pressure in a Pilot-Operated DCV
Fig. 7.10- Considerations for Proper Electrical Connection to Hydraulic Valves
Fig. 7.11- Instructions for Servo Valve Adjustments (Courtesy of Bosch Rexroth)
Fig. 7.12- Standard Test Circuit for Flow Control Valve (CFC Industrial Training)
Fig. 7.13- Circuit for Pressure Relief Valve Adjustment Test (CFC Industrial Training)
Fig. 7.14- Circuit for Pressure Relief Valve Leakage Test (CFC Industrial Training)
Fig. 7.15- Sample Test Worksheet for a Reducing Valve (CFC Industrial Training)
Fig. 7.16- Leakage Test for an Industrial Directional Valve (CFC Industrial Training)
Fig. 7.17- Standard Test Circuit for Four-Ports EH Valve (Courtesy of NFPA)
Fig. 7.18- Stationary Proportional and Servo Valve Tester (dietzautomation.com)
Fig. 7.19- Portable Servo Valve Analyzer Series F087-127 (Courtesy of Moog)

Chapter 08: Maintenance of Accumulators

Fig. 8.1- Bladder Accumulator (Courtesy of Hydac)
Fig. 8.2- Bladder Accumulator Disassembling Instructions (Courtesy of Hydac)
Fig. 8.3- Safety Base (Manifold) for an Accumulator (Courtesy of Hydac)
Fig. 8.4- Proper Mounting of Accumulators
Fig. 8.5- Accumulator Charging Kit (Courtesy of Parker)
Fig. 8.6- Accumulator Charging and Gauging Kit (Courtesy of Hydac)

Chapter 09: Maintenance of Reservoirs

Fig. 9.1- Cleaning of Hydraulic Reservoirs
Fig. 9.2- Clean versus Dirty Suction Strainer
Fig. 9.3- Best Practices for Filling Hydraulic Reservoirs

Chapter 10: Maintenance of Transmission Lines

Fig. 10.1- Bet Practices for Hydraulic Transmission Line Leakage Inspection
(www.bondfluidaire.com)
Fig. 10.2- Keep Outer Surfaces including End Joints (mac-hyd.com)
Fig. 10.3- Hose Crimping Process (Courtesy from Parker)
Fig. 10.4- Hose Crimping Machines (Courtesy from Gates)
Fig. 10.5- Best Practices for Hydraulic Hose Routing

Chapter 11: Maintenance of Heat Exchanges

Chapter 12: Maintenance of Filters

APPENDIX B: LIST OF TABLES

APPENDIX C: LIST OF STANDARD TEST METHODS

- International Organization for Standardization (ISO)
- American Society for Testing and Materials (ASTM)
- Society of Automotive Engineering (SAE)
- American National Standards Institute (ANSI)
- German Institute for Standardization (DIN)
- Iron and Steel Institute (AISI)

ASTM STANDARDS

No.	Title
D 95	Test Method for Water in Petroleum Products and Bituminous Materials by Distillation
D 96	Test Method for Water and Sediment in Crude Oil by Centrifuge Method
D 97	Test Method for Pour Point of Petroleum Products
D 130	Test Method for Determination of Copper Corrosion from Petroleum Products by the Copper Strip Tarnish Test
D 445	Test Method for Kinematic Viscosity of Transparent and Opaque Liquids (the Calculation of Dynamic Viscosity)
D 446	Specifications and Operating Instructions for Glass Capillary Kinematic Viscometers
D 471	Test Method for Rubber Property–Effect of Liquids
D 664	Test Method for Acid Number of Petroleum Products by Potentiometric Titration
D 665	Test Method of Rust-Preventing Characteristics of Inhibited Mineral Oil in the Presence of Water
D 892	Test Method for Foaming Characteristics of Lubricating Oils
D 943	Test Method for Oxidation Characteristics of Inhibited Mineral Oils
D 1401	Test Method for Water Separability of Petroleum Oils and Synthetic Fluids
D 1744	Test Method for Determination of Water in Liquid Petroleum Products by Karl Fischer Reagent
D 2070	Test Method for Thermal Stability of Hydraulic Oils
D 2270	Practice for Calculating Viscosity Index from Kinematic Viscosity at 40°C and 100°C
D 2422	Classification of Industrial Fluid Lubricants by Viscosity System
D 2619	Test Method for Hydrolytic Stability of Hydraulic Fluids (Beverage Bottle Method)
D 2717	Test Method for Thermal Conductivity of Liquids
D 2766	Test Method for Specific Heat of Liquids and Solids
D 2783	Test Method for Measurement of Extreme-Pressure Properties of

	Lubricating Fluids (Four-Ball Method)
D 2983	Test Method for Low-Temperature Viscosity of Automotive Fluid Lubricants Measured by Brookfield Viscometer
D 3339	Test Method for Acid Number of Petroleum Products by Semi-Micro Color Indicator Titration
D 3427	Test Method for Air Release Properties of Petroleum Oils
D 3603	Test Method for Rust-Preventing Characteristics of Steam Turbine Oils in the Presence of Water (Horizontal Disk Method)
D 3707	Test Method for Storage Stability of Water-in-Oil Emulsions by the Oven Test Method
D 3709	Test Method for Stability of Water-in-Oil Emulsions Under Low to Ambient Temperature Cycling Conditions
D 4172	Test Method for Wear Preventive Characteristics of Lubricating Fluid (Four-Ball Method)
D 4310	Test method for Determination of the Sludging and Corrosion Tendencies of Inhibited Mineral Oils
D 4684	Test Method for Determination of Yield Stress and Apparent Viscosity of Engine Oils at Low Temperatures
D 5133	Test Method for Low temperature, Low Shear Rate, Viscosity/Temperature Dependence of Lubricating Oils Using a Temperature Scanning Technique
D 5182	Test Method for Evaluating the Scuffing Load Capacity of Oils (FZG Visual Method)
D 5306	Standard Test Method for Linear Flame Propagation Rate of Lubricating Oils and Hydraulic Fluids
D 5534	Test Method for Vapor-Phase Rust-Preventing Characteristics of Hydraulic Fluids
D 5621	Test method for Sonic Shear Stability of Hydraulic Fluid
D 5770	Test method for Semiquantitative Micro Determination of Acid Number of Lubricating Oils During Oxidation Testing
D 6006	Guide for Assessing Biodegradability of Hydraulic Fluids
D 6046	Classification of Hydraulic Fluids for Environmental Impact
D 6080	Practice for Defining the Viscosity Characteristics of Hydraulic Fluids
D 6158	Specification for Mineral Hydraulic Oils
D 6278	Test Methods for Shear Stability of Polymer Containing Fluids Using a European Diesel Injector Apparatus
D 6351	Test Method for Determination of Low Temperature Fluidity and Appearance of Hydraulic Fluids
D 6546	Test Methods for and Suggested Limits for Determining Compatibility of Elastomer Seals for Indus0trial Hydraulic Fluid Applications
D 6547	Test Method for Corrosiveness of a Lubricating Fluid to a Bi-Metallic Couple
D 6793	Test Method for Determination of Isothermal Secant and Tangent Bulk

	Modulus
D 6973	Test Method for Indicating Wear Characteristics of Petroleum Hydraulic Fluids in a High Pressure Constant Volume Vane Pump
D 7043	Test Method for Indicating Wear Characteristics of Non-Petroleum and Petroleum Hydraulic Fluids in a Constant Volume Vane Pump
D 7044	Specification for Biodegradable Fire Resistant Hydraulic Fluids
D 7216	Test Method for Determining Automotive Engine Oil Compatibility with Typical Seal Elastomers
D 7596	Test Method for Automatic Particle Counting and Particle Shape Classification of Oils Using a Direct Imaging Integrated Tester
D 7647	Test Method for Automatic Particle Counting of Lubricating and Hydraulic Fluids Using Dilution Techniques to Eliminate the Contribution of Water and Interfering Soft Particles by Light Extinction
D 7670	Practice for Processing In-Service Fluid Samples for Particle Contamination Using Membrane Filters
D 7684	Standard Guide for Microscopic Characterization of Particles from In-Service Lubricants
D 7721	Standard Practice for Determining the Effect of Fluid Selection on Hydraulic System or Component Efficiency
D 7752	Practice for Evaluating Compatibility of Mixtures of Hydraulic Fluids
D7843	Method for Measurement of Lubricant Generated Insoluble Color Bodies in In-Service Turbine Oils Using Membrane Patch Colorimetry
D 7873	Method for Determination of Oxidation Stability and Insoluble Formation of Inhibited Turbine Oils at 120 °C Without the Inclusion of Water (Dry TOST Method)
OTHER STANDARDS	
ISO 3448	Industrial Liquid Lubricants — ISO Viscosity Classification
ISO 4263-3	Petroleum and related products — Determination of the ageing behavior of inhibited oils and fluids using the TOST test — Part 3: Anhydrous procedure for synthetic hydraulic fluids
ISO 4392-1	Hydraulic fluid power — Determination of characteristics of motors — Part 1: At constant low speed and constant pressure, International Organization for Standardization, Geneva, Switzerland
ISO 4406	Hydraulic fluid power — Fluids — Method for coding the level of contamination by solid particles, International Organization for Standardization, Geneva, Switzerland (1999)
ISO 4407	Hydraulic fluid power — Fluid contamination — Determination of particulate contamination by the counting method using an optical microscope
ISO 4409	Hydraulic fluid power – Positive displacement pumps, motors and integral transmissions – Methods of testing and presenting basic steady state performance, International Organization for Standardization, Geneva, Switzerland

ISO 6072	Rubber - Compatibility Between Hydraulic Fluids and Standard Elastomeric Materials Test, International Organization for Standardization, Geneva, Switzerland
ISO 6743/4	Part 4: Family H (Hydraulic Systems), Lubricants, Industrial Oils and Related Products (Class L): Classification Part 4: Family H (Hydraulic Systems)
ISO 7120	Petroleum products and lubricants — Petroleum oils and other fluids — Determination of rust-preventing characteristics in the presence of water
ISO 7745	Hydraulic fluid power – Fire resistant fluids – Guidelines for use
ISO 11158	Lubricants, industrial oils and related products (class L) — Family H (hydraulic systems) — Specifications for categories HH, HL, HM, HV and HG
ISO 11171	Hydraulic fluid power – Calibration of automatic particle counters for liquids
ISO 11943	Hydraulic fluid power — On-line automatic particle-counting systems for liquids — Methods of calibration and validation
ISO 12922	Lubricants, Industrial Oils, and Related Products (Class L)—Family H (Hydraulic systems)—Specifications for categories HFAE, HFAS, HFB, HFC, HFDR and HFDU
ISO 13357-1	Petroleum products — Determination of the filterability of lubricating oils — Part 1: Procedure for oils in the presence of water
ISO 13357-2	Petroleum products — Determination of the filterability of lubricating oils — Part 2: Procedure for dry oils
ISO 15380	Lubricants, Industrial Fluids and Related Procedures (Class L), Family H (Hydraulic Systems)–Specifications for Categories HETG, HEPG, HEES and HEPR
ISO 21018-1	Hydraulic fluid power — Monitoring the level of particulate contamination of the fluid — Part 1: General Principles
MIL-PRF-5606	HYDRAULIC FLUID, PETROLEUM BASE; AIRCRAFT, MISSILE, AND ORDNANCE
MIL-PRF-46170	HYDRAULIC FLUID, RUST INHIBITED, FIRE RESISTANT, SYNTHETIC HYDROCARBON BASE, NATO CODE NO. H-544
MIL-PRF-83282	HYDRAULIC FLUID, FIRE RESISTANT, SYNTHETIC HYDROCARBON BASE, METRIC, NATO CODE NUMBER H-537

APPENDIX D: LIST OF REFERENCES

Hydraulic Systems Volume 1- Introduction to Hydraulics for Industry Professionals
Author: Dr. Medhat Kamel Bahr Khalil, 2016.
Publisher: Compudraulic, USA.
ISBN 978-0-692-62236-0

Hydraulic Systems Volume 2- Electro-Hydraulic Components and Systems
Author: Dr. Medhat Kamel Bahr Khalil, 2016.
Publisher: Compudraulic, USA.
ISBN: 978-0-9977634-2-3

Hydraulic Systems Volume 3- Hydraulic Fluids and Contamination Control
Author: Dr. Medhat Kamel Bahr Khalil, 2016.
Publisher: Compudraulic, USA.
ISBN: 978-0-9977816-3-2

Hydraulic Systems Volume 6- Troubleshooting and Failure Analysis
Author: Dr. Medhat Kamel Bahr Khalil, 2022.
Publisher: Compudraulic, USA.
ISBN: 978-0-9977634-6-1

Hydraulic Systems Volume 7- Modeling and Simulation for Application Engineers
Author: Dr. Medhat Kamel Bahr Khalil, 2016.
Publisher: Compudraulic, USA.
ISBN: 978-0-9977816-3-2

R01- Basic Electronics for Hydraulic Motion Control
Author: Jack L. Johnson, PE 1992.
Publisher: Penton Publishing Inc. 1100 Superior Avenue. Cleveland, OH 44114.
ISBN No. 0-932905-07-2.

R02- Closed Loop Electro-hydraulics Systems Manual
Author: Vickers/Eaton.
Publisher: Vickers Inc. 1992.
Training Center, 2730 Research Drive, Rochester Hills, MI 48309-3570.
ISBN 0-9634162-1-9

R03- Bosch Automation Technology
Author: Werner Gotz, Steffen Haack, Ralph Mertlick.
Publisher: Bosch.
ISBN 3-933698-05-7.

R04- Electrohydraulic Proportional and Control Systems
Publisher: Bosch Automation 1999.
ISBN   0-7680-0538-8.

R05- Proportional and Servo Valve Technology – The Hydraulic Trainer Volume 2
Author: R. Edwards, J. Hunter, D. Kretz, F. Liedhegener, W. Schenkel, A. Schmitt.
Publisher: Mannesman Rexroth AG 1988. D-8770 Lohr a. Main. ISBN   3-8023-0266-4.

R06- Proportional Hydraulics
Author: D. Scholz.
Publisher: Festo Didactic KG, Esslingen, Germany.

R07- Electricity, Fluid Power, and Mechanical Systems for Industrial Maintenance
Author: Thomas Kissell.
Publisher: Prentice Hall, Inc. 1999, Upper Saddle River, NJ 07458.
ISBN 0-13-896473-4.

R08- Fluid Power in Plant and Field – First Edition
Author: Charles S. Hedges, R.C. Womack.
Publisher: Womack Machine Supply Co. 1968.
Womack Educational Publication, 2010 Shea Road, Dallas, TX 75235.
ISBN   68-22573 (Library of Congress Card Catalog No.).
R09- Hydraulics, Fundamentals of Service
Author: Deere and Company.
Publisher: John Deere Publishing 1999.
Almon TIAC Bldg. Suite 104, 1300-19th Street, East Moline, IL 61244.
ISBN 0-86691-265-7.

R10- Industrial Hydraulics Troubleshooting
Author: James E. Anders, Sr.
Publisher: McGraw-Hill, Inc.
ISBN 0-07-001592-9.

R11- Power Hydraulics
Author: John Ashby.
Publisher: Prentice Hall 1989. Prentice Hall International, (UK) Ltd.
66 Wood Lane End, Hemel Hempstead, Hertfordshire, HP2 4RG.
ISBN 0-13-687443-6.

R12- Fluid Power with Application
Author: Anthony Esposito.
Publisher: Prentice Hall.
ISBN 0-13-060899-8.

R13- Hydraulic Component Design and Selection
Author: E.C. Fitch.
Publisher: BarDyne Inc. 5111 North Perkins Rd. Stillwater, OK 74075.
ISBN 0-9705922-3-X.

R14- Planning and Design of Hydraulic Power Systems – The Hydraulic Trainer, Vol. 3
Author: Mannesmann Rexroth GmbH.
Publisher: Mannesman Rexroth AG 1988.
D-97813 Lhr a. Main, Jahnsrtrabe 3-5 D-97816 Lohr a. Main.
ISBN 3-8023-0266-4.

R15- Logic Element Technology: Hydraulic Trainer, Volume 4
Author: Mannesmann Rexroth GmbH.
Publisher: Mannesmann Rexroth GmbH 1989.
.Postfach 340, D 8770 Lohr am Main, Telefon (09352) 180.
ISBN 3-8023-0291-5.

R16- Hydrostatic Drives with Control of the Secondary Unit. The Hydraulic Trainer, Volume 6
Author: Dr. Alfred Feuser, Rolf Kordak, Gerold Liebler.
Publisher: Mannesmann Rexroth GmbH 1989.
Postfach 340, D 8770 Lohr am Main.

R17- Control Strategies for Dynamic Systems: Design and Implementation
Author: John H. Lumkes, Jr.
Publisher: Marcel Dekker, Inc. 2002.
Marcel Dekker, Inc. 270 Madison Avenue, New York, NY 10016.
ISBN 0-8247-0661-7.

R18- Feedback Control Of Dynamic Systems
Author: Gene F. Franklin, J. David Powell, Abbas Emami-Naeini.
Publisher: Prentice-Hall, Inc.
Upper Saddle River, New Jersey.
ISBN 0-13-032393-4.

R19- Modeling and Analysis of Dynamic Systems
Author: Charles M. Close, Dean. Frederick
Rensselaer Polytechnic Institute
Publisher: John Wiley & Sons, Inc.
ISBN 0-471-12517-2.

R20- Design of Electrohydraulic Systems for Industrial Motion Control
Author: Jack L. Johnson, PE.
Milwaukee School of Engineering.
Publisher: Parker.
Copyright © Jack L. Johnson, PE 1991.

R21- Basic Pneumatics
Author: Kjell Evensen & Jul Ruud.
Publisher: AB Mecmann Stockholm 1991.
S-125 81 Stockholm, Sweden.
ISBN 91-85800*21-X.

R22- Basic Pneumatics: The Pneumatic Trainer, Volume 1
Author: Ing. –Buro J.P. Hasebrink.
D7761 Moos.
Editor: Mannesmann Rexroth Pneumatik GmbH.
Bartweg 13, W 3000 Hannover 91.

R23- Electro-Pneumatics: The Pneumatic Trainer, Volume 2
Author: Rolf Balla.
Publisher: Mannesmann Rexroth 1990, Pneumatik GmbH.
Publication No: RE 00 262/01.92.

R24- Pneumatics Theory and Applications
Author: Bosch Automation.
Publisher: Robert Bosch GmbH 1998.
Automation Technology Division, Training (AT/VSZ)
ISBN 1-85226-135-8.

R25- Fluid Power Engineering
Author: M. Galal Rabie.
Publisher: McGraw-Hill.
ISBN 978-0-07-162246-2.

R26- Air Motors Ideas with Air
Author: GAST Mfg. Co.
Publisher: GAST Mfg. Co. 1978.
P.O. Box 97, Benton Harbor, MI 49022.
Book No: Booklet #100.

R27- Air Motor Handbook
Author: GAST Mfg. Co.
Publisher: GAST Mfg. Co. 1978.
P.O. Box 117, Benton Harbor, MI 49022.

R28- Troubleshooting Hydraulic Components: Using Leakage Path Analysis Methods
Author: Rory S. McLaren.
Publisher: Rory McLaren Fluid Power Training 1993.
562 East 7200 South, Salt Lake City, UT 84171.
ISBN No. 0-9639619-1-8.

R29- Hydraulics Theory and Application From Bosch
Author: Werner Gotz.
Publisher: Robert Bosch GmbH.
Hydraulics Division K6, Postfach 30 02 40, D-7000 Stuttgart 30.
Federal Republic of Germany, Technical Publications Department, K6/VKD2.

R30- A Complete Guide to ISO and ANSI Fluid Power Symbols
Author: Fluid Power Training Institute.
Publisher: Fluid Power Training Institute 200.
562 East Fort Union Boulevard, Midvale, Utah 84047.

R31- How to Work Safely with Hydraulics
Author: Fluid Power Training Institute.
Publisher: Fluid Power Training Institute 2004.
562 East7200 South, Midvale, Utah 84047.

R32- How to Interpret Fluid Power Symbols
Author: Rory S. McLaren.
Publisher: Fluid Power Training Institute.
Rory S. McLaren 1995.
ISBN 0-9639619-2-6.

R33- Safe Hydraulics
Editor: Gates Rubber Company.
Copyright 1995.
Denver, CO 80217.

R34- Electronically Controlled Proportional Valves. Selection and Application
Author: Michael J. Tonyan.
Publisher: Marcel Dekker, Inc. 1985.
Marcel Dekker, Inc., 270 Madison Avenue, New York, NY 10016.
ISBN 0-8247-7431-0.

R35- Introduction to Closed-Loop Oil Systems
Author: Rory S. McLaren.
Publisher: Rory McLaren Fluid Power Training Institute.
7050 Cherry Tree Lane, P.O. Box 711201, Salt Lake City, UT 84171.

R36- Industrial Hydraulic Technology, Second Edition
Author: Parker Hannifin Corporation.
Publisher: Parker Hannifin Corporation 1997.
6035 Parkland Blvd, Cleveland, OH 44124-4141.
Publication No: Bulletin 0231-B1.

R37- Basic Principle and Components of Fluid Technology – The Hydraulic Trainer, Volume 1
Author: Mannesman Rexroth.
Publisher: Mannesman Rexroth AG 1988.
D-97813 Lhr a. Main, Jahnsrtrabe 3-5 D-97816 Lohr a. Main. ISBN 3-8023-0266-4.

R38- Safe-T-Bleed Corporation Catalog
Publisher: Safe-T-Bleed Corporation 2001.
Catalog No. STB-PC-1201-1

R39- Industrial Hydraulics Manual – EATON
Publisher: Eaton Fluid Power Training.
ISBN: 0-9788022-0-9.

R40- Vickers-Mobile Hydraulic Manual – Fourth Edition 1998
Author: Vickers.
Publisher: Vickers Inc. 1999.
Training Center, 2730 Research Drive, Rochester Hills, MI 48309-3570.
ISBN No. 0-9634162-5-1.

R41- Industrial Fluid Power Text, Volume 2
Author: Charles S. Hedges, R.C. Womack.
Publisher: Womack Machine Supply Company 1972.
Womack Educational Publications, 2010 Shea Road, Dallas, TX 75235.
ISBN 66-28254 (Library of Congress Card Catalog No.).

R42- Fluid Power Hydraulics and Pneumatics
Author: R. Daines.
Publisher: The Good-heart Willcox Company, Inc.

R43- Hydraulics in Industrial and Mobile Applications
Publisher: ASSOFLUID, Italian Association of Manufacturing and Trading Companies in Fluid Power Equipment and Components

R44- Fluid Power in Plant and Field – Second Edition
Author: Charles S. Hedges, R.C. Womack.
Publisher: Womack Machine Supply Co. 1968.
Womack Educational Publication, 2010 Shea Road, Dallas, TX 75235.
ISBN 68-22573.

R45- Mobile Hydraulics Manual
Author: Eaton.
Publisher: Eaton Corporation Training.
Eden Prairie, Minnesota.
ISBN 0-9634162-5-1.

R46- EH Control Systems
Author: F.D. Norvelle.

R47- Fluid Power Journal
Publisher: International Fluid Power Society.

R48- Fundamentals of Industrial Controls and Automation
Author: Lonnie L. Smith and Mike J. Rowlett.
Publisher: Womack Educational Publications.
Dallas, Texas.
ISBN: 0-943719-04-6.

R49- Lightning Reference Handbook
Publisher: Berendsen Fluid Power.

R50- Pneumatics Basic Level
Author: P. Croser, F. Ebel.
Publisher: Festo Didactic GmbH & Co.

R51- Electro-pneumatics Basic Level
Author: F. Ebel, G. Prede, D. Scholz.
Publisher: Festo Didactic GmbH & Co.

R52- Mechanical System Components
Author: James F. Thorpe.
Publisher: Allyn and Bacon.
Needham Heights, Massachusetts.
ISBN: 0-205-11713-9.

R53- Electrical Motor Controls for Integrated Systems, Third Edition
Author: Gary J. Rockis, Glen A. Mazur.
Publisher: American Technical Publishers, Inc.
ISBN: 0-8269-1207-9.

R54- Instrumentation, Fourth Edition
Author Franklyn W. Kirk, Thomas A. Weedon, Philip Kirk.
Publisher American Technical Publishers, Inc.
ISBN: 0-8269-3423-4.

R55- Introduction to Mechatronics and Measurement Systems, Second Edition
Author David G. Alciatore, Michael B. Histand.
Publisher McGraw-Hill, Inc.
ISBN: 0-07-240241-5.

R56- Study Guides for IFPS Certification

R57- Work Books from Coastal Training Technologies

R58- Industrial Hydraulic Manual – Fourth Edition 1999
Author: Vickers.
Publisher: Vickers Inc. 1999.
Training center, 2730 Research Drive, Rochester hills, Michigan 48309-3570.
ISBN 0-9634162-0-0.

R59- Industrial Automation and Process Control
Author: John Stenerson.
Publisher: Prentice Hall.
ISBN 0-13-033030-2.

R60- Industrial Automated Systems
Author: Terry Bartelt.
Publisher: Delmar Cengage Learning.
ISBN: 10-1-4354-888-1.

R61- Introduction to Fluid Power
Author: James L. Johnson.
Publisher: Delmar Cengage Learning.
ISBN: 10-0-7668-2365-2.

R62- Summary for Engineers
Author: Dr. Abdel Nasser Zayed.
Publisher: Dr. Abdel Nasser Zayed .
ISBN: 977-03-0647-9.

R63- Mechanics of Materials
Author: Ferdinand P.Beer, E. Russell Johnston Jr., John T DeWolf.
Publisher: McGraw Hill Publishing .
ISBN: 0-07-365935-5.

R64- Oil Hydraulic System, Principles and Maintenance
Author: S. R. Majumdar.
Publisher: McGraw Hill.
ISBN 10: -0-07-140669-7.

R65- Contamination Control in Hydraulic and Lubricating Systems
Publisher: Pall

R66- Diagnosing Hydraulic Pump Failure
Publisher: Caterpillar.

R67- Oil Service Products Catalog
Publisher: Schroder Industries.

R68- Industrial Fluid Power Volume 1
Author: Charles S. Hedges.
Publisher: Womack Educational Publication.
ISBN: 0-9605644-5-4.

R69- Industrial Fluid Power Volume 2
Author: Charles S. Hedges.
Publisher: Womack Educational Publication.
ISBN: 0-943719-01-1.

R70- Industrial Fluid Power Volume 3
Author: Charles S. Hedges.
Publisher: Womack Educational Publication.
ISBN: 0-943719-00-3.

R71- Electrical Control of Fluid Power
Author: Charles S. Hedges.
Publisher: Womack Educational Publication.
ISBN 0-9605644-9-7.

R72- Hydraulic Cartridge Valve Technology
Author: John J. Pippenger, P.E.
Publisher: Amalgam Publishing Company.
Post Office Box 617, Jenks, OK 74037 USA.
ISBN: 0-929276-01-9.

R73- Noise Control of Hydraulic Machinery
Author: Stan Skaistis.
Publisher: Marcel Dekker, 270 Madison Avenue, New York, NY 10016.
ISBN: 0-8247-7934-7.

R74-Solenoid Valves
Author: Hydraforce

R75-HF Proportional Valve Manual
Author: Hydraforce

R76-Automatic Control for Mechanical Engineers
Author: M. Galal Rabie, Professor of Mechanical Engineering
ISBN: 977-17-9869-3,2010.

R77-Fluid Power System Dynamics
Author: W. Durfee, Z. Sun

Index

T

U

V

W

Wireless, 149
wireless transmitter, 152
working pressures, 190

Z

Zinc Anodes, 312